A PHILOSOPHER'S APPRENTICE

SERIES IN THE PHILOSOPHY
OF KARL R. POPPER AND CRITICAL
RATIONALISM

Edited by Kurt Salamun

VOLUME V

A PHILOSOPHER'S APPRENTICE

In Karl Popper's Workshop
Revised, Extended and Annotated Edition

Joseph Agassi

Amsterdam - New York, NY 2008

The paper on which this book is printed meets the requirements of
'ISO 9706: 1994, Information and documentation - Paper for documents -
Requirements for permanence'.

ISBN: 978-90-420-2434-2
Editions Rodopi B.V., Amsterdam - New York, NY 2008
Printed in The Netherlands

Dedicated
to the Memory
of my Brothers
Aaron and Judah
and my daughter
Tirzah

Boast not thyself of to morrow; for thou knowest not what a day may bring forth.
Proverbs, 27:1

And above all, my friends, let us not forget the duty of hope.
Herbert Samuel

Some of Popper's Research Assistants

Jeremy Shearmur, Arne Petersen, David Miller,
Alan Musgrave, Ian Jarvie, Joseph Agassi
(Photo by Pam Shearmur ©)

TABLE OF CONTENTS

Abstract..9
To the Second Edition..11
Acknowledgement..16
Prologue..17
Appendix: Correspondence with Professor Sir Ernst Gombrich............25
Chapter One: Prelude...29
Chapter Two: Entering Karl Popper's Famous Seminar....................41
Chapter Three: The Master's Class..................................75
Chapter Four: At the Feet of the Great Thinker...........................103
Chapter Five: The Philosopher and His Friends and Peers................127
Chapter Six: The Philosopher and His Workshop.........................167
Chapter Seven: The Open Society, Its Mentality and Style...............205
Chapter Eight: The Matter of Intellectual Leadership....................247
Chapter Nine: Epilogue: The Future of Philosophy......................287
Chapter Ten: Postscript..293
Appendices...313
Appendix One: Russell on "Logical" Positivism.....................314
Appendix Two: Popper versus the "Vienna Circle"..................316
Appendix Three: Carnap and Reichenbach in Retrospect..............322
Appendix Four: The Picture "in Essentials".........................325
Appendix Five: The Heritage of "Vienna Circle"....................329
Appendix Six: Ordinary Language Philosophy......................337
Appendix Seven: Nails on the Socratic Problem......................340
Appendix Eight: Popper on Xenophanes..............................345
Letters..349
Bibliography..353
Glossary..379
Index of Names...385
Index of Subjects...395

In the Austrian College, Alpbach, Tyrol, summer 1954

In Mid-Atlantic, summer 1957:
Joseph, Tirzah, Judith,   Hennie, Karl

ABSTRACT

This is an account both of my apprenticeship with Sir Karl Popper, at the time the greatest active philosopher, and of the way I resigned myself to his rejection of me and of my output. He was prepared to recognize only what I produced under his tutelage (which tutelage he said I should acknowledge with more appreciation).

I admire Popper because of his uncompromising fallibilist philosophy, a philosophy thoroughly free of any attempt to justify views or actions except by reference to improvable criteria, employed in order to reach given, improvable goals. It is hard to reconcile fallibilism with the recognition of science as our greatest intellectual achievement. This reconciliation is the upshot of Popper's theory of learning from experience. His model for learning from experience combines Einstein's daring with his systematic application to science of the quest to find the limits of the best extant answers to the best extant questions, thus rendering science into a Socratic dialogue, and treating empirical tests as attempts to criticize, to refute the best ideas extant. Popper's philosophy is admirable, yet it is in need of revision, especially cleansing it of hostility — to pseudo-science, to metaphysics, to nationalism, and to political power. Each of these items seems to me to have a place, and so we are better off trying to control them democratically without showing any hostility to them. Come to think of it, perhaps Spinoza was right and hostility is never laudable.

From the distance of more than half-a-century, the following appears clear to me: three personal characteristics of his played a major part in his attitude towards me: his boundless dedication and goodwill, his relentless stern moralism, and his tendency to complain — first, regarding chroniclers and peers, and then regarding disciples, myself included. The first of his complaints often fits the facts well enough yet seems, I am afraid, philosophically undignified and rooted in his image of the commonwealth of learning that, I confess, I find embarrassingly naïve. His complaints regarding his disciples are a puzzle. Possibly, they comprise a mere rider, a mere after-thought to his moralism. In the last analysis, it was his ethics that alienated me. Were his side of the story of my apprenticeship to be on record, no doubt it would be very different from this melancholy account, sincerity and sensitivity on both sides notwithstanding, and I can only guess his view. Yet this is attested: he did find my ethics unpalatable and unfeasible. For my part, I view his ethics as stale and his impatience with mine as his Achilles heel.

Young Karl Popper in the Alps
Picture that Lady Popper had taken so very long ago and kindly gave us as a precious gift.

TO THE SECOND EDITION

The first edition of this book met with many favorable responses and poor sales. It fetched only two reviews in the learned press, one critical [Wilson, 1995], one friendly [Chiariello, 1996] (by a former student). It also earned a comment in a noteworthy if amusingly inaccurate[1] work [Steiner, 2003, 169-72]. The first reviewer censured me for having allowed my appraisals of ideas depend upon my appraisals of the characters of their originators. I do not deserve this censure. I concur with the appraisal of Alfred Tarski: Popper had the better argument[2], but the people on the other side were the nicer[3] [Watkins, 1997a, 215], [Hacohen, 2000, 212]. (They also differed politically.[4]) Still, the critic delighted me as he was right in principle: great ideas deserve appreciation regardless of their sources.[5] I wish this were better recognized.

Nevertheless, this book is unsatisfactory and is excessively detailed. Its story comes to explain the most unusual decision I ever made, to risk the total loss of an academic career by leaving London and the promise of a successful career as the heir and successor of my admired and beloved teacher. My explanation is this: I had the choice between submitting to his unwanted, capricious meddling in my life and fighting with him daily in increasing frustration. Both options were preposterous, given that I loved my freedom and that I was his admirer and friend and very much in his debt. This is the whole story of this book. What remains is the philosophical background, and this you can read in my other many writings. Still, there is the flavor that this initially came with. I remember

---

1 "Popper's criteria of *experimentum crucix* and verifiability" [Steiner, 2003, 175].
2 The word "argument" here becomes Popper's philosophy: this is how it grew. He wanted to use the new logic to argue that science uses only valid inferences. As the negation of a universal follows from a particular, he wanted to show that this suffices. He then considered mere conjectures all synthetic universals. He then boldly praised empirically refutable conjectures. He then demarcated scientific theories as refutable. To apply this to political theory, he demarcated as democratic any government open to peaceful criticism, including overthrow. He then developed it as a new theory of rationality as the advocacy of criticism. This led him to the marvelous idea that the critical attitude is peculiar to the pre-Socratics: he viewed them as his precursors. His contribution was re-raising the problem of rationality and replacing Plato's view on it with his own. Other critical rationalists raised the question, but they had no solution to it. Among these were Solomon Maimon, Heinrich Heine, Michael Faraday, Albert Einstein, Erwin Schrödinger, Bertrand Russell, Bernard Shaw, and Frank Knight.
3 The word "nicer" here becomes the "Vienna Circle": they had more social grace and less seriousness than Popper, so that their nice manners were limited to personal affairs (including friendship with Popper). Their philosophy was pompous, shallow and after 1935 it included too many willful distortions of his ideas. His lack of social grace was partly due to his seriousness that boosted his boundless honesty, kindness, and concern. Incidentally, Wittgenstein was more gauche and more serious than Popper.
4 The political and educational activities of the "Vienna Circle" and their impact belong to the history of politics, of religious ideology (including the Marxist), and of culture. It has little relevance to this record.
5 Russell noted [Russell, 1897] that the rise of modern geometry in the nineteenth century was a reaction against Kant. He also reports in his autobiography that once, in a discussion period following a lecture of his against Kant's theory of space, someone mentioned in defense of Kant his love for his mother. Russell expressed then refusal to believe that humanity is so vile that in it the love of mother is scarcer then a new theory of space. This way he expressed tribute to ideas that he was criticizing.

how impressed I was when Popper told me in 1955 of his first participation in a philosophical meeting in England in 1935: in the discussion period of that meeting he said, there is no induction, and won a humorous, dismissive applause: the membership understood his brief intervention as an attack on science. I had myself a similar experience when I arrived in Boston in 1965. The prestigious Boston Colloquium for the Philosophy of Science invited me to read a paper to introduce me to the Boston philosophy of science community. I complied, reading my "Science in Flux: Footnotes to Popper" that was the centerpiece of my *Science in Flux* of 1975. My commentator for that memorable evening was the lovely, judicious Judith Jarvis Thompson of MIT. She expressed unbounded amazement. Many problems beset the theory of induction, she granted, as diverse critics from Hume to Hempel have noted, so that some people deny induction all rationality; but she could not imagine science without it. She was lost for words when she bumped into my pro-science anti-induction view. Is this response still characteristic? After all, leading academics, Adolf Grünbaum, Jaakko Hintikka, and Hilary Putnam, repeatedly identify opposition to inductivism as opposition to reason and thus to science too. So let this book stand as long as so many take for granted the identification of induction with reason.

This edition includes many small alterations: stylistic changes, meant to increase readability and decrease ambiguity, as well as corrections of small errors. It also includes here and there additional words and on a rare occasion an additional sentence. I have limited bigger corrections and additional information to new items: the notes to the diverse chapters, the final chapter, Chapter Ten: Postscript, and the appendices at the end. The opening of that chapter I wrote when I received news of the death of the great art historian Ernst Gombrich, perhaps as an expression of grief.

This edition brims with many references in square brackets spelled out in my overstuffed Bibliography. (The exceptions are Popper's *Logik der Forschung*, 1935, *The Open Society and Its Enemies*, 1945, *The Logic of Scientific Discovery*, 1959, and his autobiography; they are here denoted as [*LdF*], [*OS*], [*LScD*], and [*Autobiography*] respectively.) Some notes elaborate on tangential items. Already the first edition has much that is downright philosophical, but my intention was to cut back on such material as best I could. The additions do not show this self-restraint; so you can skip the notes and the new chapter. Possibly the changes here add up to little: their value may very well be marginal. Perhaps they express my nostalgia, my lingering on old memories that my researches take me away from and that I find hard to let go. Nonetheless, for what it is worth, here is my rationale for adding them.

A friend has pointed out to me that this book presents in a poor light Leon Roth, my controversial philosophy teacher in my *Alma Mater*, the Hebrew University of Jerusalem: my story imparts an image of him that is somewhat off-putting. I regret this: regrettably, there was no room here for reports on his having been a serious scholar and a kind, concerned teacher, remarkably humane, civilized and suave. To do him justice, I should have offered a portrait of him. Here I try to rectify the wrong impression in a brief note. The same may hold for everyone mentioned here. My anecdotes serve as parts of my narrative; they are not pen portraits. Now I try to rectify this, but only to a small extent: to do so all the way is impossible, as Ernst Gombrich stresses in his letters to me that appear in the appendix to the Prologue below. The impression that an anecdote conveys is always relevant to the reasonable image of its object, and a complete image is impossible in any case; yet anecdotes often mislead, and some do so excessively. Telling anecdotes is hardly avoidable. This is true even of the justly famous intellectual biography and pen portrait that Gombrich wrote of his mentor [Gombrich, 1970] as it naturally includes quite a few anecdotes (all of them fascinating, incidentally). Also, he could not avoid offering some general observation on the character of Popper as he told about their friendship in a significant interview [Kiesewetter, 2001, 105]. Only common sense and a sense of proportion are available as tools for the amelioration of distortions that anecdotes involve. I thus could have been more wary of misleading my readers. Some additions hopefully make some amends.

Yet even on the assumption that the amends are successful, they clearly do not provide complete and correct images. So please do not jump to conclusions without trying to check them if you can, since of necessity the clues that I provide are scanty and hardly reliable. Nor is this the whole story: in the present context the pressing question is, are there any gross distortions here? I cannot judge. I can reflect on the past and examine documents.[6] I can report honestly important corrections. This I do, but I offer no finality. Consider my anecdote about the clash that Popper had with Preston King[7] (Chapter Two). I hope when you read it you will easily notice that King is a friend — a dear friend, indeed. We have not met since our ways parted so many years ago, and we hardly corres-

---

6 Popper's archives document the tremendous efforts that he invested in me and the great anger at me that he felt later. This should not have surprised me, as my portrait of him in this melancholy account should indicate. Yet it did. It enhanced my gratitude: my education required generous, intelligent, intense efforts. A graduate student of his once had a psychotic collapse at his home and this made him invest much futile work in effort to help him out — to the almost total exclusion of every other commitment. Already in New Zealand he gave private tutorials to students who wanted to attend his courses but could not.
7 Preston King is "a civil rights hero", a Scholar-in-Residence at The Leadership Center at Morehouse College, Distinguished Professor of Political Philosophy at Emory University and founding editor of *Critical Review of International Social and Political Philosophy*.

pond, but good friends we are all the same, in the most normal sense of the word. Responding to this book, he told me that he does not remember anything like what I say here about him. Since my story is that of a clash, he also reports that to the last he was on the best of terms with Popper. I stick to my story nonetheless. I remember well that Popper did remember the clash that King forgets. Thus, the historical events are sometimes at issue. Let this be a warning. No story is foolproof [Schacter, 1997, Introduction].

Similarly, Gombrich surprised me with his angry response to my report of his lecture in Popper's seminar: he had no memory of it. I reminded him of the details of the lecture, and this seems to have placated him.

Last but most important, in this edition the new material shifts the spotlight from the autobiographical story of my interaction with my teacher to my appraisal of the great thinker — personal and more so intellectual and public. Two aspects of the public Popper require repeated updating, on democracy and on democratic science policy. Let us take it from the very start. Popper always spoke of the value of criticism, and on this all reasonable people have to agree with him. Nevertheless, his views are highly controversial. This invites explanation. There is no need to discuss the demands that scientific theory be open to criticism from some conceivable observations and that democracy is a regime that allows for criticism of its policies up to and including the peaceful overthrow of its rulers; it is even obvious that science and democracy viewed this way go well together. These ideas are often thoughtlessly ascribed to Popper. What is original with him, and is so very revolutionary and so very exciting and justly controversial, is the idea that we should make do with these minimal requirements that are universally endorsed. That openness to criticism is a minimal characteristic of both science and democracy is obvious; dispute concerns the question, does this minimum suffice? Suffice for what? Suffice as solutions. Hence, without reference to problems one can present Popper's view as flat,[8] which is what Feyerabend repeatedly did. Different problems give Popper's idea different aspects, some theoretical some practical, all exciting. They also invite different ways to update his ideas.

It looks as if the "logical" positivist identification of science with meaning is less demanding. This would be true had its advocates deemed all informative statements scientific;[9] instead, they pretended that they understood only scientific statements. This pretense imposed on them the problem of demarcation of science. Wittgenstein twisted the traditional

---
8 This is equally true of Darwin. See [Agassi, 1977, pp. 48, 55].
9 [Campbell, 1921] declares scientific all agreed-upon informative sentences. Agreed by scientists, of course. Who is a scientist?

solution to it, as he replaced verification with empirical decidability: as negations of understandable sentences are understandable, he had to declare the negation of scientific sentences scientific too. Popper rejected this, viewing negations of scientific theories unscientific. They callously ascribed to him Wittgenstein's innovation and dismissed his view as inconsistent — until they returned to tradition, which they did, forgetting to beg Wittgenstein farewell. This is the mess that they left to their heirs.

I will return to this in some detail, but only in the added material, as it concerns what has happened after the period of my apprenticeship. The Popper described here, my master, was highly volatile, suspicious, and plaintive. And then, perhaps abruptly, I cannot tell since I was not there[10], he changed — radically, it seems. He then found comfort, I suppose, in the release from the duty to fight the philistines: their injustice to him was excessive and so beyond his control: his old colleagues were dead and the new generation showed preference for new tawdry stuff. He worked then on ancient Greek culture, the only province where he felt relatively at home.[11] The classical scholars are also still grossly unfair to him.[12] Nevertheless, he preferred writing for them than for the philosophers of science, as he considered them less incompetent. His last, posthumous book is unfinished. It and the letters that he wrote then that I happen to have read in the Hoover Institution Archive at Stanford University, impart the image of him as resigned to his fate and relatively relaxed — exhibiting an uncharacteristic, blissful, highly philosophical indifference to all hostility to him. My impression is, he died in peace. Injustice to him can no longer touch him; setting his record straight is now of interest only to the commonwealth of learning, above all to those present-day youths whose eyes glitter with excitement and whose hearts throb with passion and whose minds are set on intellectual adventures, especially those related to philosophical problems. His work should not stay concealed from them: scholars — especially those who happen to appreciate it — should take courage and break their silence and make their views of him public.

My last chapter discusses all this. It is thus no part of the personal story of this book. The elderly, tranquil Popper is not a hero of my story, as this book is limited to my interaction with him — to the period of that interaction and to his earlier life as it appeared to me then. The new material presents his earlier life as it appears to me now, in a small measure due to rethinking and mainly due to the information on him that came to light later on. That new information reduces the oddity of this book, especially the memoirs by Miller [Miller, 1997] and by Watkins [Watkins,

---

10 I last met Popper in August 1978; he was then still a very bitter person.
11 For Popper's posthumous work on Greek philosophy, see Appendix.
12 For the classicists' still grossly unfair attitudes to Popper, see Appendix and [Meital and Agassi, 2007].

1997], [Watkins, 1997b], Malachi Hacohen's massive biography [Hacohen, 2000], Jarvie's encyclopedia essay [Jarvie, 1998], and Peter Munz's work on Popper and Wittgenstein [Munz, 2004]. They all offer lively pen-portraits of him. The information on Popper by Munz is of the earliest period of his academic career, the early forties. My own description recalls memories from the fifties. Watkins's, Jarvie's, and Miller's are the most comprehensive. They all differ, especially from that of Munz — for various reasons, of course, but mainly because Popper changed greatly after he found the security and relative fame of his position and his personal professorship in the great University of London. Thus, for example, as described by Munz, Popper has barely a sense of humor [Munz, 2004, 22] and by my description of him he had it in abundance; my description is in agreement with that of Watkins [Watkins, 1997]: "The seriousness was lightened by touches of humour and happy improvisations", he said. I derive great encouragement from all of these works, regardless of the differences between us:[13] we all portray Popper as we experienced and interpreted his eccentricities, warts and all, yet with endless affection, profound gratitude, and tremendous admiration.

Herzliya, Israel, Winter 2008

ACKNOWLEDGEMENT

For help in preparing this edition I thank Nimrod Bar-Am, Judith Buber Agassi, Lucien Foldes, Stefano Gattei, Malachi Hacohen, Jacob Steif and Chen Yehezkely. For permission to publish Karl Popper's picture and letters to me, I am grateful to Melitta Mew, the Popper Estate, and the Hoover Institute. For permission to publish the photo of some of Popper's former research assistants, I am grateful to Pam Shearmur.

---

13 Important new information also appeared in recently published correspondence — of Popper with Albert, and of Feyerabend with Albert, with Duerr, and with Lakatos. See also [Simkin, 1993, Appendix one].

Prologue

My initial intention was to cover my seven-year apprenticeship with Sir Karl Popper, the greatest active philosopher at the time. Though an account of an interaction, this book has a central character, whom I must describe. In addition to being a great philosopher, he was also a master artisan, as can be seen clearly in his writings, since his mastery of the craft of writing is exemplary. I hope that I learned from him some philosophy as well as some of his craft.

Somehow, however, my initial intent — to write about my apprenticeship — was lost: this account got out of hand. Not, I hasten to add, because of my treatment of the great philosopher as an artisan: I do not think he would take it as a slight were he to learn that I view him this way. Thus, his autobiography, *Unended Quest*,[1] begins with a warm mention of his own master — for, as it happens, he was a master carpenter and proud of it [Munz, 2004, 29]. (The only personal acknowledgement to a master in that autobiography is to his master carpenter.[2]) Nevertheless, as I say, things did get out of hand.

The result is wild — even by my own wild standards. It puzzles me. It is much more of a scatter-brained-product than any other work of mine, yet it has a structure and a purpose. If it had written itself, if it had merely poured out of me, this would be less puzzling. In my whole writing career I have seldom experienced a piece taking over and dictating itself: I am not that sort of intuitive worker despite what my fans or detractors may say. The closest I ever came to this sort of writing experience was with two or three very brief pieces that are highly analytically structured, as well as with the essay "In Search of Rationality: A Personal Report" [Levinson, 1982, 237-248], which I dedicated years ago to this same former master of mine on the occasion of his eightieth birthday. He did not welcome that essay. Incidentally, it may count as something of a forerunner to this account, which I suppose he would also not have

---

1 Sir Karl Popper's *Unended Quest: An Intellectual Autobiography* appeared in *The Philosophy of Karl Popper*, 1974, in the prestigious *Library of Living Philosophers*. Public interest led to its many republications and translations.
2 Popper's carpentry master influenced him by a negative example: he was after a perpetual-motion machine. Another important negative example is that of Alfred Adler, whose uncritical inductivism Popper met when he was 17. He should have recognized this influence, as it was his first step towards anti-inductivism [Agassi, 1988, 482]. He did not have the great advantage of a thrilling master that I have had. (He told me he envied me this.) His pride as self-made is just even though not in detail [Hacohen, 2000]. I wish he were more proud, less in need for approval.
    Popper also acknowledges warmly [*Autobiography*, §3] the personal influence of his parents and of one friend. The influence of Karl Bühler, his *Doktorvater*, was indirect. So was the greatest, that of Einstein: they met only in 1950. Popper also acknowledged repeatedly the encouragement of members of the "Vienna Circle": he gained his sense of his worth, he said, from encounters with its leading members, Feigl, Carnap and others [Hacohen, 2000, 199 n84]. They were academics and he was a substitute schoolteacher. Speaking with them he realized he had better solutions to their problems, he told me.

17

welcomed.[3]

Unlike those pieces, this account did not write itself in the least. I struggled to write it and lost my orientation. As it is very anecdotal, it may look as though I could add a relevant anecdote here, omit one there, perhaps move one about. The advantage of the word processor is that it greatly facilitates rewriting, shifting, and otherwise experimenting with many kinds of restructuring. Let me report that I did this — endlessly. I found it ironic: the obsessive demon of my exacting, hard-working master finally entered me, and I emulated him in working on this account — notwithstanding all the protests to such a manner of working that I launched decades ago while working under his tutelage. (Dollar book Freud gives an obvious reason: I am the child finally succumbing to repeated pressure, but doing so with mischief.) And while I was thus experimenting on the word processor it usually became evident to me, I do not know how, what the best placement was for a chapter, a paragraph, an anecdote, a point, a sentence, or a word. It is the whole of it, however, that has puzzled me. It still does. I do not know why.

While I do not know what troubles me about it, I take comfort from the fact that wiser individuals often are in the same predicament.[4] An odd autobiographical remark of Bertrand Russell impressed me, which is a report that he had been drawn into the world of mathematics in search of the order and harmony that in reality he had so painfully missed. He was also seeking the certitude he had missed and never found and learned to live without, as he explains in his moving "A Free Man's Worship", a manifesto the poetic style of which embarrassed him later in life, but the contents of which remained with him to the last [Wood, 1957, 208-9].[5] Sometime during his early adolescence, Russell reports [Russell, 1967, near the end of Chapter 1]; [Russell, 1956, 14-15]; [Monk, 1996, 25], his

---

3 Popper disliked my "In Search of Rationality", but, I am pleased to report, he told me he liked my "Karl Popper: A Retrospect" [Agassi, 1988], [O'Hear, 1, 2004].
4 Predicaments are broad problem-situations or vague problems. Putting them clearly is progress: a problem well put is half-solved, as the saying goes. (The paradigm case is Laplace's theory of capillarity: his successful presentation of the problem in Newtonian terms was already a great achievement.) It is also possible to clarify problems by inventing and examining possible solutions to them, as possible solutions of problems improve the understanding of them [Bromberger, 1992]. (The paradigm here is Darwin.) Although improving the articulation of a problem is good, one should not insist on it. Medical diagnosis rightly alternates between efforts to sharpen problems and to change the range of possible solutions [Laor and Agassi, 1990, 150-4].
  All this is fairly obvious, but Wittgenstein's popularity clouds it: he said problems deserve no notice unless worded exactly. Exact wording, he promised, would make them fully soluble [Wittgenstein, 1922, 6.5]. His famous concluding adage, "Whereof one cannot speak thereof one should keep silent" [Wittgenstein, 1922, 7] forbids the stutter that precedes proper speech. Having viewed his work as a ladder to "throw away ... after having climbed", he was in a position worse than that of one who has painted oneself into a corner, while admitting the value of stutter. Friedrich Waismann accused Wittgenstein of obscurantism for having left this position [Shanker, 1986, 50-51] – the one that Russell had declared mystical.
5 The most available version of Russell's "Free man's Worship" is in [Russell, 1917]. [Russell, 1985] includes it and his diverse comments on it. Incidentally, Popper endorsed it [OS, i, Ch. 5, note 4].

Prologue 19

elder brother taught him geometry and demanded that he first take its axioms on faith or else they could not proceed. He accepted his brother's demand, not knowing what else he could do to receive the much-coveted instruction, but it troubled him for a long time.[6] At first, he did not know why. And, indeed, few people did know: the situation was new, and due to a very modern innovation that Russell grasped only later and his brother possibly never did. For two millennia, the axioms of geometry were as secure as anything; geometry was the paradigm of rationality; and then came new kinds of geometry (years later he studied them in college) and with it came a choice, a freedom; and with freedom came a new kind of insecurity[7] and with it the quest for the lost order and harmony.[8] Success in the quest for order and harmony required faith[9] of him, no less. And he understandably resented this requirement.

The quest for harmony informs the arts and the sciences, whose structures emerge in this quest for it and express the yearning for it. This way these feigned structures can offer us the meaning that we miss in reality: they belong to dream worlds;[10] and dreams are two-edged: they are forward looking and backward looking. The forward gaze — dreams as nascent plans — needs no mention here: they breed and propagate the tension in science between the dream of seductive theory and the awakening by rude facts. The backward gaze may serve different functions. It can be the search for redemption: in our dreams we insistently alter God's fixed scripts and we try to correct them. Alternatively, dreaming may be a soul-searching act of moral stocktaking. This is very much in accord with my Jewish heritage. In my

---

6 Russell, "Liberal Decalogue" and Postscript [Russell, 1969, 60 and 220].
7 The study of attitudes to freedom is wanting. We need more than the (obviously terrific) expressions "the fear of freedom" of Russell and "The Unbearable Lightness of Being" of Milan Kundera.
8 "Harmony" is a vague word that denotes many things — also security. Wittgenstein was silent on it and on his longing for harmony that he identified as the love of clarity.
9 The demand to take axioms on faith did not trouble 12-year-old Einstein [Schilpp, 1949, 9]. It does not matter how one feels, as long as one perseveres. Mathematicians have my admiration for their ability to enter a lecture in the middle, allow for a while whatever the speaker says, to suspend disbelief and criticism, and take it from there. Philosophers, psychologists, and educationists ignore the importance for critical thinking of this ability — the ability to suspend criticism — although it is vital for growth, both scientific and artistic. They also ignore the even greater need for the opposite, for hasty dismissals. The impossibility to demarcate between these extremes is the famous insoluble Socratic paradox of learning [Agassi, 1981, 228]. The paradigm case is young Galileo who heard of an itinerant lecturer coming to Florence to speak on the Copernican hypothesis and showed no interest: afterwards friends corrected his mistaken lack of interest [Santillana, 1953, 143]. They made history.
10 Was the transition from credulity (magic) to criticism (science) gradual? Neither: both are common (Sir Edward Evans-Pritchard) and mixed (Frances Yates).
    Magic allows for no accident (Evans-Pritchard) and deems throwing lots forcing Fate to show her hand. Following the Cabbala, Bacon viewed discovery as accidental in the sense that its finder suspends judgment: only the pure of heart deserves the assurance of success. This shields his theory from refutation. Lagrange, Pasteur, and others rendered the pure heart into the prepared mind.
    Allowing for the mysterious, traditional mystics expelled it from language. Boyle and Newton expelled it from science; Wittgenstein returned to tradition. Neurath denied its very existence [Neurath, 1973, 326] and (following Frazer) he viewed it as pseudo-technology.

vanity, I would readily consider this melancholy account stocktaking; except that my Jewish heritage discourages vanity.

I am embarrassed: no matter in what way I re-wrote this account, I could not erase the impression that it displays an attitude that is as alien as possible to what is known as the philosophical outlook: pettiness, defensiveness, even vindictiveness. Perhaps I am defensive after all; perhaps, then, I am merely deceiving myself; perhaps I am vindictive, trying to settle accounts — perhaps with a former master for his having methodically ignored me, perhaps with some of my peers among his followers and former students; perhaps it is merely my private exorcism of the ghost of Imre Lakatos, my nemesis.

I do have a serious misgiving to report: I, too, wish to alter God's unalterable script; I, too, wish to reorder my little universe. I strongly wished that my master, teacher and hero would have treated me like other members of his profession, that he would have responded to my writings the very same way he would, had he never met me in person. I strongly wished he could break away from his commitment to hard work in order to have a moment to catch his breath and look around and have a little chat with a stranger like myself. I strongly wished he could shoulder his responsibility as a leader of his profession and see the requirement of his responsibility as a leader to treat his disciples and former students in a friendlier manner, or at least with more detachment. I strongly wished he would consent to discuss these things with me. But he never had any time for this, being so busy, working hard to the last on what always appeared to him as pressing business. And now it is all in the distant past, no longer alterable. I must leave it: my dream of altering God's fixed script is just that: a fading, dead dream. This account discloses no current dream, only events and dreams long past.

As to settling outdated accounts, it is pointless, especially since I prefer to bear no resentment; not even to Lakatos. When I report that we were friends of sorts, the informed are astonished: how can I call a friend someone who consistently and to his dying day expended much effort and great ingenuity in order to harm me?[11] I do not know, but I know we were friends nonetheless, as we talked a lot on all sorts of things, discussed ever so many problems, enjoyed each other's company, had great fun

---

11 Amusingly, Lakatos once arranged a party for me by my former students in London to coincide with the opening of a conference to insure my absence there. He bullied my friends to try to force me to yield to his will. He made people register complaints against me. He used threats to block invitations to me to participate in academic activities and to have my work published. (Organizers and editors asked me for explanations and even complained to me in strong words.) He solicited and published essays that scorn my work. He sabotaged senior academic appointments for me, including one in University College, London, that was very promising and that I would have gladly accepted. Yet he supported strongly my appointment at Tel Aviv University. Only after he died did evidence come to light on the effort he had invested spreading the rumor that I was a literally crazy charlatan.

Prologue 21

together. True, he tenaciously did the utmost to damage my career; and he aggressively ventured to rob me of my peace of mind. The worst was, he most ungratefully insulted my wife Judith.[12] More relevant to this narrative, he had his obsessive demon that drove him to connive and intrigue and put a wedge between my mentor and me. But if it takes two to tango, then it takes three to have an intrigue, so that if there is any account to settle, I, too, must pay.

What troubles me most about this account is possibly the worry that perhaps it is propelled by some demon, by some resentment. This is hard for me to judge. Perhaps I should take comfort in the thought that resentment is of little interest. And perhaps things are worse. Things may be worse: possibly this account displays nostalgia.

I used to muse about the nostalgia that governed the disputes between my teacher, Popper, and Rudolf Carnap, his greatest foil, who was once so very famous[13]. These two were greatly troubled by the output of each other; for a lifetime they attempted to settle a dispute that had begun in Vienna in the inter-war period; perhaps by doing so they maintained the illusion that the Vienna that Stefan Zweig had named *The World of Yesterday* [Zweig, 1943][14] still lived on. At least ostensibly, the dispute had concerned not Vienna but the subject matter of philosophy. Popper debated the great issues of traditional philosophy, especially the character of science and its place in our culture. Carnap advanced the thesis that traditional philosophy is dead and clung to the thesis that demonstrably philosophical utterances are inherently meaningless. The dispute received a symbolic example from Popper during a walk the two had in the Austrian Alps on one brief splendid holiday described most fondly in Popper's contribution to *The Philosophy of Rudolf Carnap* [Schilpp, 1964] (also in [Popper, 1963, 253] and [Schilpp, 1974, 968]). They considered the sentence, "this stone is now thinking about Vienna".[15] In Popper's view, it is false, in Carnap's view it is demonstrably meaningless. Intellectually, this is a farce: there can be no

---

12 Feyerabend narrates the story of one of Lakatos' insults to Judith [Duerr, 1995, 153]. He heard this with an open-mouth, he reports, and could scarcely believe his ears. (It was, incidentally, just before a kangaroo court session that Lakatos sprang on me.) This was particularly distressing, as he had stayed in our home for weeks at her invitation when he first moved to London and could not afford a hotel. But then, this was his style: his sense of gratitude always turned him nasty [Long, 1998, 299].
13 The fame of Carnap was much higher at the time than now. It will further diminish unless he is credited with a valuable idea. The search for it and the claim of analytic philosophers that philosophy is logic led to the repeated claim that he had contributed to logic [Awodey and Klein, 2004]. What?
14 In the inter-war period Viennese of Jewish descent (including the admirable Stefan Zweig and Josef Roth) advocated cosmopolitanism or at least a united Europe, or at the very least a federation of the Danube Basin [Hacohen, 1999]. Popper mentions the last option as a lost opportunity [*OS*, ii, 312].
15 Alas, in his published reports on the sojourn Popper ignored this intellectually pregnant stone. This is odd: it broke for him the spell of "logical" positivism [Hacohen, 2000, 199 n84]. Carnap publicly discussed this thoughtful stone soon after; he changed his mind about it [Bar-Hillel, 1964, 34].

debate about what is demonstrable.[16] But it was serious on the existential level.

This stone is now thinking about Vienna: even hearts made of stone still fondly, painfully, remember, and occasionally think about, Vienna; if I forget thee, O Vienna. The hypothesis that philosophy is meaninglessness is past history; it underwent modification in stages and thus slowly faded out of existence. (This Carl G. Hempel shamefacedly admitted and is generally recognized [Hempel, 1950]).[17] Yet they could not let go of the old disagreement and they could not recast the problems and debates of the Vienna that is no more. And here I am, discussing events that for contemporary readers are also a part of a vanished world, and concern not the debates in once glamorous Vienna on the place of science and philosophy in God's scheme of things, but small tiffs between a great master and his youthful, unruly, devoted disciples.

No doubt we — Sir Karl Popper and his close associates, past and present — make for a wild bunch. Why then bother with us? Because we are the lively bright kids around the block, the crowd of that great thinker, the greatest philosopher of the mid-century — of the post-Russell era — one among the handful who count at all, who robustly investigated philosophical problems that signify.[18] Of course, this assessment will gain a hearty dismissal from the majority of established philosophers (the analytic and the linguistic and the phenomenological and the existentialist and the fideist and above all the Thomist and the Marxist).[19] They are herewith cordially invited to save their time by putting aside this account. They will miss very little, since it is not meant to enlighten on matters philosophical. I am distressed to admit that it contains much philosophy, and readers who think this is too high a price for stale gossip, they too should skip it — and wait perhaps till extracts from it circulate as gossip proper, accurate or not, like raisins picked out of a cake. (Those who

---

16 Indeed, Carnap's change of mind won him Bar-Hillel's censure (*loc. cit.*). No matter how often he changed his mind, radicalism made him declare his assertions demonstrated and Wittgenstein made him declare them tautologies. Radicalism was important and challenging in the scientific revolution, but Einstein had superseded it. Carnap admired Einstein but he admired Wittgenstein more.
17 Hempel worked repeatedly on the demise of the doctrine of meaning and could not let go of it.
18 Gathered in a corner in a conference we would find ourselves surrounded by envy directed at the tremendous fun that we had together.
   Once in a San Francisco conference Bartley drove Feyerabend, Jarvie and me to his palatial home in the Berkeley Hills. In the car we poked fun at his attachment to gadgets: he felt at home only in over-equipped offices American-style. We then noted that Jarvie felt at home anywhere in the British Commonwealth. I added I felt at home almost everywhere and Feyerabend confessed he was at home nowhere. We expressed surprise, as he always projected bonhomie so well. He said it was contrived. Incidentally, Popper shared with Kuhn, Feyerabend, and Lakatos the disposition for tremendous, rapid swings between bitterness and joy.
19 This is my division of current philosophy to broad traditions. In the analytic trend I include British empiricism, America pragmatism, positivism and ordinary language analysis in oversight of hybrids such as the Frankfurt "critical" school, and the Thomist existentialists, not to mention the postmodernists.

continue regardless may wish to use the glossary appended below.) While I have expurgated all the philosophy that could be left out without detracting from the comprehension of its gossip, no reading is free of charge. When one wants to grasp gossip about any craft-guild, one needs to have some familiarity with their craft, or to acquire some. This cannot be helped. (I did ask friends and colleagues to suggest omissions from the draft manuscript.)

Naturally, I will feel amply rewarded should anyone better comprehend contemporary critical philosophy from reading this account, though it is no more than the story of how fortunate I was to have met the philosopher and to become his student and assistant and to work under his tutelage, and the story of how (needlessly) agonizing all this was. Naturally, I would be delighted were this account to convey the suggestion that delay in the recognition of great ideas (such as those of my teacher and mentor) is due to fear and anxiety and decidophobia among the leadership of the commonwealth of learning.[20] This sort of delay becomes increasingly expensive in our ever faster changing world. Our democratic society would do well to try to shorten it drastically. If, in addition, readers should consider this account an invitation to join the open society, then they will have well understood it. But all the same, I do not pretend that this is why I have written it. I do not know why I did: possibly, I simply like to retell piquant gossip. I hope this account of much human weakness entertains my readers without arousing malice; hopefully it will stir sympathy.

Herzliya, Summer 1987, Fall 2006;
North York, Fall 1990.

P. S. I regret I have two points to add, one procedural, one stylistic.

Procedural: I do not mean to say that I had no complaint. I was treated by Popper and by some of his disciples as badly as his peers have treated him, namely by stonewalling, by meeting the critic with utter silence.[21] Now some criticism deserves to be met by silence. Which? Where is the forum to discuss this matter? Of course, as the debate is made public the stonewall collapses, and when the debate takes place in

---

20 For the term "decidophobia" see [Kaufmann, 1973] and [Buber Agassi and Agassi, 1985]. The Romantic view of autonomy is anti-democratic élitism. Kaufmann rejected élitism yet, following the Chicago school, he distrusted democracy and this left him vulnerable to Romanticism.

Decidophobia intertwines with the wish for stability and of fear of experimentation – the strongest conservative arguments. Conservative political leaders tend to suffer from it. Conservative military leaders tend to seek technical and organizational innovations and love to test them in action. Both tendencies are undemocratic. To counteract them, Popper recommended replacing the traditional search for stability with the search for efficient democratic controls. This is now urgent as global political problems are mounting and as political stability is gone for good.

21 I am very glad to rescind this complaint, as my views were discussed critically, especially in the two volumes in my honor [Jarvie and Laor, 2001] and as I was invited to speak to the Popper 2002 Centenary Congress in Vienna.

private, critics from the outside are prevented from commenting on it. Why does it matter, however, that one is or is not noticed? I will return to this later on when discussing some of the complaints about stonewalling that I used to hear daily throughout my apprenticeship days.

Stylistic: I do not know how to refer to my erstwhile master: I cannot refer to him as Sir Karl, and even Popper sounds funny to me. I tried "the Master" in earlier drafts and had strongly negative responses from readers. So I decided to use "the philosopher", though not consistently. Personally, I speak of the philosopher's conduct, as well as of his relations with other professors, with his Viennese peers, including his responses to their publications, and with his entourage, including his attitudes to them. Intellectually, I speak of Popper and of other philosophers, including some of his predecessors and disciples, and of his attitudes to them, and of his publications and his published ideas.

I enclose here a correspondence I had with Sir Ernst Gombrich, Sir Karl Popper's closest friend. He has my most sincere gratitude for his permission to publish it. A few readers told me that it is better omitted, as it is out of tune, perhaps merely out of synch, with the rest of this account. This may be true, but all the more reason for including it here: as my account is unavoidably some sort of fiction, let the following correspondence be a token presentation of reality; whether the representation is typical or not, and if yes, of what, is not for me to judge. Perhaps I should say that the greatest incentive I have for sending it to the press is that however differently readers responded, they all reported having enjoyed reading it. So you, too, enjoy.

P. P. S. Having put this manuscript in its final stage in 1987, I left it lying around, as is my habit, especially since it was harder than usual to get accepted, and I kept tinkering with it. The death of W.W. Bartley III, Popper's heir apparent and official biographer[22] reminded me of how confusing can be the revising of one's responses to old events in the light of recent ones and I decided to make no new additions.

---

22 Bartley died much too early, leaving only one chapter finished. He proofread his last publication on his deathbed [Bartley, 1990, Stephen Kresge's Preface].

APPENDIX: CORRESPONDENCE WITH PROFESSOR SIR ERNST GOMBRICH

London, 9 July 1988

… … I am most grateful for your letter … … and for your concern which I certainly share. I am all the more sorry to say that I do not think that it is within my power to intervene in this most distressing affair. I believe Professor Agassi has written a similar unpleasant account in Hebrew many years ago and I cannot but agree with you that this kind of personal polemic can only do harm to him. Personally I do not think that it can do any harm to Professor Popper in the long run, because I am sure that the accusations are baseless, though they may well give passing pleasure to Sir Karl's opponents. In any case, my attitude is a very simple one: Having been very close to Karl Popper during all the years in question I know how very much Professor Agassi owes to his teacher's care, effort and tolerance. It is my considered view that as a recipient of all these immense benefits Professor Agassi is very ill advised to air any grievance he may think he has in public. Such rank disloyalty and ingratitude must inevitably reflect on the writer.

Let me explain therefore, why I feel unable to intervene: I am still as close to Sir Karl as I ever was. We talk three or four times a week on the phone, and it would be quite impossible for me to hide from him any communication on this matter. But if I cannot hide it, I must tell him, knowing that it would deeply hurt and upset him. Though he is approaching his 88th birthday, he is still working every day from morning till night with incredible intensity on a variety of important problems. In addition he receives, of course, a very large amount of letters, of manuscripts, books, and invitations and often grudges the time he has to spend on his answers. To engage him in an exchange with Agassi might set him back for weeks if not months and I would not want to be responsible for such a crisis which can lead to no good.

I am 100% sure that Karl Popper never had anything in mind but the intellectual and material welfare of Agassi, and that any impression to the contrary Agassi may have had must be due to misunderstanding. These are matters which cannot be argued, let alone demonstrated. I cannot see what I could do to persuade Professor Agassi that he is wrong and that he should drop his charges both in his own interest and in that of Sir Karl's peace of mind.

I would have no objection to your showing this letter to Professor Agassi.

Yours Sincerely,

(E. H. Gombrich)

Appendix

As from The Department of Philosophy,
York University, CANADA, 1988.08.14
Professor Sir Ernst H. Gombrich, ...
Dear Ernst,
A copy of your thoughtful letter ... of the 9th last has just reached me (with your kind prior consent). I always envy people, like you, whose even small personal letter to a stranger on a complex painful matter shows excellence in quite a few ways. It has been a long time, and I realize I am taking liberty pestering you in addition to him. My apology for that; I will address myself in this letter solely to its content; I do hope it will prove less embarrassing to you than what has preceded it.

Allow me to concur with all you say in your letter ... . except to suggest that a certain allegation it refers to, reporting some misconduct on my part, is somewhat inaccurate. If I knew how to show you that my gratitude and admiration for Karl are still strong and unqualified, I would gladly do it. For, though not particularly known for astuteness, ... X was right in divining that I deem very weighty any opinion of yours regarding any matter at all.

Perhaps I should begin with the allegation — of some expression in Hebrew in print of "disloyalty and ingratitude" to Karl. I do not know what you are referring to, but I do promise nonetheless and quite unconditionally that I will publicly repudiate any such expression of mine or in my name, and with all the fanfare I can mobilize to that end, if only I were shown the text referred to.

Perhaps I should also ask your permission to quote your letter, in full or in part, in my humble memoir in which my debt and gratitude to Karl are described in detail. I suppose this view of my memoir does not exactly tally with the impression ... X has given you, but I never attempted to meddle in his conduct, nor need I comment on his letters to you. With your kind permission allow me to add that the memoir is unfinished nor is any part of it in final draft.

I was very glad to hear about Karl from you — however indirectly — and it warms my heart to learn that you too do not think that any smear campaign — by myself, heaven forbid, or by anyone else — can do his reputation much harm.

With great respect and with my very best wishes,
    Very sincerely,
                                    (Joske)
                                Joseph Agassi

27 August 1988

Dear Joske,

Thank you for your letter of 14th August ...

I suspect there must be a misunderstanding somewhere: My reference to an article in Hebrew was based on a memory I had that many years ago you published an article in Hebrew in which you offered something like a caricature of Karl and also such certain allegations against him. If such an article never existed, so much the better, and I apologize for this allegation.

As you may remember, I have spent some time on the study of caricatures[1] and I perfectly realized that anybody can be shown in an unfavorable light. As a friend of Karl's I know how much depends on interpretation. But precisely because this is so, I feel doubly strongly that anyone who owes as much as you do to Karl (and I was glad to read that your gratitude to him is "still strong and unqualified") should not and must not indulge in this easy game! Obviously I can give you permission to quote my words; like a certain Pontius Pilate I can only say, *"Quod scripsi, scripsi"*.[2]

I fear if you really want to write about Karl, you will soon come to realize, as I have often found, that nothing is more difficult than to describe a person — and the more unusual and outstanding that person, the more difficult it becomes, because *"Individuum est ineffabile"*.[3]

There is no more I can say. Karl now lives rather far from us, but every time we visit him, he reveals another wholly unexpected aspect of his interests and personality, the range of his reading is prodigious and the intensity of his engagement unchanged.

With the best wishes,

Yours,

Ernst

---

1 [Gombrich and Kris, 1940].
2 *John, 19:22*.
3 See [Pieper, 2003, 728].

York University ..... 8 October, 1988
Professor Sir Ernst H. Gombrich ... ...
Dear Ernst,
Your letter of the 27th of August ... is most gracious....

I much appreciate your friendly attitude, and, of course, I fully agree to what you say. Yes, I do intend to write my memoir, and I shall call it *Apprentice*, to stress that it is but one memory/view of my experiences, and only mine: others who have shared it with me will doubtless have memories and views of them different from mine. This is not to get off cheaply. As you invoke your study of caricatures, allow me to brag that I remember the part of it I have read, and you quote there the question, when does an inaccuracy become a caricature? Insofar as the answer is obvious, intention to be funny at others' expense, I have no fear: I have no such intention, least of all when talking about Karl. But insofar as a caricature may be a result of ineptitude beyond the usual difficulty to appreciate and describe greatness, I, for one, do not trust myself. I will consult others and ask them to purge the manuscript of all caricaturing and all cheap scoring of points. It is to that end that I have asked for your permission to publish this correspondence and, I assure you, if I do, it will be in a prominent place: as an appendix to the preface.

I do not presume to impose my manuscript on you, certainly not in its very poor preparatory stage. But do allow me to assure you that I will gladly mail you a copy in case you wish to glance at it so as to insure that I do not misuse your name.

I am grateful for your news about Karl. It is only my respect for his wish to avoid contact with me that I restrain my wish to communicate with him and do so minimally.

Please accept and convey to your family greetings and sincere good wishes — my own and those of my wife Judith.

Joske

P. S. Let me add to this sad correspondence one piece of information: Popper refused to look at the manuscript, expressing the wish not "to be involved" in any publication of mine (letter from his secretary Melitta Mew to me, May 4, 1990).

November 1991.

Chapter One: Prelude

And if by chance we find each other,
it's beautiful. If not, it can't be helped.
(Fritz Perls)[1]

Philosophy is a harsh mistress, said Sir Isaac Newton, half proudly, half plaintively, intending by the noun what we would call today the combination of science and scientific philosophy (or scientific *Weltanschauung* or *Weltauffassung*), namely, the scientific outlook[2] plus scientific research[3]. And he should have known, as he was (or so they say) both the greatest and the most intuitive. This remark sounds very deep, especially coming from the pen of such an authority. It is objectionable all the same. Harshness need never be sanctioned — not even from a master or a mistress, not even from the most beloved and well-intended and noblest, be it concrete or abstract, be it real or imagined: we prefer sweet reasonableness to harshness these days and lovers to masters and mistresses.

If one needs a master or a mistress, then perhaps philosophy, namely science, is the best. If one needs an authority, then, admittedly, there are good and strong arguments in favor of the view that science is the best candidate for the post. Yet there are also arguments against this view, ones expressed already in the seventeenth century by St. Roberto, Cardinal Bellarmino, the grand inquisitor who was responsible for the legal murder of daring thinker Giordano Bruno, and who threatened the life of wise, brave and noble Galileo. I suppose Galileo was of two minds about the matter: he clearly opted for total individual autonomy, yet he equally clearly submitted to the authority of the Catholic Church on matters of faith and to the authority of science on matters of reason — hoping to bring harmony between the two authorities. All his followers were convinced that the arguments of Bellarmino are answerable, but they cared

---

1 [Perls, 1969, opening paragraph].
2 The "Vienna Circle" manifesto is the best-known pro-science document. (Only Heidegger's Rector's Address is better known.) The manifesto supports the input of Einstein, Russell, and Wittgenstein, and expresses the desire to combine them as up-to-date science and up-to-date logic. It exudes contempt for metaphysics — taking no stock of its own faith in rationalist metaphysics, namely, in the faith in the siblinghood of humanity, in the ability to reason, and in the right and duty to do so.
3 Newton was ambivalent [Manuel, 1968]. In a letter to Halley (20 June, 1686) he said,
"philosophy is such an impertinently litigious lady that a man has as good be engaged in law suits as have to do with her."
Faraday said [Bence Jones, 1870, 1:54] [Gladstone, 1872, 5],
"My desire to ... enter into the service of Science, which I imagined made its pursuers amiable and liberal, induced me ... to take the bold and simple step of writing to Sir H. Davy, expressing my wishes, and hope that ... he would favour my views ... At the same time that he gratified my desires as to scientific employment, he still advised me [against it] ..., telling me that Science was a harsh mistress ... He smiled at my notion of the superior moral feelings of the philosophic men, and said he would leave me to the experience of a few years to set me right on that matter".

for autonomy more than for faith. They disagreed with him about reason: they said proof is possible. Real autonomy, they also said, belongs to those who recognize proofs and submit without compromise to all and only the verdicts of their own reason; autonomy will thus never clash with science, with proof; rather, science and autonomy are one. Galileo was in doubt, but his followers were optimistic: the authority of science lies in proof, in Reason. But proof is hard to come by. Philosophy, namely, science, is then harsh. Indeed.

That is how things stood until Albert Einstein and Karl Popper preached freedom from all authority. All. A new age has come. From now on, those who wish to follow Reason can do so out of love,[4] not out of submission, not out of acceptance of a harsh regime. Life is harsh enough as it is; we should dismiss the educators who make things worse by their imposition of authority. Science is no longer a master; those attracted to science are now invited to a love affair. There is no proof[5] in science and no need for proof. Nature, Einstein mused, can say no, and She can say maybe; She never says yes: science never imposes. Since experience can sometimes say no,[6] Popper argued, scientific research can continue and

---

4 Einstein said [Fösling, 1997, 26], [Calaprice, 2000, 14], "love is a better teacher than a sense of duty", and [Oberdorfer, 1995, 127]
    "never regard study as a duty, but as the enviable opportunity to learn to know the liberating influence of beauty in the realm of the spirit for your own personal joy and to the profit of the community to which your later work belongs".

5 The greatest asset of the new rationalist philosophy is its recognition of the freedom that reason enjoys, especially within science. Many philosophers deem this freedom a defect [Agassi, 1999b]: freedom and proof seem to clash. (Hence, suggested sagacious Heinrich Heine already in 1852, we should not let proof dazzle us so much as to block criticism and deprive us of our "inalienable right to admit error".) Rationality is freedom of engagement in critical debates in the quest for the truth rather than for proof-surrogates. Expressions of this idea surfaced repeatedly. David Brewster expressed it in his life of Newton (1831) and William Whewell expressed in all of his writings on science (1837-60). It needed a background of non-defeatist version of skepticism, and this Einstein and Popper provided with the theory of science as solving problems in series of approximations. Without reference to problems, these ideas are dull. Richard Robinson notes in the end of a favorable review of Popper's *The Open Society and Its Enemies* Vol. I [Robinson, 1951], that Popper's moral precepts are trite; he praises him for having stated them but he can scarcely justify this. The reason Popper made them is that he was a minimalist. His minimalism is questionable, and in the study I repeatedly criticize the corollary he drew from it, namely that ethics is trivial. The triteness of Popper's methodology justly made John Ziman ask in his review of Popper's *Logic of Scientific Discovery* of 1959 that he referred to much later [Ziman, 2000, Preface], what merit is there to the stress on the refutability of theories that sticks out like a sore thumb? With equal justice, Eccles had the opposite attitude to the same refutability: beset by problems and worried about his own refutations of his ideas, he was relieved to hear from Popper that refuting hypotheses is a virtue, not a vice [Eccles, 1970, 105-6], [Popper, 1999, 12-13]. The theory of rationality as proof-surrogate is popular because it leads to agreement. Ziman came to object to it [Ziman, 2000, 254]: "Nobody expects a group of lawyers, politicians, theologians, or doctors to have identical expert views. But any outward sign of disagreement among scientists is taken as a grave weakness." This could have sent him to reconsider his review of Popper. Sadly, he died soon afterwards.

6 [Dukas and Hoffmann, 1979, 18-19]; [Einstein, 1981, 19-20]:
    "The scientific theorist is not to be envied. For Nature, or more precisely experiment, is an inexorable and not very friendly judge of his work. It never says 'Yes' to a theory. In the most favorable cases it says 'Maybe,' and in the great majority of cases simply 'No.' If an experiment agrees with a theory it means for the latter 'Maybe,' and if it does not agree it means 'No.' Probably every theory will some day experience its 'No' — most theories, soon after conception."

be exciting and worthwhile. Science, in the view both held, is kinder and gentler than Newton had imagined.[7]

The greatest lesson from this exciting development is that criticism is a friendly act, not an act of aggression. When informed colleagues hear this statement, they are quick to notice that it is one that Plato has explicitly and unreservedly made in his *Gorgias*. They are right: it is, I suppose, the thesis of that marvelous dialogue, rather than the garbled up thesis usually attributed to it, namely, that sophism is intellectually not serious enough, since it does not live up to the ideal of truth. Popper begrudges the fact that modern languages force "sophism" to denote mock-criticism; he admired the sophists; most of them were democrats, he reminds us [*OS*, Chapter 6, note 53; Chapter 7, iv]. He took as his example of mock-criticism an ancient Hebrew example rather than an ancient Greek one: false prophecy. That is a suitable word, even though it has religious or, worse, anti-religious overtones. Perhaps mock-criticism is the best word: the truly free person always prefers frank criticism to mock-criticism, not to mention flattery and intellectual submission. This, then, is the lesson we learn from Einstein and Popper: in Plato's message as stated in his *Gorgias* (and elsewhere) lies the key to liberation: the authority of science is extraneous: science as conjectures and refutations is exciting enough and interesting enough. Plato's own message, the view of science Plato himself propounded, or is presumed to have propounded, is not that science has no authority: rather, most historians of philosophy and most Plato scholars agree, he claimed that true science has a supreme authority. Commentators seldom read the passages from Plato's *Gorgias* that deliver the said anti-authoritarian message as fully anti-authoritarian; rather they view them as opposed to the authority of the sophists. It is easy to speak against some authority, against the authority of this or that oracle/prophet/sophist/scientist; the autonomous bow to no authority of any magic/prophecy/philosophy/science. People are impatient when they hear that, as they are sure that they know this. Do they?

As the new attitude is free of psychology, it leaves this question open (to empirical investigation). The picture of science emerging from this new attitude to it is, then, at least in one sense, very disturbing: with no recourse to psychology, it pits autonomy against scientific authority. According to the new picture, there is the authority of the standard science textbook, the advocacy of its authority, the defensive presentation of science, and so on, all on one side; and then there is on the other side the scientific attitude (the critical attitude) the anti-authoritarian and un-

---

[7] Hermann Weyl echoed the passage cited from Einstein in the previous note [Weyl, 1931, xx]; Popper quoted him [*LScD*, 280]. Most philosophers of science center on the obligation to agree. Einstein, Weyl and Popper centered on the practice of admitting error that they considered commonsense and commoner in science than elsewhere.

compromisingly non-defensive. Now this is bound to be very surprising in the light of the fact that scientists usually and systematically defend their theories, and, more to the point, they defend the authoritarian view of science.[8] Michael Polanyi and Thomas S. Kuhn have observed this approvingly. So much the worse for the facts, of course; yet the facts will not vanish by themselves. Their abolition requires much work, including a revolution to make the standard science textbooks user-friendly. How?

Many problems beset the new view, of course. As a social phenomenon, science is bound to have many problematic aspects. I had once thought that writing his Preface to Galileo's great dialogue [Drake, 1953, vi-xx] Einstein described as excessive Galileo's attempt to convince the authorities of the Church of Rome of the truth of the Copernican hypothesis. I was in error. His aim there was to refute the idea [Crombie, 1952, 232] that general relativity allows to turn the wheel backwards (xiii) and to deny that Galileo had used the inductive method. Einstein notices that Galileo's significance is in that he had refused to base science on authority. They both spoke against the intrusion of politics into science as a threat to the freedom of thought. The problem remains: what is the right attitude to it? Should we defend science? Not as the right answers. How then? Not as very good but merely as the very best we have thus far.

To defend science at Galileo's difficult time was necessary. In Einstein's view, to defend science today is superfluous and distasteful to the aware of the immense power of the contemporary scientific community and of the heavy responsibility that they should shoulder but scarcely

---

[8] The defects of the standard science textbooks are too many to list, but let me make a stab at it. Textbooks are unreadable. This way they support the popular view of learning as boring and arduous. As a model for writing, they encourage suppressing problems and expressing contempt for criticism. It makes writing defensive and aggressive and thus reader-hostile. This way they support the popular view of criticism as hostile. As they are inductivist, they suggest that everyone can contribute to science. When Sir Francis Bacon said so, he encouraged amateurs to do research; he thus helped bring about the scientific revolution, as it instituted ways of encouraging and helping novices; when science teachers and textbooks do so today, they discourage and dishearten, since too little guidance is available to high-school students, since the Baconian ethos is gone.

Standard science textbooks present competing theories as true, as, for example, the series of competing versions of the gas law. To admit openly that competing theories are not all true is to reject Bacon's demand that all falsehood be mercilessly erased. This demand is untenable: some refuted theories are most important. Classical mechanics is the paradigm here. The combination of Bacon's demand to discard falsehoods with the teaching and applications of classical mechanics is disastrous: it amounts to the claim that refuted scientific theories are true. This is a silly version of the intelligent suggestion that we should rescue refuted theories from their refutations by either increasing vagueness or adding *ad hoc* limitations. Duhem and Poincaré made this proposal; Kuhn, Feyerabend and Lakatos presented the proposal as a fact. The former were reasonable though mistaken; the latter were simply frivolous. What is necessary here is the admission that some falsehoods are too significant to forget. This admission raises an interesting question: which falsehoods are significant? The answer may provide a list of important characteristics. There are diverse reasons for appreciating and remembering old defunct ideas. Einstein said, a scientific theory that approximates an extant theory is memorable. There is more. The ideas of Aristotle are of no interest for science; those of Democritus, for example, are. And ideas of Plato are of interest for the foundations of mathematics. This still leaves much for further discussion. Descartes' ideas are of importance for many fields. So is the idea of the siblinghood of humanity.

do.[9] Yet the persistence of the defense of science poses the following important problem (unless one dares assume that there is nothing more to the contemporary defense of science that so many philosophers still engage in than the pleasure of playing it safe). Why is the defense of science still so popular? Of course, one might say, because the hostility to science is still so popular. Now, admittedly, obscurantism is still popular and it is anti-scientific by definition. Yet nobody attacks science these days, not even the most militant preachers of obscurantism — except in the darkest parts of the globe, and even there merely for internal consumption with no intent to have effect elsewhere. Why, then, do philosophers of science and even intelligent scientists defend it still? Perhaps they do not know that there is no more need to defend science.

Science is better off not defended but severely criticized (by frank admirers). This is not easy to discover. It appealed to me greatly when I learned it, but learn it I had to, and from the philosopher Karl R. Popper.

My greatest fortune (next to my marital fortune) was to have met the philosopher Karl Popper, and to have become his student, research assistant, associate, close friend and admiring colleague — beginning in January 1953 and up to August 1960.

After that we met a few times, once in a while, but most of the time he was not on speaking terms with me. As far as I can tell, we both agreed on this point: despite all attempts to patch things up between us and despite some bright moments, some of them wonderful, the cutoff point for him was in spring 1957, when one day to his utmost surprise I declared my period of apprenticeship over and said I was his pupil no longer (though, of course, I still was very much his disciple) — and even more so when he realized that I was not going to accept the job offer he kindly procured for me and settle down in London (though he accepted my decision understandingly and even helped me very generously to get a job elsewhere). For me the cutoff was in spring 1962, when he graciously agreed to be guest examiner in the department of philosophy and psychology in the University of Hong Kong, where I was in charge of that department for a brief period. On that occasion I took him once for a car-ride around the breathtakingly beautiful island, and we eagerly engaged in a conversation. He then invited my comment on something he had just published and was naturally proud of. As it happened I strongly disliked it (I still do) and said so. In response he called me a liar. The incident was not important; his error (as I will explain later on) was understandable;

---

9 Einstein said (*Reporter*, 18 Nov., 1954),
"If I would be a young man again and had to decide how to make my living, I would not try to become a scientist or scholar or teacher. I would rather choose to be a plumber or a peddler in the hope to find that modest degree of independence still available under present circumstances."
Already at the age of 16 he spoke of "a certain independence in the scientific profession that greatly pleases me" [Dukas and Hoffmann, 1979, 13].

my outward reaction was as conciliatory as I knew how: if memory serves, I simply did not react at all, then or thereafter, and I never allowed the incident to dictate my conduct in any way (until now, of course). But inside me something snapped and I instantly went into a state of mourning that took many years to get over.

One of the few bright moments I spent with the philosopher some years afterwards, was during a visit to him that I made while passing through England. As we were strolling in his large garden on his lovely lawn, I said to him, I would not ask him to publish critical comments on any of my publications but would take it to be the greatest honor he could possibly bestow on me were he to do so. He liked my sentiment and said it was well put. I now think he meant he was not going to grant me the honor and was glad to learn I could live with this.[10] Likewise, naturally, he could live without my honoring him with my criticism. But it is different for me: perhaps he had no use for my criticism, but I (and my readers, hopefully) do: without it I simply could not write, since for decades I took his philosophy as my point of departure, and I often still do, depending on the situation and nothing else.

It is thus not surprising that he hardly mentioned me in his writings, and that his (almost obligatory) response to me in *The Philosophy of*

---

10 Does my dissent from Popper vacate my discipleship? With no measure of distance between ideas, it is generally moot. Berkeley and Hume saw their ideas as modifications of the philosophy of Locke. This is very nice. So is Nancy Cartwright's expression [Cartwright, 1989, 4] of her (misguided) intention to police science in the tradition that goes "from Francis Bacon and Joseph Glanvil to Karl Popper and the Vienna Circle" — even though as it happens Glanvil succumbed to Bacon-style policing only most reluctantly and Popper was a true liberal who insisted on the legitimacy of metaphysics no matter how poor. Cartwright's mention here of Popper and the "Vienna Circle" in one breath is fine. Not so Schlick's assertion that Popper's views resemble his own, since it is non-specific and meant as an insult. (He said, Popper's motive for distancing himself from the "Circle" was self-aggrandizement.) A few have repeated this verdict of Schlick — even Hempel (in his review of *Logik der Forschung*) — but no one discussed it except Popper, who in amazing generosity attributed it to the refusal of the members of the "Circle" to take seriously his defense of metaphysics. This is a scarcely credible story, but to my surprise I can corroborate it. When Philipp Frank passed once through London and invited Popper to see him in his hotel, Popper took the opportunity and introduced Bartley and me to him. He kindly took interest in our studies. Bartley reported that he was working on theology and science and I that I was working on metaphysics and science. He took this as a great joke and, assuring us that he was really interested, he asked us very nicely to give him our true answers. He refused to accept our assurances that we were serious. This amazed Bartley.

With no measure of similarity, at times commonsense helps adjudicate such matters. It should thus also suffice to dismiss Feyerabend's claim, "Popper's ideas bore a surprising similarity to the ideas of Wittgenstein" [Feyerabend, 1979, 203-4], [Feyerabend, 1991b, 50] and Lakatos' distasteful assertion that I follow Neurath and Hempel [Lakatos and Musgrave, 1970, 113n]. I still wince at this, and more so at Popper's 1959 preface to his *Logic of Scientific Discovery* that addresses the analytic school as fellow rationalists. His conciliatory mood is lovely, of course, but rationalists they never were. His former students, on the other hand, surely are; you would have expected him to acknowledge that much. He did not. Can I call myself his disciple nonetheless? I do not know, but I do so anyway. It is surprising that Feyerabend did not. His reason was strange. He once asked me, Joske, will you really not mind if you will not be the greatest philosopher on earth? I then called him Dr. Columbus and he signed his letters to me with the diminutive "Paulchen" written in Greek letters.

Karl Popper [Schilpp, 1974, 1114-7] is minimal[11]. In our last public encounter, in the discussion after a public lecture of his in the Austrian College, Alpbach, Tyrol, summer of 1978, he flatly and quite unpleasantly refused to respond to my (not important) critical comment. He was determined, as he said many times before and since, not to respond to criticisms from any of his former students.

The day after that embarrassing public encounter, late in the evening, we had a private meeting at his instigation. It was lengthy, tense, and grim. It also was our last. He hoped for reconciliation, it was clear, but since I never demanded anything of him, I had nothing to contribute to any negotiation he might have had in mind, and he had nothing to propose to me. He had a demand from me, or so it seems to me to my regret, namely, that I desist from publicly criticizing him; but if so, then he did not say so then. (He alternatively affirmed and denied that a few times.) In any case, I do not remember that he made any demand or any proposal then. The conversation had nowhere to go. He realized that. He probably realized that there and then; if not then, then the next morning when I approached the breakfast table he and Lady Popper were occupying at the hotel where they were staying and bade them a very friendly farewell. (He could scarcely hide his surprise and disappointment at my polite departure.) Yet on the previous night, just before our grim conversation was over, the philosopher made a request: just at the end of our very last conversation ever, he asked me to ensure that the embarrassment that had taken place in the public meeting a day earlier would not happen again. It was very easy for me to respond with an immediate consent — to his displayed surprise. He did not ask me to explain, or else I would have told him: his public conduct towards me on that day had made me decide that I should avoid meeting him publicly again, come what may (except at his express behest, of course). So I never made anything of that public encounter and my promise to him inhibited my writing freely about him (as a person, not as an author, of course). But something had changed.

Things began to intensify, I heard — he and I naturally regularly heard about each other, of course — and what I heard that he was saying about me depressed me somewhat, perhaps wrongly so. Once he said into a tape recorder[12] something rather unsavory about some little concession he once had made very long ago (before he broke off with me for the first time), concerning the possible validity of some of my criticism. (I do not know if the criticism is important enough to discuss, and I regret that due

---

11 In Popper's response to my view [Schilpp, 1974, 1117] he characterizes it as "ism-ism". Gilbert Ryle has coined this expression as pejorative [Ryle, 1937] and Richard Robinson echoed it [Robinson, 1954, 597]. Both viewed Popper's philosophy this way. He found this agreeable; when he responded to my note he used it somewhat differently and disapprovingly. He did not explain.
12 The tape belongs to Mark Notturno.

to that uninvited and unpleasant and irrelevant concession it is my best-known criticism of his output; so perhaps I shall discuss it later on. As an aide memoir let me mention that the concession is known among the experts by the funny name he gave it: the whiff of inductivism.) What Popper is on record as having said on that concession, taken literally, cannot be true, as it is a self-incrimination: he suggested that possibly his concession was not sincere — as he said that it was the mere outcome of my alleged wish (in later versions the wish became a demand) that he further my career by some such concession. Now, the philosopher told me many a time, and I think in truth, the following story: his colleagues, the members of the "Vienna Circle" in the twenties and the thirties, demanded a concession from him as a condition for inviting him to their company. Had he made any concession to them, had he merely found a way to praise Ludwig Wittgenstein, their mentor and patron saint, then their attitude to him would have been much friendlier. They made this very clear to him. Indeed, they were specific enough: they wanted from him in exchange for some recognition some statement to the effect that Wittgenstein had been an important philosopher. It is not only his testimony that I go by. The acknowledged leader and the theoretician of the "Vienna Circle", Moritz Schlick and Rudolf Carnap, accused him of having overemphasized differences. How is it possible that Karl R. Popper, a person who insisted on his utter intellectual integrity at such a very high cost, could compromise it merely because (allegedly or in truth) one of his young, unruly former students wanted him to compromise? This is strange; I do not know how to take it.

Things stay confusing and even worsen. The philosopher also said (a few times, I am told) that he abstains from criticizing the publications of his former students out of sheer kindliness: he wished them no harm.[13]

---

[13] Admittedly, the view that criticism is beneficial demands qualification, as it may have side effects. Although criticism is respectful, its exposure need not be: it can harm a candidate for office by public exposure of incompetence. In such cases, it is advisable to seek private critics before going public: private and public criticisms differ this way, having different targets. Criticism is respectful and beneficial to all, but in different ways. The best is to criticize worthy ideas, but even to criticize racism, sexism or other follies is respectful as the recognition of their potent social influence: to recognize any idea as influential is to acknowledge its strength in some sense. Finally, unfortunately it may be hard to separate the shabby from the admirable, since at times poor ideas appear as significant innovations and it may be hard and unpleasant to unmask them [Agassi, 1975, Ch. 3; 2003, 97].

More generally, as all discourse is context-dependent, one comment may have different effects in different contexts. Critics may succeed because they use powerful platforms or rely on superstition. An example is the notorious *Jensen Report* [Jensen, 1969] that is a criticism of the anti-racist reform of the American system of education. Its racism is a superstition and its statistics sham as all statistics that rest on superstition must be: he could get similar results with redheads (as Asimov noted) as the demarcation by color is conventional. The source of the public attention that it received is the powerful platform that has launched it, the overrated *Harvard Educational Review*. [Agassi, 1972].

Pseudo-criticism or fake criticism may be different. Some agreement may be put as criticism. Critics may distort the ideas that they criticize and even attribute to their targets ideas that they repeatedly repudiate. This is as related to criticism as any deceitful act to its honest model. One need not agree with a repudiation of an idea: when Buber ascribed to Jung ideas that Jung claimed to have re-

I take him to have meant this in earnest, but I refuse this kindness as I view any criticism as a mark of recognition. I do not demand it, but I certainly have no wish to see any criticism of my output suppressed. The philosopher was certainly kind enough not to wish anyone harm, certainly not his former students. I do not see this as a motive that should have prevented him from criticizing their works. Did Feyerabend's vicious attacks on him reduce his reputation any?[14] Did the refutation of his theory of approximations to the truth (we do not have to discuss it) harm him in any way? He knew that the truth goes the other way; he knew that were he to criticize my publications, validly or not, he would not have harmed me or my reputation: probably it would thereby increase. So what was going on? I do not know.

In the conspicuous case in which Popper did respond publicly to a former student — Bill Bartley, it was — his conduct was very distressing. It was in a highly publicized international symposium. He was expected to respond to Bartley's critical discussion of his view — a discussion which I, for one, found both interesting and expressed in an eminently respectful manner. He was invited, then, to parry the criticism. Instead, he offered a discussion of the question, whence Bartley's hostility to his kind, dedicated teacher?[15] The teacher's kindness and dedication are beyond doubt. But how could the great master of critical philosophy consider hostile any criticism of his ideas, whether valid or invalid, whether enlightening or obfuscating, whether to the point or merely side-tracking, whether in the service of the truth or self-serving, whether well-meant or ill-meant, whether by friend or by foe, how could he, of all people, how could he take any critic as hostile?

Bartley and I met after that disastrous meeting. He (and I) lamented the conduct of the philosopher: we would have much preferred to see him

---

jected, Buber argued back; he never concealed this claim of Jung. This is fine. The critically minded authors often mention possible criticisms of their views and answer them. At times some critics repeat these criticisms as their own and as unanswerable. Deception of this kind happened to Marx and to Popper.

14 Feyerabend's work is still interesting [Agassi, 2003, 351-4]. See [Farrell, 2000, abstract],
    "I ... trace the development of Feyerabend's philosophy in terms of a commitment to the central Popperian themes of criticism and critical explanatory progress. This commitment led Feyerabend to reject Popper's specific methodology in favour of a pluralistic methodology, but the commitment to the central values of criticism and critical explanatory progress remained. ... Feyerabend was not a disappointed Popperian, but, in many respects, a die-hard pluralistic Popperian."
Regrettably he also said, "the Popperian colony in Toronto" need explanations "three times in baby language" [Feyerabend, 1979, 88n], ascribed to me the proposal to leave education to experts [*ibid.*, 113n], and posed as a follower of Wittgenstein, rather than as a "radical critic of logical Empiricism" [Broad and Wade, 1982, 133].

15 The Bartley-Popper exchange appeared in [Lakatos and Musgrave, 1968], the proceedings of that 1965 symposium. Popper presents a complaint about Bartley [Popper, 1994a, xii-xiii] and a response to his criticism; most regrettably he did not mention the target of his comment. In an essay on Lorenzo Ghiberti, Gombrich discusses Ghiberti use of perspective, using the same evidence as Richard Krautheimer while ignoring his different view. (Later in that paper Gombrich refers to a paper of Krautheimer on a different matter, not his famous book on Ghiberti; all this is sadly distasteful.)

conduct his performance a bit more philosophically and perhaps also a bit more by the book.[16] But I do not remember that Bartley resented it or felt harmed by it; regret was what he (and I) felt. Bartley did indeed get hurt — not so much by our joint guide and mentor, as by Lakatos, who maliciously spread ugly rumors about him, and by John Watkins, Popper's heir and successor in the London School of Economics, who had spent so much time correcting the faulty English of my doctoral dissertation and who had published very nice things about me [Watkins, 1958], when (in January, 1967, as I was passing through England) he told me in all seriousness that I could not stay his friend and Bartley's too. He accused me then of having incited Bartley to attack our admired master. It was at a dinner in a posh restaurant, and I was surrounded by friends and colleagues — only to find myself subject to a kangaroo court that Lakatos presided over, but with Watkins taking the more active part of the proceedings. It ended in a stalemate: kangaroo courts are not such that the accused can ever be pronounced innocent: the accused can break down, denounce the self-appointed court, or play the whole thing down. My own determination was the latter: I wanted to contain the damage this session was causing, come what may. Kangaroo court proceedings are hard to bring to conclusion if the accused plays it down rather than breaks down or breaks it up. In my case the Kangaroo court fizzled out as we moved from highly emotional and gossipy exchanges to philosophy proper. We spent some time discussing the criticisms I had been directing at Popper, as Watkins was very earnest and wanted to judge for himself that I was not just raising dust. When we parted, it was already well past midnight: we had moved to a hotel bar but had to leave there too and anyhow he was too tired to continue. As we parted company we made a breakfast appointment for a few hours later for a continued philosophical conversation. When I turned up at the agreed time and place Lakatos was there too. Seeing me surprised at his uninvited presence, he gave me a triumphant grin. I got the hint and gave up: I left it to Watkins to decide whether to reopen the discussion that he had begun with me the previous night. In any case, it was he who had initiated it as he had wished to decide for himself how serious was my criticism of the ideas of our joint and beloved teacher.

He did not reopen the discussion — then or on any other occasion. He never took it up again.[17] Most of the time he was on closer terms with Bartley than I happened to be — or so I understand: I am not very familiar

---

16 Upholding the value of criticism, Shaw says [Shaw, 1965], "the difficulty is to distinguish between the critic and the criminal or lunatic ..." Popper viewed his critics among his former students as not serious, as having ulterior motives. His negative view of us troubled us less than his refusal to hear us criticize his judgment of our seriousness, his love of criticism notwithstanding.
17 Watkins' evasion of the issue was brief. He returned to it later on.

Prelude 39

with later personal developments. Afterwards Bartley found himself isolated. Eventually this got to him. When I once told the philosopher he had ruined Bartley, tears rolled down his cheeks. Fortunately, my assessment was erroneous: Bartley did recover — though the friendships were never fully recovered to be as free and easy and delightful as they once had been; as he died, in 1990, he was on reasonably good terms with the philosopher as well as with a number of associates. Lakatos died suddenly in 1974, estranged to the last from the peers with whom he had the strongest bondage and for whom he had the deepest love and respect.[18] His posthumous contribution to *The Philosophy of Karl Popper* of 1974 says of Popper that he had not contributed anything to the philosophy of science. Popper's response to him was very forceful and, presumably, it was one of the chief causes of the rapid decline of the status of Lakatos as a philosopher of science among the more informed: as usual his ignorance and extravagance were exposed, yet this time he was no longer there to dismiss the incident with a joke, as he usually did so well, and with so much success. I am glad to see, however, that in the meanwhile his reputation as a leading philosopher of mathematics has become reasonably secure.[19] I know he would have approved of this change, as he always knew that only his contributions to the philosophy of mathematics were of great value.[20] Soon four volumes on his alleged contributions to the philosophy of science appeared, which were proceedings of two conferences he had arranged in his own honor (one of which took place after his death).[21] Watkins appears there as a supporter of the philosophies of science of both Popper and Lakatos. Bartley's contribution to the volume in memory of Lakatos says that Lakatos had added nothing to this field, except for some catchy expressions. Watkins defended Lakatos' alleged philosophy of science not by answering Bartley or by restating Lakatos' contributions to the field, but by the use, in a volume devoted to Lakatos' alleged contributions, of the doubtful epithet "the London School of Economics position". My review of that volume [Agassi, 1988, 341-4] shows that there is no such thing: Watkins was constrained by the limitation to the possible loyalty to two masters and friends, even though he finally learned to be friendlier.

All is well that ends well. My review of Watkins's *Science and*

---

18 Ample material in Lakatos' archives in the LSE testifies to his *modus operandi*.
19 Popper deemed Lakatos' contribution to the philosophy of mathematics the greatest since Gödel's [Watkins, 1977a]. The contributions of Paul Cohen and of Abraham Robinson are no less significant.
20 [Davies and Hersh, 1980] have forcefully exposed the ideas of Lakatos on mathematics. So did others. By contrast, James Franklin says [Franklin, 2000], "Lakatos's project is fundamentally dishonest." Charges of dishonesty require justification. Someone should force Franklin to justify or withdraw.
21 The proceedings of Lakatos' two conferences are one volume edited by Latsis [Latsis, 1976], one by Howson [Howson, 1976] and two by Gerard Radnitzky and Gunnar Andersson [Radnitzky and Andersson, 1978 and 1979]. See Lionel Robbins' friendly review of the Latsis volume [Robbins, 1979]. See also [Agassi, 1988].

Skepticism [Agassi, 1988, 469-78] was friendly though in dissent. I made only one light-hearted comment — on a passage that, out of loyalty or in plain error, takes seriously a remark of Lakatos. Unfortunately, that book, which has somewhat patched up things between Watkins and myself, has cost him dearly: its publication lost him the friendship and support of our mutual teacher and mentor [Watkins, 1977a].

I should tell my story as I remember it and as I perceive it. Perhaps it is of little consequence. At my age — I have passed my eightieth birthday — I do not much mind minor tiffs that happened to me before I was thirty, as I do not see myself as a grumpy old fogy, perhaps due to some self-delusion. I am old enough to look fondly on what remains of my youthful memories and notice the calm with which I now remember these somewhat turbulent days. I hope the story is interesting enough to let readers and future commentators judge it as they wish: let the critics sharpen their pencils or put new floppy disks into their word processors.

CHAPTER TWO: ENTERING KARL POPPER'S FAMOUS SEMINAR

Sworn allegiance to no master,
wherever with her the wind me carries,
a visitor I travel.
(Horace)[1]

Let me begin with a brief self-portrait and some autobiographic observations. These should serve as background to my first encounter with my master, Professor Karl R. Popper. Whether you endorse them or not matters little (all self-portraits are inevitably self-serving): you may find them not boring and somewhat helpful in developing your own perspective on what comes later.

Right or wrong, perhaps even dead wrong, I happen to consider myself a good philosopher — even a very good one. I have been of that opinion even before I graduated (nearing the age of thirty) as I always considered myself an independent thinker. My self-esteem derived solely from my view, possibly equally dead wrong, that I was independent and that my independence itself had made me an unusual philosopher. Independent thinkers, regrettably, are still very scarce,[2] especially in philosophy. Yet I do not delude myself on one matter: I know I am not inspired: I am lucky to have certain gifts, some degree of wisdom, a broad, though rather superficial background-knowledge in many fields, and I have some reputation for glitter on account of my good synoptic view of things: I boast a sure sense of priorities and I care about the welfare of my readers, students, colleagues, collaborators, friends and kin, and I always seek the right encouraging word to say. But for all that, I am not inspired: I will never be a Quixote, at best a Sancho, a Boswell of sorts. An avid commentator, an anecdotist, I am a frustrated artist promoting intellectual freedom, while playing the Jewish-traditional-style maggid who, between

---

1 Horace, *Epistles*, I.i.14.
2 The sense of proportion requires noting that before the French Revolution both autonomy and literacy were for the privileged. The frequency of literacy and of autonomy in the population at large was low and of autonomy among the literate relatively high. The rise of popular education should count as progress although it lowered the relative frequency of autonomy: education for autonomy is still rare, but any education is better than none.
    Shaw said repeatedly (e.g., in *Fanny's First Play*) that the aristocracy and the working class but not the middle classes are autonomous. (He ignored the peasantry.) It seems he romanticized the workers (e.g., Preface to *Major Barbara*).
    Nietzsche derided heteronomy as "slave morality" and identified autonomy with nihilism (or allowed his readers to do so). In response to this Dostoevsky said, one can reject morality without gaining autonomy, describing a fan of Nietzsche this way (*Crime and Punishment*). The "Frankfurt" or "critical" school self-styled Marxism is a mask for their following Nietzsche in expressing contempt for mass culture in the greatest contrast to Marx's sympathy with them even while explaining their shortcomings. Nor is the "Frankfurt" or "critical" school quite Nietzschean: they bemoan alienation in urban society, paying Nietzsche lip service. Complaining about increased freedom, they display their lack of moral fiber. Popper rightly viewed their doctrine as expression of desperate escape from politics into the art world.

the end of the day and the beginning of the evening, entertains with the light side of scholarship the captive audience in the synagogue in the few minutes they pass idly till the sun set before the evening prayer may commence.

I do not know who enriched my life most, which way, how, or why: but I can tell stories about them, stories with a touch of a moral in them, perhaps, but hopefully more interesting than instructive, and hopefully more useful and conciliatory than demanding or provoking. I say hopefully: I know I am often quite unsuccessful: all too often I am told, much to my frank displeasure, that I am just a be-provocative-and-love-it sort of person. My response to those who tell me this is usually to request that they try to ignore me and my writings and my lectures and so on. No: it does not work: they come back at me and tell me in a great spectrum of ways that of course I am still very much a be-provocative-and-love-it sort of person: they are convinced that I just love to hear this repeatedly, that only matters of style prevent me from saying so out loud; why else would I be so provocative?

This is a good question: when it is meant literally, that is; usually it is meant rhetorically. My answer to the question is, I do not know how to avoid provoking people. Goodness knows how I envy those who avoid provoking people while doing things rightly and honestly. Or do they? I do not know.

I do not know who enriched my life most broadly, but I take it that I am not called upon to mention all those who have morally influenced me — I could mention them if I tell some stories about my early life[3] or if I come to tell of the role my wife Judith plays in my daily life and in my major decisions and in my thinking process and in my regular work: she has been a constant major influence over me in many ways, ever since I met her (long before I courted her). But I better skip this, since it appears here merely as background to my apprenticeship. Similarly let me skip the story of my intellectual progress prior to the start of my apprenticeship (when I was 24) — except for a few significant details, perhaps. These may surface later on.

I do not know who has enriched my life most deeply. I do know that I have changed my view on this question now and then. This does not matter here much, since my chief aim here is to tell that — my early life and my wife, Judith, aside — it was undoubtedly the philosopher, Professor Sir Karl Popper, who influenced my life in depth more than anyone else, and who influenced me intellectually more than anyone else: I will mention later on a few others to whom I am also in great debt, though to a

---

3 I have recently published a few words about my parents [Agassi, 2003, xxii-xxiv].

much lesser extent, especially Judith and Hillel Kook.[4]
I do not even know who influenced my intellectual makeup most, but I suppose that it was to a large extent formed in early childhood by close relatives, neighbors, school-teachers, rabbinical tutors, in some cases also nameless passers by, remembered with gratitude. Yet I know for sure that by the time I entered university I had no use for my rabbinical training and that I was a failure by then. Since adolescence my central interests had been philosophical, perhaps because of loss of faith — I had a fairly strict religious upbringing, though my education was never as narrow as required by late-medieval-East-European Jewish tradition that my grandparents shared and my parents did not. It still sounds strange to me to hear so often that it is easier to be irreligious than religious. I take this to be merely a lame explanation of the decline of established religion in the modern world, a lame explanation that is semi-official due to the deplorable defensiveness of its elders. At least my own personal experience was very different: my loss of faith came after a great struggle,[5] one that made me a philosopher just as I emerged from my rather ludicrous mid-adolescence-identity-crisis. (This is the case with many individuals, except that few choose philosophy as a vocation; otherwise, most people around would be philosophers: there is, after all, room for only a few philosophy professors in our society, which regrettably takes for granted the demand that only philosophy professors should be philosophers. Things were different in the Age of Reason.[6]) The Jewish-style love of learning remained with me, but I became alien to the subject matter of Jewish Law: only with efforts could I develop a broader and more sympathetic attitude to it [Agassi, 1974] or to anything else that I was taught before entering university; at first I deeply disliked it all. (I still am bad at doing

---

4 Hillel Kook, alias Peter Bergson, was once very famous as he organized in the United States during World War II public committees for the rescue of the Jews of Europe and for a Hebrew Republic of Palestine. He was the only member of Israel's Constituent Assembly who protested against its illegal self-dissolution. His advocacy of separation of church and state is still very unpopular in Israel. I have dedicated to his memory my *Liberal Nationalism for Israel* [Agassi, 1999].
5 George Orwell's *A Clergyman's Daughter* (1935) sums up her failure, saying, one day you wake up and realize that your faith is gone and you can do nothing about it. Her story is about her failed efforts to do that. She is then ready to seek better challenges.
6 In the Age of Reason, most intellectuals were philosophers engaged in science. This changed with the Reaction that followed the failure of the French Revolution. In Russia the change came later as Mikhail Lermontov indicates in his *A Hero of Our times* (1840) that describes an odd philosopher as a charming figure of fun. Obviously, Lermontov had a penchant, but no appreciation, for the Enlightenment.
 Having views in a wide variety of fields is unavoidable. The philosophy of life is more diverse than philosophy proper. Regrettably, only a few keep up active critical interest in it; fewer of these are philosophers, professional or amateur. It is still the monopoly of novelists and dramatists and teams of TV soap opera writers. A once popular text on English philosophy [Warnock, 1958] ends with a complaint about the public expectation from philosophy professors to transcend their narrow specialty. Freud too complained: every mother deems herself a psychologist ("On the Question of Lay Analysis"). Sidney Hook disliked Einstein's readiness to meddle in philosophy and in politics [Hook, 1987, 461]. Fortunately, Hook did not practice what he preached and efficiently defended democracy as an amateur politician.

sums; I do it a bit better in English than in my native Hebrew.)

Entering university I registered not for philosophy but for physics — under the influence of Sir Arthur Stanley Eddington's superb *The Nature of the Physical World* (1927): it made me realize that philosophy requires scientific training. Most of my classmates were high-school graduates in possession of a smattering of calculus and physics. Having spent my adolescence partly in theological school and partly working for a living and mostly in depression, I was unprepared: I could not do the simple exercises and regrettably I developed then a permanent distaste for computations coupled with an impatience for any sustained exercises — an impatience that I overcame only during my period of the intense work of my apprenticeship, as I shall soon narrate in detail. In my first year in the university, however, I had no one to consult, and I was left behind almost at once. I was poorly prepared for classes, and more so for student social life, including the intense courting of female students (they were then more scarce than now, especially in the exact sciences) and the even more intense political activity (that was the norm then, in the exciting and turbulent last days of the incompetent, treacherous British rule in Mandatory Palestine). I broke down.

I went into what is technically known as ambulatory or sub-clinical severe depression — hebephrenic, I suppose. Simple daily activities required tremendous efforts, exhausting my resources and leaving no energy for anything else. In particular, no matter how hard I tried, I could not study. Fortunately, I did not know what had hit me. What saved me was the reading of and about music and the study of psychology — by myself. (I do not know what gave me the wisdom to ignore urgent required studies and center instead on music, on some biographies of composers, and on Flugel's lovely, pluralist *A Hundred Years of Psychology* [Flugel, 1935, 1970].

Finally, I emerged with a strong political interest and read Marx and Engels and Lenin and even a bit of Stalin. This last should have triggered the alarm bell,[7] but I foolishly managed to postpone the alarm for a while, since I had meanwhile joined a minuscule political-educational group, mainly students, that was busy helping bring about the world-revolution and even attempting to cure the rabid anti-Semitism of the Soviet Communist Party no less. It also made sincere efforts to overcome the Soviet hostility to the Jewish settlement in Palestine. (Soviet official

---

[7] Reading Stalin's "Dizzy from Success" should have made me reject Marxism. He blames contemptuously nameless party-functionaries for their alleged over-enthusiastic implementation of his plan to control agriculture; without admitting failure, he claimed that its cause was this over-enthusiasm. (He gave no explanation and no example.) Surprisingly, some scholars still praise his policies, although he caused famine that killed vast populations and led to popular refusal to fight during the early stages of the Nazi occupation — until its limitless cruelty raised public wrath (Mikhail Sholokhov, "The Science of Hatred", 1942).

propaganda portrayed the Zionists as imperialist colonialists rather than as idealist settlers who preferred to avoid exploitation of the local population. The exploitation and worse came much later, decades after Independence.) Unlike other left-wing groups, the one I joined was not Zionist yet it was brave enough to declare open and emphatic dissent from the official Communist hard-line hostility to nationalism, including Zionism. This forced the group to dissolve: one cannot be a Marxist for long without losing one's autonomy.[8] The group's attempts to modify attitudes imposed by Stalin proved, indeed, that it was naïve to the utmost. The possibility that we were naïve did not occur to me: I was charmed by the openness, frankness, and sincerity of the group's young members, and shocked by the discovery that these qualities were not fully shared by its older members.[9]

My inspiration was the group's intellectual leader, Shmuel Ettinger, who later became a distinguished professor of Jewish history in Jerusalem and a foremost authority on the vast history of anti-Judaism (anti-Semitism included). He greatly influenced me, while secretly agonizing over his loss of faith in Marx. Soon Judith and I lost faith too — and we took active part in the disbanding of the group. (We had the option of joining other groups, also to enter professional politics. But, we realized, this would require compromising our intellectual integrity[10] as it would force us to advocate political views that we were not ready to advocate as they stood, though we could do so with some measure of *ad hoc* adjustments.[11] The detestation of *ad hoc* measures has stayed with me and has

---

8 Those who cling to party discipline after having lost faith often appeal to loyalty and denounce frank defection. The appeal to loyalty is redundant when it rests on proper moral grounds and wrong otherwise [Agassi, 1974b].
9 Clingers often excuse their weakness by clinging to the view that struggle for reform must come from within.
10 This is not to speak against political compromise but against the demand to lie that is strong. The assertion that politicians must lie is false, but its refutations are regrettably all too scarce.
11 One reason I was happy to write about Faraday is his distaste for *ad hoc* hypotheses. Concerning weaknesses of an otherwise satisfactory hypothesis he said ("On the Conservation of Force" quoted in [Agassi, 1971, 159-60]),
   "the place of deficiency or opposition should be marked as the most important for examination; for there lies the hope of a discovery of new laws ... The deficiency should never be accepted as satisfactory, but be remembered and used as a stimulant for further inquiry."
The popularity of the recommendation to employ *ad hoc* hypotheses rests on the empirical argument that many researchers do so. David Bohm parried this argument, observing that the practice is reasonably limited: peers of defensive researchers grant them periods of grace. Sticking to old ideas without making their *ad hoc* hypotheses attractive, they push themselves into isolation. Bohm's observation is empirical: *ad hoc* hypotheses endure only when they gain corroboration or follow from new valuable theories. Bohm's observation is a refutation of current sociology of science.
   Duhem and Poincaré legitimized the ancient instrumentalist license to use *ad hoc* hypotheses in order to rescue good scientific systems. Einstein offered an alternative: he rejected the demand to forget all past errors, observing that they remain as approximations [Agassi, 1990]. He made light of the proscription of *ad hoc* measures despite his aversion to them. He felt unease about his *ad hoc* addition of his λ (lambda) to his gravitational equations to account for the expansion of the universe and was greatly relieved to learn from Alexander Friedman that he could dispense with it.

prevented me from ever seeing the allure of the philosophy of Pierre Duhem and Henri Poincaré, all their strong arguments in its favor notwithstanding, not to mention Thomas S. Kuhn, Imre Lakatos, and the cohorts of their fans who approve of defensive *ad hoc* measures *en bloc*.)

This blanket approval is vulgar. Its popular appeal is due to its tolerance of clinging to poor ideas.[12] Being no temptation to me, it still is a challenge for those who do not disregard the strongest argument in its favor: it tallies with the ubiquity of such measures in science. I knew all along that the vulgar are defensive on behalf of science, though it is the most powerful ideology ever. (Ironically, the ideology conflicts with the partiality towards *ad hoc* measures, as this partiality is defensive whereas the ideology of science is critical, as I learned later from Popper.) And I knew all along that they looked for a method of argumentation that will render the adherence to science obligatory (which, as I learned later from Whewell and Duhem, is quite pointless). All this is insufficient: if the scientific spirit is so averse to defensive conduct, why is the conduct that many scientists display so very defensive? I do not know, but I am still working on this problem.[13]

My brief encounter with intense political activity together with my brief military service distracted me significantly from my studies: if I ever had any prospects of becoming a decent physicist, these factors wiped them out, especially since I was handicapped from the start.[14] Perhaps I was simply not cut out for science. Ettinger was always concerned with my studies and even taught me what he could, including philosophy, in between political meetings: only later did I notice that he had taught me more than my professors of philosophy, though he did so in an undisciplined way at a time when I needed most to develop strict discipline. This task was left to the philosopher, as it happened.

The problem of discipline had an ideological aspect to it: discipline in science is a favorite of positivism (I will soon tell all about it, perhaps more than the reader wants to know) and I always had a strong distaste

---

12 Lakatos first opposed all *ad hoc* measures and then he advocated them all. Once he had amassed power, he promised me, he would right this wrong and more. One never amasses enough power, especially if one dies young — as, regrettably, he did.
13 Public relations demand support for *ad hoc* hypotheses, forcing researchers to be publicly defensive even as they privately try to do better [Agassi, 2003, 152-3]. The beautiful starting point of Lakatos was his effort to make sense of the up-to-date handbook's barely comprehensible, tortured definition of the polyhedron by searching for the counterexamples excluded by qualifications added in the course of the centuries to the obvious, initial definition. Beginning with the Descartes-Euler version of the theorem, he reconstructed brilliantly a series of versions plus simple counterexamples to disprove each. This is his splendid *Proofs and Refutations* that leads dialectically stepwise from the first conjecture to algebraic topology (Poincaré). He called this process concept-formation although it is concept-revision. To go beyond Poincaré, he needed a different approach. His search for it petered out, yielding first to his politicking and then to his early demise.
14 I suppose this reference to distractions is a mere excuse on my part: my classmates suffered them too, yet most of them became decent physicists.

for positivism. Errors often gain confirmation from fortuitous facts, and my erroneous attribution of a pro-discipline attitude to positivism has gained such confirmation: my science teachers supported positivism and discipline, some of them with a vengeance,[15] and my acquaintances opposed positivism and discipline out of (silly yet popular) Romantic prejudices.

Ettinger's reputation as a scholar is that of a meticulous pedant: not so his reputation when I joined his political activities: his brilliance served then as an argument against the "bourgeois" conception of scholarship. I was confused then; perhaps I still am. On the one hand I still think the ideas that most academics preach concerning scholarship are too scholastic: they encourage heaping a series of small *ad hoc* modifications to some unspecified theories in efforts to circumvent some unspecified criticism: this makes much of the standard curriculum unnecessarily opaque and its significance obscure; also, academics often use scholarship shamelessly as a mere means of intimidation for the sake of keeping the academy a closed guild. On the other hand I always admired the mathematical proficiency of true-blue scientists, the golden hands of precise technicians, and the artistic proficiency of veritable artists. I needed competence and I did not know how to acquire it. I could not even identify it. I had the choice between an artistic career (in music) and a scientific career (in physics) and I viewed myself disqualified as one who had no technical proficiency and no patience needed to acquire some. Ettinger taught me how to write some political propaganda, but he rightly found this increasingly distasteful and, anyway, he was always limited in his ability to express himself in writing, perhaps because he did not need it, as he could be satisfied with his charismatic presence and great verbal facility in conversation, in lecturing, and in speech-making alike. I admired him and was unable to imagine that I could ever acquire a skill he did not possess. I felt lost.

My fortune in having met the philosopher was my lucky break: he was the first anti-positivist I encountered who saw himself as a proud artisan and who valued competence; and he patiently taught me to improve my reading and my writing and my ability to assess a problem-situation in any field of study and, above all, he taught me how to persevere. He was an incredibly exacting and eccentric master, and I do not know if I could have taken it from him despite his generosity and concern, were he not such a staunch critic of the standard tough-and-no-nonsense attitude

---

15 I noticed exceptions, of course: the famous Abraham Halevi Fraenkel and applied mathematician Menahem Max Schiffer. Both spoke against the nuisance of excessive precision. Colleagues aptly described Schiffer (*Stanford Report*, 13 February 2002) as
"an outstanding mathematical stylist, always writing, by his own testimony, with the reader in mind. He sought always to convey the joy of discovery and the deep satisfaction in the unity of the subject."

of the positivists (now known as "the shut up and calculate approach"), a stern opponent of their promotion of competence-for-its-own-sake and of their precision-for-its-own-sake, who was justly proud of his abilities to outdo them in these very respects, and who tried to pass on his abilities to all of his apprentices, enjoining them to use their skills for good causes of their choice.

Positivism is a (rationalized) distaste for philosophy. What (if anything) is wrong with positivism? This question has engaged me for most of my research activity. I could not guess this then: it is not and was not to my liking to work on things I dislike. My dislike for positivism was both a poor excuse for my poor training and a healthy immediate response: after all, why hate philosophy? Why hate? No doubt, ignoring philosophy is acceptable. (In my youth I considered this, too, a kind of barely excusable superficiality: perhaps because of my religious yearning transformed to a philosophical quest, perhaps because of my mere mixing of philosophical quest and religious yearning. Though this is a barely charted territory, I was here sharing the idea that religion answers philosophical-metaphysical questions. This idea is fairly popular among all sorts of students of religion. In any case, as an apprentice to a friend of so many positivists I learned to compromise that much with positivism: I no longer consider thinkers who shun philosophy to be necessarily superficial.) Still, what can possibly make philosophy so offensive? Amazingly, Eddington had judged his own views of the universe purely private because they were philosophical and thus metaphysical. What makes speculating about the nature of things so unfit for public consumption that Eddington viewed it a private vice? What makes them so distasteful, that my physics teachers should have hated it so? And why do they find doubts about the validation of science so frustrating?

I found the answer[16] to this question only after I graduated, when I was working on my doctoral dissertation: traditional positivism condemns all speculations, and traditional metaphysics is speculative. This is why Francis Bacon asserted that science will culminate in a grand metaphysical system: he opposed only unscientific metaphysics, as he considered it a competitor and thus an impediment to scientific research. Research that starts with speculations, he said, goes astray; proper research starts with observations and ends with metaphysics, having no use for speculation.

---

16 Popper observed early that the usefulness of some metaphysical speculations refutes "logical" positivism [*LScD*, conclusion]. In response, Carnap said, he opposed only the metaphysics that impedes science. In this he confused "logical" with traditional positivism. He remained famous as the apostle of clarity nonetheless.

Traditional positivism is the better and the wiser of the two. The possible goodness of some metaphysical speculations does not refute it. Bacon said, as adherence to error is so dangerous, it is more prudent to start afresh despite the loss that this move might incur: speculations may accelerate progress, but renounce them and you will gain much more and even quickly retrieve all losses. (All proselytizers share this advice.)

The view of science as speculations put to empirical tests invalidates this argument against metaphysics. (The then popular "logical" positivist movement condemned metaphysics for all eternity on the principle that the speculative-in-principle has no meaning and on the supposition that metaphysics is speculative-in-principle. They did not mind that this idea is speculative-in-principle, expressing unjust, unbecoming hostility.)[17]

Positivism enjoyed immense popularity after World War II, in centers of learning and in the Jerusalem periphery.[18] My impatience with it clinched my isolation there: hardly any of my professors there influenced me at all. (I will tell you about the exceptions soon.) I then learned that Einstein was an anti-positivist, open-minded and conspicuously kind and tolerant of poor performance by students: at that time a leading physics journal dedicated an issue to his 70th birthday and I read there in a pen-portrait by Philipp Frank [Frank, 1949, 350] that he was tolerant of one particular student[19] whom everyone else considered a pest on account of his habit of asking awkward questions. I identified with that person instantly: it was for me a source of immense consolation. I felt that if only I could meet Einstein in person and ask him to be my teacher, then I would regain hope and my life would acquire some meaning. I did not have the means to travel to the United States. When I finally graduated somehow, I was lucky to scrape up barely enough money to get to England.[20] When the United States finally became accessible to me, he was dead. (I considered his death a severe personal loss.) I never had the fortune to see him, and it is a great regret of my life. Thanks to the philosopher I had the good fortune of meeting the admirable Erwin Schrödinger, though, and

---

17 The "Vienna Circle" had no theory of meaning to speak of [Quine, 1977]. Hence, it never had reason, let alone proof, for its dismissal of metaphysics, much less for its (traditional) demand to dismiss the unproven. Alas, even their opponents took them seriously.
18 My isolation in my student days was enhanced by the prevalence then of positivism, operationalism, behaviorism and similar popular debris of Enlightenment philosophy that never attracted me despite their noble ancestry. As I wrote my dissertation, I read Laplace and began to see what in positivism attracted my physics professors. This was a relief, but nothing like the relief I felt when I read Bacon. Positivism then began to make sense to me. As to "logical" positivism, I knew nothing of it, although I met Bar-Hillel in my student days and even read his doctoral dissertation, as I could not distinguish between it and traditional positivism. Even reading Russell's *Principles* did not alert me, and reading Carnap made no sense to me. I had to master Wittgenstein's *Tractatus Logico-Philosophicus* to begin to understand "logical" positivism. I never found any value in it and read it because the philosopher took my familiarity with it for granted. I still have no idea how Wittgenstein impressed some of the best logicians of his day. But then this is not the worst. Otto Weininger's immeasurably worse *Sex and Character*, 1907, impressed many intelligent people who should have known better.
19 This is an error on my part. It was not a student, but "a certain physicist"; and Einstein liked not the awkward questions but the disposition to delve ever deeper into problems, although this leads more to frustration than to success. Frank reports that Einstein admired that physicist and thought poorly of those who preferred minor challenges. He cites then Einstein's obituary on Mach to say, the ablest researchers are concerned with the theory of knowledge. (Einstein generously overlooked Mach's shortcomings.) This view is the exact contrary to the instrumentalism that my physics teachers used to justify their preference for minor challenges.
20 As Judith's promised grant was sabotaged, we were lucky to get a family loan that enabled us to go to England for one year and to take it from there.

that made me very happy. Of philosophers I wished to meet only Bertrand Russell (as I knew nothing of Popper until the time I met him). I met Russell for an instant, when he came to talk to the graduate students' society in the London School of Economics about nuclear disarmament (these were his Campaign for Nuclear Disarmament days). He was just terrific, but he was rushed off to meet important individuals and then to the (so-so) lecture before a packed hall and then away.

Let me tell you how much Einstein meant to me. My studies included some specialization, and I chose to study his general theory of relativity, as it was the most coherent piece of contemporary physics, and anti-positivist to boot.[21] I read as much Einstein as I could. His "Replies to My Critics" (in the celebrated *Albert Einstein: Philosopher-Scientist* [Schilpp, 1949]) granted me three or four lasting impressions. He says [Schilpp, 1949, 682], quantum theory is not a matter of the smallness of size, but of the smallness of Planck's constant. Thus, very slow billiard balls will behave in a wave-like fashion and exhibit a quantum interference pattern when crossing a suitable grid. Years later I forgot that I had read this. (The leading sociologist of science, Robert K. Merton, says [Merton, 1976, 448], this phenomenon is so common, it deserves a name: cryptomnesia[22].) I wrote a brief paper elaborating on it. The paper was rejected as silly, once on the recommendation of a referee with very high scientific credentials.[23] When it finally was in print [Agassi, 1963b], it raised a sneer. My slip in not having acknowledged Einstein's idea to him thus offered me an insight into the immense conformism that holds back even very competent physicists. In the same "Replies" Einstein says, to his regret he has to deny the charge that he has a [quasi-]classical theory: it is only a program for one (the very idea that the laws of physics are all expressible as partial differential equations invariant to broad sets of

---

21 According to general relativity finding the local metric and local measurements of distances are interdependent: knowledge of the one is necessary for finding the other and *vice versa*. A few writers view this as a serious problem and even a paradox [McVittie, 1956]. As interdependence is common in physics, and met with bootstrap operations [Agassi, 1975, 169, 172], I failed to comprehend the difficulty until I read Einstein's dismissal of it as resting on the positivist demand for testability for each assertion (see below). This was his cavalier dismissal of both the operationalism of Bridgman and the verificationism of the "Vienna Circle". Bridgman then weakened his operationalism by allowing what he called paper-and-pencil operations. He then required of them testability. See his posthumous contribution to the Popper *Festschrift* [Bunge, 1964].

22 "Cryptomnesia" is a term that Carl Gustav Jung had used to denote mock-memories that happen to be true. Merton's term denotes unconscious borrowing of ideas, ideas assimilated, forgotten, and rediscovered. In a sense it is ubiquitous: rereading a good old text may be an impressive experience that stirs a sense of gratitude. (See Russell's *History*, chapter on *The City of God*.) But as the scientific tradition does not require acknowledging general ideas, these involve no theft and so no cryptomnesia.

23 The anonymity of readers for *Nature* prevented direct contact. The editors informed me that the reader of my paper was an eminent physicist. I asked them to pass my reply to that reader — who kindly answered me in detail. The new reply wrongly posed as no more than an elaboration on the older one. The editors understandably put an end to the correspondence. Before I learned that Einstein is the source of my idea, a rejection could make me put it aside; the two comments of that reader and the disparity between them made me decide to submit it again elsewhere.

transformations [Schilpp, 1949, 675]). This led me to develop in my doctoral dissertation (again with no acknowledgement, alas, but this time less inexcusably) the view that metaphysics may generate research programs, fruitful on some blessed occasions.[24] Years later, to my horror, Lakatos managed to engage Popper in a priority dispute[25] regarding the invention of this idea — which is really quite traditional. There was some novelty in the critical way in which I worked it out, and indeed the source of this novelty is in the writings of Einstein and of Popper. The conclusion of Einstein's "Replies" says, one quarrels only with brothers and good friends; strangers are too alien for that. Somewhere in that essay, full to the brim with innovations, Einstein repeated the commonplace statement, a theory is scientific only if it is verifiable; and he added to this platitude a most remarkable parenthetic aside, appending to the word "verifiable" the expression "(viz. refutable)" [Schilpp, 1949, 676, line 3]; see also [Schilpp, 1949, 666, last line]).[26] This is half of the revolutionary critical philosophy of science of Karl Popper put in one brief aside. I noticed it and was deeply moved by it: when I met it again in the original version in the works of Karl Popper while he was my guide and mentor, it was like meeting an old friend. Decades later, Robert S. Cohen gave an interesting presidential address to the international Philosophy of Science Association. His topic was Einstein and the [professional] philosophers, and he

---

24 Einstein and Russell have impressed me quite generally and more greatly than I can possibly say and in ways that I can hardly describe. I take comfort in the fact that this has happened to Popper, too. He could say little about his early years and what he read then that shaped him as he was. I heard that he once acknowledged in a conversation with a Marxist the profound influence that Karl Marx had had a on him. I also heard that once he acknowledged in a conversation with a neo-Kantian the same about Hermann Cohen. In his autobiography he refers to Otto Selz as an influence, but only in psychology, a field that he deserted. (He hated a book [Berkson and Wettersten, 1984] that discusses the influence of his psychology on his philosophy.) He indicated to me that Samuel Butler was such an influence too. I think he learned of Butler from Shaw, and I do not know when he learned about Shaw, so I cannot say how much he was in their debt, except that it was significant. He read them avidly and he discussed them with me on quite a few occasions (never as a literary critic, though: he limited his literary discussions with me to the German classics).
   Our greatest debt is to tradition as such (see "Towards a Rational Theory of Tradition" [Popper, 1963]). Example: Whewell greatly influenced Oswald Külpe, the teacher of Bühler, the teacher of Popper. His influence came through tradition: Popper read him only in his fifties.
25 The brash, hilarious claim for priority that Lakatos made for the hackneyed idea that [at times] metaphysics serves science, displays his incredible ignorance of the history of philosophy. (Regrettably, Popper responded, proving his independence of Lakatos. Regrettably, Bartley did so too [Popper, 1982, 32n].) To confuse the heathens, Lakatos introduced three versions of Popper's ideas, and he did this in three different ways, so that his aficionados now have their hands full sorting him out. The long and the short of it is that he defended ad hoc blocking of refutations; earlier he had explained that it is foolish, but by then he was in search of popularity. (He also foolishly deemed common the very rare case in which the refutation of a refutation is rehabilitation. A commentator deemed this an endorsement of Hegel's rule of negation of negation, whatever exactly this rule is.)
26 It took Einstein no time to learn Popper's criticism of the "Vienna Circle". He probably viewed it as commonsense and dismissed their "basic principle (meaning = verifiability)" off-hand, observing that testability is required of theories, not of single assertions. "It is hardly necessary for me to enter this ticklish problem" of formulating precisely the idea of what makes a set of statements a theory, he added, "inasmuch as it is not likely that there exists any essential difference of opinion at this point" [Einstein, 1949, 678-9]. This is a powerful if gentle dismissal.

noted the agreement between the two giants but missed this little treasure. His audience, incidentally, left the lecture hall in a state of shock. Evidently they soon got over it: their journal continues to offer the mixture as before.

Nor is this all. In a famous passage there Einstein said, researchers are opportunists: they try every conceivable option from every conceivable viewpoint. This is a supreme expression of freedom.[27] Einstein had practiced this idea in 1905, when he offered two theories of light, one as corpuscles and one as waves. I used it in my "Sensationalism" [Agassi, 1975] and elsewhere. It disposes of a problem that beset Robert Boyle and Karl Popper: how should researchers choose between theory and observation report when they conflict? They said, researchers should prefer the observation over the theory that it conflicts with. (Popper wisely added a proviso: unless and until the observation is refuted.) They found it hard to stomach their own demand, since the observations are generalized and so their status is that of hypotheses. Duhem and Poincaré said, and Kuhn and Lakatos echoed them, researchers should resolve the conflict (and rescue the theory) by *ad hoc* measures. I said, researchers may try any option — of course tentatively, perhaps also playfully.

Let me also tell you how much Russell meant to me. The first work of his that I read was his *The Conquest of Happiness*. A number of respected professors of philosophy have told me since that this book is a potboiler (true) containing slight pep-talk (true) and nothing else (false), so that no serious philosopher can take it seriously (hilarious). This qualifies me as a non-serious philosopher. The book is doubtless readable as mere pep talk, but all-the-same it is a profound contribution to moral philosophy. Its advice to lose one's own problems in attempts to help others, especially others whose problems are more real, made a great difference to me in my lonely and miserable adolescence. It was a great thrill for me to learn that its author was a philosopher, and it encouraged me to study logic though I had had no occasion to take a course in it or otherwise study it with a teacher. I read Russell's famous but neglected[28] *Principles of Mathematics*, not knowing that it was unpopular because it had been left uncompleted to be replaced by Whitehead's and Russell's classic *Principia Mathematica*. I was lucky that the available part of the university library had only the first edition of the *Principles* (since most of its treasures were in its campus that was under siege then), as the preface to

---

27 Alfred Landé declared something of a scandal the prevalence of instrumentalism among young physicists [Landé, 1973, Preface]. Disagreeable as instrumentalism is, however, it is often advocated because of its freedom from inductivism, the freedom to "let the imagination go" (as Faraday has put it). It is regrettable that its advocates find the need for a justification for freedom, even more so that they find it in instrumentalism. All the same, their love of freedom is admirable.

28 The neglect of Russell's *Principles* may be a thing of the past: there is currently an increased interest in the literature on the rise of modern logic.

the second edition clarifies the situation, and it would have sent me to the *Principia*: I could somehow struggle through the older book but not through the newer one — not without help, I mean.

After having read Russell (without having understood him well enough, of course), I asked for advice. I was told to read Carnap — by people who had not studied him but knew of his very high prestige. I spent many hours poring over Carnap's diverse books and was surprised that anyone could take seriously such presumptuous, anti-intellectual, and apologetic material. Only later, during my apprenticeship, did I realize that I should have paid no more attention to public assessments of philosophical idols than to artistic, cultural, religious or political ones. (I will later narrate the single event that cured me somewhat of my naïve view of all "real" philosophers as progressive.)

Carnap did have sincere and informed followers, of course. His most enthusiastic disciple ever was Yehoshua Bar-Hillel, of the Fraenkel and Bar-Hillel justly celebrated text on abstract set theory [Fraenkel and Bar-Hillel, 1958]. He stopped me in the street somewhere in Jerusalem one day. I was stunned by his having heard of me, though I was a mere student, and a poor one at that, while he was my senior by more than a decade and then one of the very few Israeli philosophers of international repute. (At the time Buber was already very famous, but he was not considered a philosopher. I concurred then.) On that chance encounter Bar-Hillel spoke to me with the enthusiasm characteristic of one covertly eager to proselytize, which I easily recognized from my political days. I was amused by his ability to buttonhole me, glare with excitement, and say that positivism is right, that there is nothing in philosophy to get excited about. (Among philosophers this is known as a pragmatic paradox.[29]) Though we seldom met, we became friends at once and remained so until the day he died. It was an eerie feeling to play host to him in London, as we were two Israelis continuing a senseless feud that had begun in the Vienna of another era: he was then defending Carnap against Popper's onslaught. The onslaught succeeded, but it took two to three decades before the Establishment conceded that Popper had a point, and he was bitter about it — understandably yet regrettably.

Bar-Hillel could not but be friendly even though the philosopher's response to his attacks was (fittingly) scathing. When he passed through London, he wanted to visit the philosopher and pay him homage. The phi-

---

29 The term "pragmatic paradox" is the invention of G. E. Moore. His example is "p and I do not believe p", where "p" is shorthand for any proposition ("Today is Tuesday and I do not believe that today is Tuesday"). It does not seem to me paradoxical at all, as it follows from the following: "as the Bible tells us, p; but I cannot bring myself to believe it" where "p" is shorthand for some incredible biblical factual report. Norman Malcolm made the concept popular [Egidi, 1955, 195-205]. Its paradigm case is, "I am asleep now". It has a true instance, though: somnambulist actor Jerry van Dyke taught himself to say it during surprise visits that he made in his pajamas and that understandably perplexed his hosts.

losopher could not face him and asked me to play host instead. I was truly surprised to learn on that occasion of a variance between Bar-Hillel's written and spoken expressions: in conversation he expressed much agreement with the philosopher, but not in his published attacks. This variance bespeaks bad faith. I cannot possibly ascribe any bad faith to Bar-Hillel personally, but bad faith clearly was there; the result was that most of his published work is simply worthless, famous at the time but now all but forgotten. I find his pioneering work on machine translation a valuable morsel and think it should be studied.[30] He was one of the friendliest and most honest and moral people I ever met, but we had little in common and so we spent little time together until he was utterly disappointed in Carnap. It was, he told me in a moving conversation just before he died, only after Carnap's death that he could assess the worth of Carnap's output, and he declared it a bankruptcy. He was deeply disappointed in himself, then, and he thought his health was then deteriorating rapidly because of his loss of self-esteem. I did not know what to say: he moved me greatly although, of course, he had learned already in our first encounter that I had no good word for Carnap's output regardless of his great repute.

I should say something about the paradoxical claim made here, that bad faith can be impersonal. Whether this paradox can be explained away or not, it is real. It is not just my prejudice in favor of Bar-Hillel. I find the same bad faith in the writings of Carnap. I always regarded him as a very decent person although I never met him. (He bravely declared himself a socialist during the high tide of McCarthyism, when it was very dangerous to do so.) Moreover, it is the same bad faith as I found in Hempel, as I will narrate and explain later on, and Hempel is reputed — probably rightly — as the nicest individual in the western academic world.[31] The paradoxical bad faith is an almost inevitable outcome of two general facts. First: a statement pulled out of context and translated into a different context is bound to undergo distortion. Second: "logical" positivists refuse to recognize any philosophical context, viewing it indescribable (Wittgenstein) or nonexistent (Carnap). They thus demanded that every sentence be judged on its own.[32] And so they could not acknowledge that they were engaged in translation, much less that their translation

---

30 [Bar-Hillel, 1960], [Bar-Hillel, 1964, Chapter 10, The state of machine translation in 1951; Chapter 11, Aims and methods of machine translation]. Happily, his work on this matter enjoys a renewed interest [Hutchins, 2000, 299-312]. He is "the logician who might be called the theoretical father of machine translation" [Lehmann, 1982].
31 See Chapter Five for Hempel's attitude and Chapter Ten for its background.
32 Positivism and "logical" positivism share hostility to non-science; the "logical" version is the pretence that it rests on a theory of meaning (that of young Wittgenstein or that of young Carnap). This pretense survived as Otto Neurath, the chief propagandist of the "Vienna Circle", fended off all criticism. Arne Naess said, Neurath repeatedly belittled new criticism, suggesting that it was already answered [Naess, 1968, 13, n13]. Popper agreed [*Autobiography*, n114].

was distortion. To some measure this is always so. This is a profound general thesis, known as "Quine's radical untranslatability thesis"; discussion of it here is superfluous.

Let me also explain to you my hardly complimentary attitude to most philosophers of the previous generation and of my own. Suppose you meet a physicist who does not know twentieth-century physics, or an economist whose knowledge is likewise limited. What will you think of such a person? My reaction is always to try to find some explanation. If the explanation is trivial, say, the persons in question are ignoramuses with no idea as to how ignorant they are, I try not to discuss with them matters from their own fields of expertise. You may think those cases are exceptional. And you will be quite right. So let me take a field where this is the rule rather than the exception. Suppose you meet a person concerned with educational reform who has not heard of Homer Lane, Bertrand Russell, Janusz Korczak, Anton Makarenko, A. S. Neill and Carl Rogers, of their demand to democratize education, of what that demand means and of how it can possibly be met. In my experience this is the rule. One learns to live with it, and even with the fact (of which I have first-hand experience) that ignored ideas are barred from most of the relevant scholarly literature by the process of peer review or even by editors' frank censorship (say, of *Harvard Educational Review* or *The Journal of the History of Medicine and Allied Sciences*, from whom I have letters to that effect). For those not professionally engaged in the field in question, it is rather easy to avoid the company of deliberately ignorant practitioners. It is difficult to do so when the deliberately ignorant are in one's own department.

Many if not most philosophy professors these days do not even know elementary modern logic.[33] The rest are only slightly better equipped. Until recently most of the better trained philosophers I had met knew nothing of Popper's ideas[34] or knew only some garbled versions of them. Worst of all, when I disputed their wordings of his views they could not see the difference between their versions of his views and mine. (I will come to all this later on in some detail.) I confess it is hard for me to learn from such individuals — although I know of some very notable exceptions, mostly people who are active in those areas of philosophy which Popper left unstudied or concerning which he had some blind

---

33 Ignorance of logic is more understandable than ignorance of the literature on the democratization of education, as even some able scholars cannot master elementary logic.
34 David Hollinger says [Hollinger, 1996, 169], Popper "did most of his work in the era of World War II but suddenly became part of the action in the early 1960s". This sudden exposure was the Lakatos 1965 conference, of which more later. Hollinger refers here to Kuhn's works and mine. This is incorrect: the best-known reference to Popper till 1970 was that of Carnap [Carnap, 1936], which is the distortion discussed here.

spot[35].

Two experiences stand out in my memory concerning my failure to make any peace with positivism: one as a student, one as a colleague. When I studied at the London School of Economics, I naturally looked around in other places for different teachers. I visited the seminar of A. J. Ayer in University College up the road. I liked him tremendously. I was immensely complimented when, on some later occasions, he remembered me and we had a friendly conversation.[36] But in his seminar[37] I found him discussing very uninteresting and unimportant matters. In these discussions he showed ignorance of all the technical details somehow involved in his discourse. (In his autobiography Sir Solly Zuckerman suggests that Ayer's ignorance of the natural sciences was deliberate if not even feigned.[38]) I left soon enough. Years later a positivist colleague who was a neighbor was unable to comprehend Popper's ideas — from dogmatism, not from ineptness, though ultimately the two amount to the same thing. He wanted to discuss them with me, especially since he knew I was always ready to enter a friendly dispute. I tried very hard to accommodate him, as I appreciated his good will. I gave up. I regret that I offended him but I had to forgo: I saw no other way: dispute requires disagreement and disagreement requires familiarity (with the issue involved, with the question-and-answers), and familiarity requires comprehension, and comprehension requires cerebral flexibility; not much but some. My neighbor was so intent on taking the task of philosophy to be the justification of be-

---

35 His most obvious blind spots are his oversights of the absence of repeatability in psychoanalysis and of the pretext in the compromise between Russell and Wittgenstein in Carnap's work, of the utopian character of the categorical imperative, of Descartes' texts on explanation that dispose of his priority dispute with Hempel and of Descartes' and Kant's ones on metaphysics as a regulative idea for science that dispose of his priority dispute with Lakatos.

36 My friendly conversation with Ayer happened during an Oxford job interview. The chair stopped us impatiently. The committee members presumably knew that the interview was a charade: efforts to overcome the uneasy atmosphere failed.

37 Ayer made his intent clear in print soon afterwards [Ayer, 1957]. He advocated then commonsense knowledge — including some scientific knowledge. This only shows how variable commonsense is. See the final paragraph of J. O. Wisdom's terrific review of this book [Agassi and Jarvie, 1987, 50].

Ayer's 1957 book won him a fan letter from Russell [Russell, 1969, 130], no less, expressing agreement with him, although with some reservations, while ignoring Wittgenstein's influence and while playing down inaccuracies. Russell always respected common sense and science [Schilpp, 1944, 700], and his view receives a magnificent expression in his [Russell, 1944, Replies] and more so in the first chapters of his *Human Knowledge, Its Scope and Limits* [Russell, 1948] (that Ayer's book obviously echoes). Russell felt regularly misunderstood. He expressed the wish that those who see a conflict between his ontology and his epistemology should prefer his ontology. Commentators usually did the opposite. Ayer seemed to him to have respected this wish of his. More likely, Ayer simply saw no conflict here. Oddly, Quine did, and he did respect Russell's wish. This is no small matter: even sensitive Watkins came to view Russell as an idealist [Jarvie and Laor, 2001]. Commentators have the right to disregard an author's wish, of course, but the complaint still stands: courtesy requires that they should inform their readers of the situation.

38 The first volume of Solly Zuckerman's autobiography exposed the pro-science attitude that Ayer expressed in his first book as somewhat spurious, since Ayer showed no interest in science, not even in the impact of scientific technology on society [Zuckerman, 1978, 92]. He also says there, I do not remember where exactly, Ayer's ignorance of science was often feigned.

liefs, that he would not admit the possibility that some philosophers may refuse to justify their refusal to justify beliefs; he insisted on looking for Popper's justification of his refusal to justify beliefs, and he dismissed his philosophy anyway since, evidently, the attempt to justify the refusal to justify is inconsistent. (Erudite Stanley Jaki blames me in print of this very inconsistency.[39]) But I should be speaking now of the time prior to my apprenticeship.

I would love to claim that I was intellectually influenced by Abraham Halevi Fraenkel, the great Jerusalem mathematician, co-author of the famous Zermelo-Fraenkel-axiom-system, but, regretfully, I doubt it. He was the teacher there whom I most admired. He was an excellent teacher and a model gentleman; he was the only person in the Hebrew University who helped me — even though I hardly was his student — and he did so in many ways. His positivism was always tempered by a luminous tolerance that sprang from his profound religiosity — and this in a society in which most religious people were given in some measure to obscurantism or to self-deception, and the rest were given to a strong anti-religious intellectual and political sentiment. (This is no longer so: obscurantism and self-deception are on the increase in Israel and criticism of religion gave way there to hostility to religious political parties.)

Fraenkel shared with the positivist philosophers the view that scientific theories are (not assertions about the world but) mere means of computations. I found it strange that coming from him this did not trouble me, even though I found it offensive when I heard it from other professors. Did I accept it from him because I liked him? No: I did not agree with that view no matter who was asserting it. But why did I find it inoffensive when coming from him? Because the doctrine has a different import when the tough-and-no-nonsense scientific positivists advocate it than when religious people do, I later found out. In the first case it has the import of hostility toward metaphysics and the consequent loss of a comprehensive view of the world; it is then a crass utilitarian view of science; in the second case, when religious people advocate the same assertion, it does not import hostility to metaphysics even though it still is the assertion that science has nothing to do with it. They mean not to defend

---

39 Stanley Jaki ascribes to me the view that "there is no absolute truth" and asks me rhetorically, "how can one be sure that the big mistakes (in science or anywhere else) are gradually being replaced by smaller ones?" [Jaki, 1978, 421]. As he allowed that I speak of mistakes, what he meant to ascribe to me is not the view that the absolute truth is nonexistent but that it is unattainable. Even this I did not say categorically; yet understood me to say it apodictically although at most I consider certain only theorems proven *a priori*. My answer to his rhetorical question then is, we have no apodictic knowledge: we can never be sure of anything: we suppose that science approximates the truth; this is a supposition, not a proof, even though it is commonsense. At least as long as science lasts, says Popper rightly, its challenge is wonderful.

Jaki also ascribes to me [Jaki, 1978, 333], a "Popperian rejection of metaphysics" and calls Popper "an 'informal' member of the Vienna Circle" [Jaki, 1989, 252]. Sad.

science from metaphysics but the other way round: their intent was to prevent science from conflicting with the speculations that are part-and-parcel of their religious doctrine, with the speculations that portray God and His world and that they (mis)understood their religion to impose on them. The positivism of most scientific positivists, then, is a pro-science attitude that becomes anti-metaphysics through aggressiveness, whereas that of religious scientists is their way of separating science and religion so as to keep the truce between them. This clearly does not work, but it is not aggressive in the least, and so it is not offensive.

Having left religion in my late adolescence I assumed it would never interest me again. I was in error, though I was never tempted to re-convert or even to rethink: my loss of faith came with a loss of interest in theology, and this loss was final. But I developed a new kind of interest in religion as a historical and social phenomenon. This was possible because I never developed hostility to religion. Fraenkel's personal example taught me to respect religion and to observe the possibility that traditional religion and enlightened toleration may go hand-in-hand. Incidentally, I found what Karl Popper had to say on this greatest disappointment in him: a religious agnostic *par excellence* preaching a modern positivist version of Christianity as if it were the genuine original is too hard for me to stomach any day; reading his masterpiece *The Open Society and Its Enemies* soon after I met him, was a most exciting experience for me and I still remember it this way and I have experienced the excitement anew when I worked as editor on a Hebrew translation of this stunning work. Yet now I speak about his attitude to religion that was then, and still is now, the source of my greatest disappointment in him: I found very unsatisfying his version of secular religion, more precisely his version of secular puritanism or protestant ethic so-called. It rendered him unable to appreciate the sincerity and importance of any other sort of religious impulse, despite his declared efforts. His ethical convictions make these great efforts sufficient evidence of his reasonableness. The conviction that the limit of what can be reasonably required of anyone is hard work and great efforts is a conviction that leads to dogmatism; it is also a reinforced dogmatism,[40] since hard work and great efforts cannot overcome

---

40 Popper's observation on reinforced dogmatism is philosophical. Read as psychological, it is irrelevant. Its philosophical interest is this: dogma may depend on the refusal to listen to criticism; reinforced dogma does not: it is a trap, like the assertion that doubt comes from the devil. One may free oneself of it by realizing that it is a trap, perhaps also by considering alternative traps with the same mechanism (the way the claim that doubt comes from the devil works in competing religions). The traditional idea that research must steer a course between dogmatism and skepticism is psychological and obsolete: methodology should be free of psychology, as Popper has suggested [*LScD*, §20]: treating research as a game Popper-style dispenses with it. Possibly some doubt is good for it [Einstein, 1949, 13]; possibly some dogma is [Popper, 1970]. Methodology should make do with the wish to play, namely, with active curiosity, perhaps boosted by the hope that the game is fruitful [*LScD*, §§38, 80]. As Einstein said, researchers are opportunists.

it, yet it prevents its adherent from following a different and possibly more useful course.[41] (The claim that what is not attainable by hard work and great efforts is not attainable at all is clearly refuted by whatever is attainable only playfully, for example, or lovingly. Examples of these modes of progress abound. Falling in love and making friends are most conspicuous. But also juggling and tight-rope walking are.)

This happens to have been the philosopher's root strategy, the one that I find now in all of his early writings and that I do not appreciate: he polarized ideas into acceptable and objectionable and he proposed that one should try as hard as possible (a) to render every idea objectionable and declare them acceptable only after one fails to render them objectionable despite hard work and great efforts and ingenuity, (b) to render all acceptable ideas accepted by all reasonable people, and (c) to render all objectionable ones so unpalatable as to have everyone reject them no matter how reluctantly. This strategy has led him to work very hard and repeatedly at his efforts to dissuade his peers from their positivism and from their inductivism. It also led him to some extraordinary and very interesting attempts to make some metaphysical ideas, including some (admittedly rather thin) version of Christianity, agreeable to the most ardent positivists.[42] This way he only managed to empty Christianity of religion: it is preferable to be more honest and admit that one is or is not religious, to oppose some religion as immoral and to respect at a distance some other religion just as it is. In addition to the vulgarity of Popper's effort to appear Christian in some sense, it made him endorse a kind of anti-Judaism[43] peculiar to Vienna of his early days, I understand (I will return to this later), with a slight flavor of Max Weber.

The matter of attitudes to religion was one of the earliest topics of debate that I regularly tried to engage the philosopher with, once we got close enough to have debates in private. He explained to me then that his remarks on religion in his classic book were conciliatory and he declined to discuss this with me. He only agreed to omit the most offensive anti-Jewish expressions from later editions of the book. This, of course,

---

41 Assigning place for dogmatism in empirical studies — psychological, social, political, or educational — is taking science as an institution rather than a game. It reopens the door to the study of proposals like that of Einstein or of Popper mentioned above. And opening a traditional philosophical problem to empirical study is success — in line with Russell's program to render philosophy scientific. This offers a version of Popper's view of work that is refutable – and refuted.
42 The positivist reading of Christianity that Popper adopted is quite common. See [Freyne, 2002, first chapter],
... according to the standard account, the triumph of Christianity was the result of its ability to shed its Jewish past and embrace wholeheartedly the universalist, Hellenistic spirit of the age. Such a portrayal requires a critical evaluation, since it operates with stereotypes of both Judaism and Hellenism that ignore the complexity of the relationship.
43 Anti-Semitism is secularized anti-Judaism, except that since secularization is never full, it gets easily complicated and uncontrollable, especially as it is vulgar and anti-intellectual. Its thrust was the reaction against the emancipation of Jews as the recognition of their status as equal co-nationals.

looked to me very reasonable and very flattering. Since I was not interested in religion then, I let things rest there.

Otherwise, my interest looked to me surprisingly close to the interest of the philosopher. I was passionately interested in science and, moreover, in science as an intellectual activity, not as a success story and not in any way as a surrogate religion, much less as a surrogate magic. This may sound trite these days, even if most philosophers of science still cling to the attitude to science as a surrogate and discuss almost exclusively science as a system of rational beliefs and potent practices; in my student days in Israel I had lost all hope of finding a teacher. Gershom Scholem was typical:[44] he was angry when I said very briefly at some philosophical meeting that scientific views are not obligatory matters of faith. We never met, but as a friend of Judith's family, after the meeting he scolded her all the way home. Later on, after we left for England, he declared on the strength of that expression of mine that I would return to Jerusalem only over his dead body. He was a true believer in science and in his classic studies on Jewish mysticism he express contempt for it. He was caught in its net all the same and died endorsing the most pathetic faith: the irreligious, chauvinist mysticism against which Heine, Nietzsche and Weber cautioned their readers.[45]

It was a tremendous relief to meet the philosopher and savor his attitude to science as intellectual admiration with no intellectual servility.

There are two sides to the comparison between myself before and after my apprenticeship. The one is that I considered myself a failure before — even though far from feeling desperate — and I considered myself a success after, when I parted company with my mentor; and the success was thanks to his patient training. The other is that from the very start I had very unpopular philosophical attitudes and intellectual tastes, and these posed for me many difficult problems that I had hardly anybody to share with, and then I met a philosopher who was just ready to help me in my very position with my very attitudes and with my distinctive tastes. We shared so much, it was a bonanza. Consequently, it is no surprise that whatever differences in attitudes and in tastes that we had came forth later on and that he found them most annoying.

The majority of philosophers of science and of scientists, especially physicists, were at the time (and still are) authoritarian about science; I

---

44 Scholem was devoted to the study of the cabbala that he deemed downright superstitious, idolatrous, and a revival of Gnostic mythology [Scholem, 1965, 181, 183 *et passim*]. In his zeal to defend science he belittled the parallels between mediaeval astronomy and the cabbala (*ibid.*, 168). Its influence on both Copernicus and Kepler did not interest him. Nevertheless, like many early Zionists, he was a mix of Enlightenment and Romanticism. See next note.

45 Gershon Weiler, author of the classical *Jewish Theocracy*, 1988, notes in the Hebrew press in his review of Scholem's last collection of Hebrew essays that he finally succumbed to irrationalist chauvinism.

was not. There were a few causes for that. One was that I had developed an allergy for apologetics and *ad hoc* tinkering, as I had a mouthful of it in my rabbinical education and in my short period of political activity. And when my physics professors displayed apologetic attitudes, they equaled my rabbis. Yet for some reason — I was disturbed by not knowing which — I refused to declare science bankrupt. Perhaps it was because of my most impressive experience. This was my half-a-year stay in hospital when I was five years old — especially in view of the fact that since then and till recently I was never sick in bed for more than one afternoon and evening. In hospital I learned the charms of reading a fairy-tale, and there I met death: though infant mortality had been greatly reduced by hygiene, nutrition, and immunology, this was before the days of antibiotics; dying and dead children were part of my environment for six months. I also was myself near death — I was treated for malaria when I had a different tropical disease that is rare in the Middle East and known, if at all, as visceral leishmaniasis or by many other names, including *kala azar* and *dumdum* fever. It is an incurable and often fatal disease that is very difficult to diagnose. I remember my delirium with my mother sitting late nights next to my bed waiting patiently. Finally she raised hell. A young physician, probably a resident, showed initiative and so I passed to the care of the leading parasitologist, Saul Adler of Jerusalem, the head of the Royal Commission for the study of *kala azar*. He was expounding views opposed to received empirical findings.[46] Yes, you are getting the gist of my tale at once, I can see: scientific heretic Adler saved my life.

Decades later I met Adler in London in some public to do. I introduced myself to him. He was a shy person and then on the point of death from cancer. To my great regret we had no occasion to spin a conversation. I much regretted this, since my childhood memory of him was radiant. I remember that once he came with his hands behind his back, hiding a chocolate bar. Later they told me it was my turning point, when he thought there was a hope that I would recover. His widow and I once met by chance in the street somewhere in Jerusalem, and she recognized me even though before that she only saw me as a child. She then told me a few details of my case. For my part I told her I am not bad at writing a scientific biography (having written one about Faraday) and would gladly

---

46 [Hay et al., 1997]:
"... visceral leishmaniasis, or kala-azar, is caused by Leishmania donovani, a life-threatening parasite which invades and destroys immune system cells often associated with the liver and spleen ... . Untreated visceral leishmaniasis usually results in death and it is estimated that 350 million people are at risk worldwide, with approximately 12 million people infected at any one time (WHO 1990)."
The obvious diagnostic sign of the disease gave it the name "tropical splenomegaly". (I have a photo of myself with the size of my hugely enlarged spleen drawn on my belly. See my WebPages. It was prepared for Adler's lecture.) Diagnosticians have to find the parasite *in situ* [Adler and Theodor, 1931], as diagnosis is difficult: leukemia and other diseases accompany it and mask it.

write one about her late husband. Nothing came of it, I do not know why, but I regret it, since he was as near the ideal normal scientist as I could imagine — to the exclusion of individuals like Pasteur or Einstein, of course. Anyway, when I was in hospital he was called to my bedside, came at once and diagnosed me. I was off quinine at once and received the usual injections (a bismuth complex) instead, but this improvement was not enough. The received observation, established in Singapore by the two discoverers of the disease, was that the reservoir of the illness was not dog. It was. He found the carcass of the dog that was the source of my illness. He developed a serum and cured me. I was taken one night to the medical meeting where he described his findings, to serve as an exhibit; I was disappointed that I was not presented to the medical audience. His serum, I understand, is not in use: it hardly ever works (it works differently for different species of the parasite, of which at least twenty are known [Shortt, 1967, 23-6]). I have read that even in Singapore dogs were found to be the reservoir for the disease.[47] But I also read in other journals of tropical medicine that in diverse parts of the globe other animals function as reservoirs for the same disease. I will not trouble you with the added complications of the story (concerning the vector, for example): its moral is clear: Adler was a pioneer, a brave person, a great thinker; he discovered an error and he was himself in error.[48] Condemn science for its errors if you can. Declare science the body of empirically confirmed theories if you can. I am alive to tell a different tale.

But I violate chronology again, yet I only wish to report how fortunate I was that I became a close student, follower, and associate of the great philosopher. I decided to go and see him as soon as I learned about his ideas: I made this decision before I finished reading one essay of his,

---

[47] The discoverers of the Leishmania parasite in Singapore refuted the assertion that its reservoir is dog (at least there). Adler questioned the refutation and corroborated on me the previously rejected hypothesis. Later his refutation was corroborated even in Singapore, as reported in a scientific paper that I read — in *The Chinese Medical Journal*, 1960, if I am not mistaken.

[48] The normal sense of the term "normal scientist" as used here is the opposite of Kuhn's sense: in the normal sense scientific research is autonomous; in Kuhn's sense it is performed in obedience. Kuhn said, scientists are normally professionals; with a handful of exceptions they were all amateurs until the French Revolution. When writing philosophy Kuhn did not care for the social history of science and *vice versa*. Adler admirably encouraged amateurs to explore and publish. Kuhn describes leading scientists as pathfinders who are also leaders in the organization of research. Compare Bernard and Pasteur or Einstein and Bohr; all four were leading scientists but only Pasteur and Bohr were organizers. Contrary to Kuhn's teaching, path-finding researchers seldom have time and patience and ability to lead; Pasteur and Bohr were exceptional in both capacities: they excelled as path-finders and as leaders.

Adler was an outstanding researcher, an unusually gentle soul, a traveler, a polyglot, an amateur chess player, and a mathematician. He was also widely read and immersed in poetry. His research comprised meticulous experiments that are theoretically heavily biased in display of a tremendous sense of proportion. It embraced quite a few fields of biological and medical research from entomology and population dynamics and parasitology to epidemiology and internal medicine and toxicology. He had to distinguish many species and strains of the parasite and of the vector — sand flies — some of which evaded detection as they are nocturnal or twilighty. His available brief life [Shortt, 1967] is rather conventional but usefully detailed.

on the nature of philosophical problems (of 1950; reprinted in [Popper, 1963, 66-96]), which moved me greatly, apart from its having constituted an attack on positivism. I went to him with the clear intent of making him my teacher. I was a failure and I knew it and I needed a teacher badly and I knew this too. (I had requested Fraenkel to be my teacher, of course. He hesitated and finally refused: he was too humble to think he could teach anything but mathematics; even his self-appraisal as a mathematician was astonishingly below his professional reputation. He was very glad to see me in London after I became a student of the philosopher, showed interest in my studies, and offered help yet again. I visited him again when I was on my way to Hong Kong for my first appointment away from my mentor.)

The transition was dramatic. Before my apprenticeship I had nothing to show but a poor master's degree in physics that entitled me to be a high-school teacher and disqualified me for further studies at home. I was in a foreign country without a labor permit and with a wife and a child to support, yet with money for no more than a year at the outside — about half of it already consumed with nothing to show for it. Whereas Judith had found a good adviser and begun working on her dissertation almost at once,[49] I had landed in a miserable department; I knew at once I had no future there. Most of the undergraduate courses I had taken in Jerusalem were poor, but none came close in poverty to the courses — and the whole miserable higher degree program — in the philosophy and the history of science in the great and exciting University of London. I was troubled (not by being registered for a second master's degree but) by the role of the course in the philosophy of science as mere embellishment to the courses in the history of science, and by the unspeakable dullness of the whole program. One lecture there stands out in my memory. It consisted of the lecturer reading off the blackboard a list of Latin names of books by Aristotle, which he had carefully inscribed in clear capital letters before the lecture, and making a few rather flat comments on some of them — and the students were supposed to take notes, and they all did except for me. On another occasion that lecturer suddenly interrupted his lecture to explain to the class at some length the great importance of note-taking for the success in the ensuing exam (claiming that the information he was imparting was not available in books), darting a glance now and then at me, as I was sitting there, arms folded and fountain-pen conspicuously in my jacket's breast-pocket. Other lecturers were worse: they flooded us with incoherent collections of information that even I, never a student of

---

49 At first Judith's progress was faster than mine. Soon she had a misfortune: her adviser left for research abroad, and two advisers replaced her who had no interest in her problem and who shared a dislike both for her decentralist views on democracy and for her lifestyle as a student and a mother of a small child. Regrettably, due to their discouragement her excellent and pioneering dissertation is still unpublished.

the history of science, could and often did correct. I know how glad they were to see me leave. When I was established in London years later the leading historian of science there repeatedly told mutual acquaintances how inconsiderate I had been when correcting the poor lecturers' errors in class. (When I described the profession — not mentioning that department, of course — commentators declared my description a caricature though it was less damning than the truth.)[50] But all's well that ends well: it was the then department's head, Herbert Dingle, who helped me out. He was as narrow a positivist as any, and he really had little patience for my concerns, but he was frightfully humane and I was very happy that in the course of the years he became increasingly friendlier to me. At the time he barely knew me and he was very clear in his expression of reluctance to instruct me[51], yet he earned my warm gratitude for his deletion of my registration in his department as a technical error: but for this fib I would disqualify for a degree in the University of London — which understandably forbids switching colleges in mid-course.

I really should not have received a degree from the University of London. Not only did I switch colleges, in clear violation of regulations; I fibbed about my grade in Jerusalem, saying there were no grades for the master's degrees there (and concealing the fact that the graduate school there had refused me entry to their doctoral program). In addition, I simply refused to sit for the qualifying exam required by the university by-laws as I had switched from physics to philosophy. (Consequently my doctorate is in "general science: logic and scientific method", not in philosophy.) It was lucky for me. Later on, as I was already teaching there, a very talented and intense woman came over from the United States and impressed everybody by both her stunning appearance and her quiet intelligence: she was humble and eager to learn and regrettably was diffident for no obvious reason, being, as she was, intelligent, kind, good-looking and well off. She was encouraged to register as a graduate student, though she had no need for a degree, and to sit for a qualifying examination somehow required by some regulations. The philosopher wanted her to pass the exam and casually allowed me to tell the other member of his department, John Wisdom, to see to it; I did not. She failed and left with an unpleasant feeling. Why he had decided to help me but merely allowed me to instruct my senior to cheat, I do not know; she certainly impressed

---

50 My understated description of the sad state of affairs in the field of the history of science appears in the opening sections of my *Towards an Historiography of Science*, 1963. Reviewers repeatedly alleged that it was out-of-date, exaggerated, and unfair, especially in the choice of amateur historians of science as objects of criticism. Kuhn censured me severely for my lack of loyalty to the profession.

51 Dingle's reluctance to instruct me was reasonable. When we first met and I showed him a brief statement of my intended project that was naïvely critical of traditional positivism, I was ignorant of his reputation as a keen advocate of it. He sent me to read the forthcoming second volume of E. T. Whittaker's *History* in the naïve hope that it would straighten me up.

him no less favorably than I did and was the better qualified. I am reporting all this not only as an expression of gratitude to the philosopher (and, I understand, to Ayer too, who then chaired the university committee for graduate studies in philosophy). I am saying this to exhibit my view that academic success is often quite accidental yet many who do not achieve it suffer loss in internal terms of superfluous agony due to self-reproach in addition to loss in external terms of pay and prestige. Also it may serve as a caution: regrettably students penalized because of bureaucratic technicalities usually accept defeat; the rest usually try to settle matters by themselves; it is better to seek the support of an interesting and able teacher. There is no doubt that my backers knew I was cutting corners and should be disqualified. But it pleased those who were helping me to be helpful. I try to help too, but all too often people who most need help refuse it because of some pointless scruples[52], as if morality was involved rather than sheer, pointless bureaucracy.[53]

There is documentary evidence to corroborate my claim that I should not have received a London degree: I have a letter from the University Registrar promising me that I would never receive any degree there. Fortunately, although I did not answer that letter I received another from the same person, telling me he was happy to inform me that in retrospect

I really should not have received a Jerusalem degree either. The Jerusalem campus was under siege after the Israeli War of Independence and so records were unavailable; I falsely declared that I had completed my coursework. As a veteran I was relieved of the duty to write a master's thesis. I passed my master's exams by the skin of my teeth (all except for philosophy, in which I terrified my professors with allusions to mathematics and such — without the slightest suspicion of what I was doing, of course: students seldom notice the impression they leave on faculty). This game can be played two ways, of course, and I was lucky that

---

52 Cheating bureaucrats to help a student is opposed to Kant's categorical imperative: he forbade even cheating killers to save their victims. Popper's siding with him [*OS,* i, 102, 256, ii, 238, 386] is odd, since it is Utopian and a-historical. Also, here Popper disregards Russell's criticism, which he seldom did. Russell had helped a fox by cheating foxhunters, he reported, feeling no compunction about it [Russell, 1999, 188]. This is less of a puzzle than it sounds: Popper endorsed the common confusion of Kant's categorical imperative with the golden rule of Hillel the Elder: what you hate done to you, do not do to your neighbor (*Talmud Babli, Shabbath* 13:1). Cheating bureaucrats to help a student is at least as good an example for that as cheating hunters to help a fox and it agrees with Hillel better: some nice bureaucrats helped me and my friends to fill forms incorrectly. One would expect them to be more frequent in universities as these are growing increasingly bureaucratic. Alas, I see no evidence for this and it is very hard to study empirically, particularly as it is becoming increasingly hard to know what regulation makes sense and deserves respect and what regulation is pointless and invites cheating if this is the only way to restore some fairness and come nearer to practicing equality of opportunity.
53 Sissela Bok offers a non-Kantian, commonsense attitude [Bok, 1974]. This is quite easy, once we replace the quest for certitude with the reasonable demand to test our conjectures before we act on them in full knowledge that we may be in error and in readiness to pay the price if we are. That should keep us suspicious and critical: all amendments to the categorical imperative are quite *ad hoc.*

my chief physics professor who hated my guts (nothing personal; he was a super-positivist and I was a poor, exasperating student) did not try to fail me. (He appreciated my grasp of general relativity but resented my heresies on quanta. Later on I learned that before his death he joined the new fashion led by Paul Dirac, who had hoped to see an escape from quantum mechanics.)[54] He was satisfied with insuring that I had a degree that disqualified me for further studies. (He also insured, poor soul, that the graduate school made no exception in my case even though Shmuel Hugo Bergman, the professor of philosophy, took the liberty of pleading for me.)

I should not have passed my doctoral exam. My dissertation was typed in a hurry though it was a mere draft, as the philosopher instructed me to submit it at once when he had received an invitation to go overseas[55], and I submitted to the external examiner small batches of uncorrected typescript,[56] which annoyed him no end. He was Leon Roth, a famous scholar and a former professor of philosophy in Jerusalem. I tried to dissuade the philosopher from choosing him, but he did it anyway as a token of appreciation for my *alma mater*. I did not argue against this choice beyond saying that Roth was a super-pedant and that he deemed ethics a demonstrable science. What I had against Roth, despite his having been an excellent teacher and a true liberal, relates to his conduct towards me as dean in Jerusalem before he left, just before Judith and I graduated. A professor there misinterpreted her records (the originals were unavailable due to the siege, you may remember) in spite against her grandfather (who was a leader of the unpopular peace movement; he recognized the injustice to his granddaughter but understandably let it ride). Consequently, the university withdrew its promise to her of a scholarship, a fact that had caused us untold damage. I demanded that matters should be investigated. Roth[57] was the dean of arts then. He came to see

---

54 [Dirac, 1978, 36]:
"Most physicists are very satisfied with the situation. They say 'Quantum electrodynamics is a good theory, and we do not have to worry about it any more.' I must say that I am very dissatisfied with the situation, because this so-called good theory does involve neglecting infinities which appear in its equations, neglecting them in an arbitrary way. This is just not sensible mathematics. Sensible mathematics involves neglecting a quality when it turns out to be small — not neglecting it just because it is infinitely great and you do not want it!"
55 Out of the blue Popper received in spring 1956 an invitation to run a three-week seminar in Emory University, Atlanta GA. Behind it stood Charles Hartshorne, I think.
56 Popper was more of a pedant than Roth. Fortunately, he had no time to read dissertations. This is hardly unusual. Other advisers and readers of dissertations are much less conscientious: to conceal their indifference they often limit their comment to marginal aspects of works that they have to read, often offering insignificant and random verbal alterations and corrections of spelling and punctuation.
57 Leon Roth appears here in a fairly negative light. This is sad, as he was admirable on a few counts. He was an excellent and dedicated teacher, honest with his students and a true liberal. He taught for free and left Israel soon after Independence for want of any liberalism there: he and Hillel Kook tried to organize a voluntary association for civil liberties and they failed miserably. He was a true scholar. His best known works are his edition of The Correspondence of René Descartes and Constantyn Huygens,

me, and he promised me to do something. I was tremendously flattered and did not know what to do next. He knew what he had to do and he did it: pacify me with vain promises. I resented this, but it was my luck: he consequently felt obliged to make redress. He came to the exam in a fury because of the untidiness of the dissertation and of the pressure on him to rush but also, I suspect, because he could not fail me. He spoke for half-an-hour mainly against my untidiness but also against my skeptical view of experts. He had but a few words of approval of my praise of metaphysics. I listened attentively and the philosopher took notes for me. (He promptly lost them.) I then thanked the still irate Roth. Taking this to be facetious, he exploded. It was with great effort that the philosopher mollified him, assuring him of my sincerity. They then gave me a reluctant pass, signed the document, and went off to lunch. I never saw him again.

I just told you of the formal end of my apprenticeship, in mid-June, 1956, when I intended to report its beginning, in early January, 1953. I must try to be a bit more disciplined.

It was a dull wintry afternoon. I knocked on the philosopher's door at the London School of Economics to ask his permission to participate in his seminar.[58] Dr. Wisdom answered the door. It is strange to remember

---

1635-1647, [Roth, 1926], Spinoza [Roth, 1929, 1954], and Is There a Jewish Philosophy? [Roth, 1999; Ajzenstat, 1999]. Philosophically, he was a follower of Samuel Alexander. Both clung naïvely to seventeenth-century high ideals.

58 Watkins said [Watkins, 1977, 79],

"... in this seminar what mattered was not so much the papers as the remarkable and often devastating interventions they provoked from Popper."

The following report by a colleague in a neighboring college [Maxwell, 2002] is accurate enough on reputation and slightly exaggerated on facts:

"The Department at the L.S.E. was famous for Popper's weekly seminar. Notoriously, visiting speakers rarely succeeded in concluding the announcement of the title of their talk before being interrupted by Popper. Each speaker was subjected to a devastating critical attack by Popper, almost sentence by sentence; quite often, the subject of the seminar would be continued a week later. The seminars were always dramatic, sometimes farcical, but nevertheless created an overwhelming impression of Popper's passionate determination to get at the truth, even if conventions of politeness and good manners had to be sacrificed."

Let me offer here my overview of the seminar in disregard for the demand to avoid repetition. The papers were often excellent, even though there was only one filter: he allowed everyone to read a paper as long as it was on a significant problem. There were some top-notch guest lecturers, including William Grey Walter and Ernst Gombrich; there were papers by Popper, Wisdom and Lakatos; there were also papers by Popper's students, including Bartley, Jarvie, Sabra and myself; and there were papers from some non-academic members of the seminar. Despite Popper's questionable idiosyncrasies, when speakers had good manners the seminar went smoothly. Yet it was too often too stormy. I refer below to Ettinger's just criticism of this trait: though the rule was to avoid ego-involvement, Popper broke this rule all the time, erroneously considering his ego involved with propriety than with recognition of his contributions. (He never referred to his own writings, but found ways to stake claims anyway.) He utilized the traditional prerogative of the professor to the full. He never pulled rank, inside or outside the seminar, but he behaved as a determined chairperson. When this led the seminar to a halt, Wisdom often took over inconspicuously and with a few well-chosen moves brought the seminar back on track and inconspicuously retreated. Yet it was no doubt the combination of a few factors that rendered the seminar such a miracle: the high-level papers, the experiment in etiquette that was so unconventional, the tremendous intensity of the process, and Popper's personality, as noted by Watkins in the above quoted passage: when Popper was absent for any reason the thrill was gone.

that brief first encounter: we later became close friends, but then we were utter strangers. Professor Popper is at the physician's, he said, and will join the seminar as soon as he is back; in the meantime Wisdom was going to chair the meeting. And yes, he granted me the desired permission and led me to the seminar room. The session started very unceremoniously. The seminar was in Logic and Scientific Method. I had then only a vague knowledge that this was its title, but I soon forgot it: every question of intellectual significance from any field of study the philosopher treated as proper seminar agenda, provided the speaker could explain that significance. He gave permission to anyone who wished to attend to do so and he encouraged everyone there to read a paper on any question. And he invited everyone to comment on the paper, with no respect for expertise.

This made the seminar a pioneering experiment of the first order. It therefore faced procedural problems that the commonwealth of learning seldom faces. Its sessions repeatedly illustrated to me the identity of procedure in the sense of etiquette and in the sense of research methods.

For example, as individuals were encouraged to participate, the right of speakers to speak uninterrupted was never recognized. Terrific as it is for those who can take it, for most speakers it is too hard. Much later I struggled with this hardship in the seminar and tried to develop some idea as to what should replace the generally accepted rule permitting invited speakers to bore their audiences for an hour or so, and to devise a substitute rule to insure that speakers not suffer the strain some of them suffered in Popper's seminar. Similarly, distinguished and not-so-distinguished visitors would often repeat the profundity that is at the root of inductive philosophy: responsible scholars and researchers should be able to back up what they say and avoid engaging in wild speculations. This is the heart of inductivism; the heart of Popper's critical philosophy is precisely its contrary: speculations are hard to come by and they are valuable even when untenable. This is not to say that the philosopher had no rules of responsibility. In particular, to visitors who espoused the inductivist etiquette he told in no unclear terms that they were violating etiquette when prescribing etiquette before having learned what rules are practiced where they are guests. He also had demands of those who were offering speculations: what kind of argument, he asked, would make them admit that their speculations were false? This is a terrific question, I understood at once, but it took me years to come to my present view that it is very hard on the individuals who offer speculations to demand of them that they have a ready, satisfactory idea of a possible criticism of it. This is perhaps the poorest aspect of Popper's moralist attitude to scholarship, of his protestant ethic, his demand to work very hard: he could not praise a visitor for some new idea without at once pressing hard for more, for some critical apparatus with which to meet it. This made him unreasonably hard to please. Some visitors complained; we are trend-setters, he responded to them, and it behooves us to demand of ourselves much more than the customary rules do. I never liked this.

The meetings that were very pleasant also illustrated the force of Popper's method. I remember particularly two American amateurs, one was a physician who spoke of the living organism as a self-correcting mechanism and the other was a retired Hollywood lawyer who spoke about the desirability of replacing rules to avoid disasters by changing the environment whenever possible with the intent to immunize it to those disasters.[59] Both spoke directly, welcomed interruptions, and won the philosopher's immediate, sincere appreciation. But these were exceptions. Most guests in the seminar were not as placid. Nor were their papers the most instructive. I remember these two speakers more because they befriended me outside the seminar. The philosopher befriended the retired lawyer, which I remember because it was very remarkable, as his political opinions were on the far left. More instructive were two outstanding guest lectures. One of them was by Ernst Gombrich, on the limited reliability of art experts. It included a remarkable analysis of the case of Han van Meegeren's Vermeer forgeries. The other was by William Gray Walter, who came all the way from Bristol to talk about his terrific artificial tortoises. For that occasion the seminar moved to the great lecture hall; he sat there on the stage with the philosopher, and they enjoyed a most engaging exchange.

And so, the seminar was always very exciting yet all too often also very taxing — intellectually, morally, and emotionally. Some people I knew repeatedly tried to participate in Popper's celebrated Tuesday Afternoon Seminar but could not stomach the highly intense emotional and intellectual mixture. They considered his techniques of aggressive criticism and of interrupting speakers almost at once wrong and unbearable.

I remember, in particular, Preston King. In the meantime he has published some books, and even Popper admitted to me that he had found his work impressive — not an everyday occurrence by any means. King fitted our image of the stereotypical American Southern Gentleman in every respect except that his skin was deeply and beautifully dark. The philosopher just loved him. When he read his paper in the seminar he was greatly concerned about the response from the philosopher. He was then absent: doctor's appointment again. Wisdom was absent too (he was on a sabbatical) and so I was the moderator. I found King's paper hard to comprehend and I tried as gently as possible to tell him so. The meeting

---

59 That speaker was the Hollywood attorney, Dr. Morton Garbus. At the time I found his idea rather trite, but I was in error. Applying it systematically to military hardware, for example, would significantly reduce the permissible accident rate in training.

was unexciting but also uneventful. The next week the philosopher was back and asked King to continue. They broke up almost at once. He left the seminar. I went a few times on a mission of conciliation, trying to win him back. He flatly refused and I cannot blame him, though I am still convinced that this was his loss.[60]

I often wondered why the philosopher had such a bad press among students and young colleagues. He was really so kind, so eager to help, so full of exciting ideas. I remember the first time I heard him lecture on logic. It was in the middle of the course, yet I saw at once how powerful his ideas about logic are — no, they are still neglected — and got very excited. I remember him saying to a student who had criticized him, after he had demolished the criticism, as you see your criticism was not very good today; better luck next time! and it was one of the greatest impressions of my whole life, especially since so often my Jerusalem instructors said the very opposite: as you see, your criticism is poor; now that you see that, do shut up for a while — or something to that effect. I do not deny that the philosopher's bad reputation was well-earned, yet it is puzzling all the same. Indeed, when he became a grand old man, much worse conduct that he occasionally displayed was rightly tolerated and even wrongly admired. To paraphrase Bernard Shaw (*Doctor's Dilemma*), not all rude people are so kind and so interesting. Why was his rudeness so amply punished? I still do not know. Perhaps some people find it hard enough to tolerate so much brilliance and kindness and patience regardless of the manners with which they appear, and then such people may console themselves by the pretence that the chief cause of their irritation is the poor manners, not the terrific matter.

My best memory of the philosopher's performance is a lecture he gave impromptu to the student union in a neighboring college. He sat there on a table, his feet dangling and his head slightly cocked in half friendly, half mocking pose, animating the students, inviting them to respond increasingly forcefully, wearing all the weight of learning lightly like a flower, to use an apt expression by Tennyson, and enjoying it enormously.[61] My worst memory of him is the opposite image of that meeting, one that was an extremely well prepared, carefully written paper. (It is known in the inner circle by his nickname for it: "The Problem Children"; it says two things: I am on top and I recommend that you find a problem and fall in love with it and devote yourself to it. Rather anticlimactic.) He delivered it to distinguished colleagues in the Stanford think-tank where he had spent the year 1956-7 (and where he had got me

---

60 To repeat, King reported that his relations with Popper were cordial to the last, and that he does not remembers the incident here recorded. Popper did.
61 The lecture to the University College Students' Union took place in 1954 or 1955. Its topic, I remember, was political: he spoke of Alex Weissberg-Cybulski's impressive 1951 *Conspiracy of Silence* of.

invited as his research associate). It was met with tremendous hostility. (Some participants found the opportunity years later to gleefully remind me of that horrible experience.) Yet the philosopher thought nothing of his meeting with students and he thought his Stanford performance exceedingly satisfactory. I know that, since we discussed both before and after the Stanford event. He declared the success of his Stanford performance — for he saw it as a success — proof enough that one should not improvise but work hard. I did not ask him what experience would make him think otherwise. Hard work was his religion.

I will return to discuss the philosopher's lecturing style on a later occasion. Here my point is not his lectures but his impact on audiences.

I hope you indulge me my stressing that the philosopher was just unaware of the impact he had on students. A student from Boston, where I was teaching, moved to California, where Bar-Hillel was a frequent guest, and managed to take courses in the philosophy of science with each of us. She did not know which course she preferred, or rather she was ambivalent about them, or rather about the views expressed in them. Having found her ambivalence very sincere and fresh, I sent her a dialogue between herself and Popper. It did not satisfy her, but I liked it well enough, even though the session it presented was naturally somewhat idealized. Eventually, I had my letter to her published almost unaltered [Agassi, 1975, 81-90]. On three different occasions the philosopher gave me the same response to it — though we scarcely met by then: he did not like it, as it presents him as so much less patient towards students than he really was. Discussion on this item three times ended abruptly after I reminded the philosopher of the case of King and similar ones. He repeatedly forgot: his tremendous kindness naturally blinded him to his at times rather unpleasant impact. He forgot.

But I should reach the start of my story at long last.

January, 1953. A roomful of animated young people engrossed in an animated debate, skillfully orchestrated by the department's junior member, Dr. J. O. Wisdom. The paper read then was on the difference between the natural and the social sciences. It was a poor paper, given, I later found out, by an ambitious Polish Jewish refugee. He was scraping a living in London as a consultant of sorts — these were the days of popular cybernetics, operational research, systems analysis and such — trying his best to get a degree at the London School of Economics. I now admire his courage and vitality. It is unbelievable that he knew of Popper's reputation as an ogre yet joined his seminar and even volunteered to read a paper there. Luckily for him, the philosopher was absent for a while and a free and energetic exchange flowed under Wisdom's always-superb moderation. I knew nothing about Popper the person, nothing about rules of conduct — in his seminar or elsewhere — and my English was atrocious, yet I participated. Blame it on the Israeli provincial sense of self-worth or on youthful impatience with intellectual woolliness as you wish.

Suddenly the door was flung open. The philosopher appeared. Silence. He took over, allowed the speaker to read a little from his paper and took off. No response from the audience except for one fool, too slow to notice the change of atmosphere, of the rules of the game and all. That fool offered a brief insignificant remark, a near-quote from Russell on the foundation of mathematics, I remember. Evidently the remark did not register. The philosopher reacted in a flash: you are a charlatan, he said.

These were the first words the philosopher addressed to me. After the meeting was over he approached me. We met at the door. He spoke very sweetly: who are you? he asked. I said my name and added, unasked, I need your help. He was amazed. Later he told me you could then knock him down with a feather. I really do not know why: I really did need his help. Badly.

He granted me audience the same day, though it was, by then, very late in the afternoon, an early grey depressing evening already. He said he was willing to help, of course, but how was he to know I was no charlatan? I proposed he could examine me. He agreed. We had an animated conversation. I said, we do not ever have answers, only improved questions. Though this is merely a high-sounding folly — a double or triple confusion between answers, satisfactory answers and final answers — he liked it, on account of its spirited anti-positivist sentiment, I surmise. He said, do not switch colleges or else you will be barred from receiving a degree here, but come informally. I said I was looking for a teacher, not for a degree. He evidently liked that too. He nearly missed his train home and had to rush to the High Holborn tube station, a few hundred yards away, to rush to the train station, to drive home not too late. On the way out he said he appreciated my enthusiasm and accepted me. I said — where had I acquired so much conceit? — by Jewish tradition the choice belongs to the student to pick a teacher. The philosopher smiled benignly: he was sick and in great pain, dead tired, and above all in a very great hurry; but he was not displeased. It was an exhilarating moment for me.

For anyone still reading, not distracted nor busy with anything urgent, I hope they will indulge my concluding this chapter with a brief sermonette — about this our great, lovely, beautiful achievement society: we can overdo it: we can judge people solely by their success or its absence, which is neither friendly nor wise. I should know: I have achieved and I had two utterly underachieving brothers who were my betters in many respects (one older one younger). The same goes for my daughter who was to the last a good student and an industrious free lancer, with unflinching expectation of just reward from an amorphous abstract system. Where she quarried so much faith despite the prevalence of so much heartlessness I do not know.[62] What I had that made a difference are three things: first, a spouse to sustain me; second, a lot of help from all sorts of people, some of them utter strangers — which was vital to me, especially as a poor foreign student in this prosaic but, oh so marvelously humane England — ; and, thirdly, a lot of lucky breaks. This is my deepest reason for my utter inability to make peace with positivism. Positivism tells me that I am successful because I am smart[63]. I am too smart to believe such folly. Positivism says that we should apply science to daily life whenever possible and that this must lead to success. What is this application of science to daily life? It says, I am simply lucky to have used felicitous theories, *i.e.*, scientific ones.[64] Theories seldom affect the success or failure of individual lives. It is amazing: according to positivism either science is irrelevant to personal daily affairs or science is a magic formula. The commonsense idea that science is a part of ordinary life because it is a form of trial and error, like everything else in life — only with much more intensity and regularity and advancement and sophistication, of course — this idea had to wait for twenty-five centuries to move from commonsense to the philosophy of science. The commonsense idea that the difference between success and failure even in the commonwealth of learning may very well be due to good-will, kindness, and the readiness to bend some rules, this still awaits its turn.

End of sermonette.

Only do not judge people by achievements alone, though!

P.S. At the age of eleven I had two visions concerning my adult life. One was that when I grow up I will not betray the values I had as a child, particularly religion. This, of course, I did not do: I am not religious in the sense that I do not share any established faith and belong to no religious congregation: I view my Jewish affiliation as socio-cultural

---

62 This book is dedicated to their memory.
63 This is an allusion to the silly adage "probability is a guide for life" that Carnap chose as his motto. See my comment on him in my "Contemporary Philosophy of Science as a Thinly Masked Antidemocratic Apologetics" [Agassi, 1995] that discusses ill effects of the exaggerated rationalism of the positivists. Two millennia ago Cornelius Nepos exaggerated in the opposite direction, saying,
"I am far from viewing philosophy as a guide to life or as promoting happiness; rather, I presume that no one is in more need of instruction in living than most of those engaged in teaching it."
Cornelius did not deny that philosophy could bring financial security, such as most philosophy professors have. Since happiness is elusive and rare, his remark may well be untestable. Popper said in class, life is unpredictable: it is very unlikely that I will die rich, but one cannot disregard the mere possibility that a rich person will like my books and make me rich. And this is what later did happen, and in a most opportune moment, since just then on the advice of some eminent economists in the London School of Economics Popper speculated on his pension and lost it all. He was in an awkward financial situation, from which George Soros extricated him.
64 Scientism or positivism is the rejection of all non-science. Ever since Descartes, it had to allow for commonsense — allegedly pro tem on the condition that it will not clash later on with some new scientific finds. Yet Descartes was aware of the impossibility of total adherence to scientism, He said, as he began his adventure into it he had to allow for following non-science for a while, and he took truth by convention to be second best to truth by nature. How consistent was that?

(and I deeply regret that my country deems Jewish affiliation a political matter, thus refusing to be a normal liberal nation-state). My other childhood vision concerning my adult life was that when I grow up I will not switch loyalties and in the conflict between the young and the adult I will side with the interest of the young. I am still moved by this vision, which is profound; it was not original with me: it can be found in Mark Twain.[65] It was probably first articulated by Homer Lane [Lane, 1928]; [Wills, 1964].[66] I absorbed it from Janusz Korczak's *King Matt the First* [Korczak, 1928], which was a support during my miserable childhood; I can scarcely tell how important it was for me.[67]

---

[65] The literature on Mark Twain's *Huckleberry Fin* and *Tom Sawyer* shows no concern with the intergenerational conflict: the topic is surprisingly new, although sociologists and anthropologists have studied it soon after World War I [Mead, 1928, Preface] or at least soon after World War II. Harold Bloom, for instance [Bloom, 2000, Prologue], names *King Lear* as the paradigm for it. If it comes to this, then surely Jane Austen's *Northanger Castle* describes a much more direct conflict. But all this is skirting the issue. "Beauty and the Beast" is obviously better, and much more direct, being all too realist a tale of a Father saving his skin by throwing Beauty to the Beast. (The popular pseudo-Freudian reading of the story as Beauty's revulsion to sex is crude: she moans her loss of her freedom to choose her mate.) Linnea Hendrickson's *Children's Literature: A Guide to the Criticism* [Hendrickson, 1987] seems to me characteristic: she mentions only two essays on the literature on adults and children. One of these observes that this literature idealizes adults; the other reports beginnings of notice of adults who are "less than perfect". Both ignore Huck Finn's father who is a no-good drunkard. Its best expression is obviously in migrants — as it appears in works by Ole E. Rölwaag (1927) and Wallace Thurman (1929). The outstanding screenplay by Ruth Prawer Jhabvala and James Ivory of the 1965 Merchant-Ivory movie *Shakespeare Wallah* portrays the conflict as rooted in nothing but the difference of childhood backgrounds. This factor is obviously commoner in immigrant societies. It finds a clear description of it in the neglected study of Paul G. Cressey of the 1930's [Jowett *et al.*, 2007, 136]. The works of Chaim Potok and of Bernice Rubens center on it yet with no discussion of their sociology. Robert Cormier's 1974 *The Chocolate War* returns to the intergeneration conflict the way Korczak saw it: not as closely linked to family relations as to the failure of adults to recognize the respect that is due to human beings on the pretext that they are too young to deserve it.

[66] Shaw clearly presents it already in his 1910 *Treatise on Parents and Children*:
"Between parent and child the same conflict wages and the same destruction of character ensues. Parents set themselves to bend the will of their children to their own — to break their stubborn spirit, as they call it — with the ruthlessness of Grand Inquisitors. Cunning, unscrupulous children learn all the arts of the sneak in circumventing tyranny: children of better character are cruelly distressed and more or less lamed for life by it."

Alfred Adler notices this conflict too and Allen Wheelis' Adlerian 1958 *The Quest for Identity* is a pioneering report on parental neurotic ruthlessness.

[67] Korczak's role reversal between children and adults was thoughtful and *avant-garde*. If judged as escapist, it only proves escapism valuable and crushes all contempt for it.

CHAPTER THREE: THE MASTER'S CLASS

I have learned that if you have something critical to say about a piece of scientific work, it is better to say it firmly but nicely and to preface it with praise of any good aspect of it.
(Francis Crick)[1]

My first meeting with the philosopher set me on a busy course. To begin with, I went to all his lectures and seminars and guest-lectures and I read things he discussed or his other students mentioned as interesting and I soon concurred. This way, in particular, I read Émile Meyerson, E. A. Burtt, Arthur Edward Waite and, later, R. G. Collingwood, all of whom I found splendid, not to mention Einstein and Schrödinger. By that time, I had decided not to read anything other than what I enjoyed, unless it was a part of my paid work. This gave me enormous freedom. Freedom is what most characterized my studies at the time, partly from choice, partly from the absence of any formal obligation, which was a blessing, and from the paucity of guidance in my graduate studies. Except for time spent in the terribly inviting graduate students' lounge of the London school of Economics and my increasingly frequent visits to the philosopher, I went as I chose to lectures and seminars and libraries, especially the magnificent and ever so generous British Museum Library that I loved and that still fills me with gratitude.[2]

As to my dissertation, here is its story. I first showed the philosopher a page or so which I had written in response to Max Born's lecture [Born, 1953] to the British Society for the Philosophy of Science that was a positivist, unfriendly response to Schrödinger's classic "Are There Quantum Jumps?" [Schrödinger, 1952]; [Schrödinger, 1956, 132-60]. Right away, the philosopher decided that it was my dissertation proposal. Some weeks later, I read to him a longer abstract of my proposed dissertation, comprising some thirty handwritten pages. In the course of my studies for my dissertation, I also read two papers in his seminar on material related to it. Much later, when I was submitting the blessed thing, I showed him a few typed pages of the finished product, which he corrected, as he did later with many other short pieces of mine. Had he only glanced at my dissertation, he would have found that it contained passag-

---

1 [Crick, 1990, Chapter 6]. Crick's view is timid compared with that of Plato, who presented criticism as a token of appreciation that should never give offense; yet Crick's suggestion to offer regularly praise of the targets of criticism is novel enough.
2 I should have mentioned here with equal gratitude the benefits we had as students from occasions of exposure to terrific art like the cheap theatre and opera tickets of the British Council, the free theatre tickets from the Anglo-Israeli Society, and all the wonderful free art galleries. They all helped reduce financial plight and loneliness.

es that contradict his views, perhaps even distort them. But then neither of us suspected that. Nor did Watkins, who was most helpful — I cannot begin to tell you how much, and how much indebted to him I am — in correcting with extreme patience its atrocious English and in curbing somewhat my catastrophic disposition to digress. Watkins also mentioned my dissertation in one of his famous papers on the positive role which metaphysics in the design of scientific research [Watkins, 1958], which was the point of my dissertation, though it was not new and all he borrowed from me was my example from the life work of Michael Faraday, which engaged my attention during most of my studies for my doctorate. The more experienced I am the more I learn about the rarity of acknowledgements (especially to dissertations).

My dissertation was over 800 pages long. A three-page typed appendix in it was my paper on the discovery of Boyle's Law or the gas law: the pressure of air is proportional to its density [Agassi, 1977]. It concerns the confusion caused by the incredible conduct of Robert Boyle, its celebrated discoverer and the leading founding Fellow of the Royal Society of London. He made an acknowledgement — probably too generous — to an assistant, at the time when canons for acknowledgement were not yet established. (Incidentally, they were established soon after, at his instigation.) The very paucity of acknowledgements[3] contributed to the confusion about the discovery. When I tried to publish the paper, there was so much opposition to it from learned referees that, since it is my habit to incorporate comments in my work and respond to them (favorably or not as the case may be) the paper expanded and in the final draft was over one hundred typed pages. Amusingly, when the paper was finally in print, many years later, it stopped the flood of papers on the question — either because the professional historians of science were convinced that my paper closes the debate on the question or because they feel unable to publish on it without discussing my work. For, discussing my work is still out of fashion. It surprises novices to hear, but it still is an empirical fact, that there are tacit rules about whom a learned paper in the professional literature can mention and whom not. Popper, for example, was regularly ignored; this annoyed him in an unphilosophical manner, I dare say. It also explains why he thought that his mentioning me in a passage in his *Conjectures and Refutations* must mean so much to me. (Yes, I am glad

---

3 The chatty late seventeenth-century *Brief Lives* by John Aubrey describes Boyle as very generous. It also says, "*vide* Oliver Hill's book, where he is accused of grosse plagiarisme". (I could not lay my hand on it). Although examples of his failure to acknowledge are readily available, since he wrote so much, none of them is "grosse". The reference seems to me more gossip than serious challenge, especially since there were still no canons of acknowledgement then. This fact is often overlooked, and current standards are applied to old cases. The paradigm cases are Galileo's repeated accusations of plagiary and historians' blame for Descartes for his not having mentioned Willebrord Snell's priority in wording the sine law of the refraction of light.

you remember: in that passage he lamentably concedes to me that perhaps he was guilty of a whiff of inductivism, to which I have promised you to return. I will.)[4]

Though my dissertation was largely argumentative, it is still, inherently, a student's work: lamentably its end is to convince the reader; it is not a mature scholarly work. (There was no excuse for this: my familiarity with Faraday's discussion of the matter should have prevented this: I should have learned from his repeated declaration that all he wished for was public attention to his ideas and his arguments as well as public debates on them, not public assent, since this violates the right of readers to judge for themselves [Agassi, 1971, 136, 148, 286].) I already told you of my having submitted a mere draft of a dissertation agitated the external examiner and of how lucky I was that he was charitable to me even though he found my dissertation tremendously unconvincing (he was a staunch inductivist). I am proud that my dissertation is referred to in some books and is still read by some scholars, but nevertheless I think it is a mere draft. And it hardly shows the mighty hand of the philosopher: only his marvelously exciting and stimulating presence: even an attempt to do justice to some of his ideas gives a tremendous impetus to any work which includes that attempt, and in my dissertation I used some of his ideas to beat with them the staple positivist arguments that had plagued me as a student of physics. Yet in so doing I did not notice that I was contradicting some ideas in Popper's early masterpieces. At the time he was developing magnificently his realist stance and thus his way out of classical positivism, and with gusto that greatly affected me, in close encounters but, to begin with, in his public performances, which were most unusually exciting lectures.

Let me tell you about his terrific lectures.[5]

Let me start this with an observation on the academic settings. Though at the time the philosopher was a professor, his chair was personal, as was the custom in the University of London, as distinct from an established chair at a component college: there was no department of philosophy proper in the London School of Economics at the time; even

---

4 This material is scarcely known: only the Popper circle has some familiarity with it, simply because it is problem-oriented. All problem-orientation is out of fashion, even Popper's. Compare his case with that of Faraday. To recognize Faraday's criticism requires that we desert the view of science as proof. The philosophy Establishment has worse trouble with Popper. Bartley said [Bartley, 1984, 269],
"Popper is not really a part of the contemporary philosophical debate. On the contrary, he has ruined that debate": the distance between him and them is "as great as that between astronomy and astrology. ... If Popper is right, most professional philosophers the world over have wasted or are wasting their intellectual careers."
Physics witnessed repeated efforts to cope with Faraday's criticism, and this led to the triumph of his ideas. Physics before the scientific revolution and philosophy today are different. Those who wish to see today's philosophy follow pre-classical physics and give way to a better alternative may hope to learn from the study of the social aspects of the scientific revolution.
5 For more a detailed discussion of Popper's teaching methods see my short piece in [Michalos, 2003].

when I left, in 1960, the School had only a combined philosophy and economics program, not its present fully fledged philosophy department. When I arrived there it was only a service department with two members, Professor Popper and Dr. Wisdom. (Luckily for me, that did not prevent them from having doctoral students, as doctorates were university matters, distinct from college matters.) Watkins was a disciple too and was always around, but he belonged to the government department then; the philosophy and economics program was initiated later on, and this gave him occasion for a promotion and a switch of departments so as to inaugurate energetically a course in the history of philosophy which broke away from the traditional mold and was in line with his teacher's new and revolutionary philosophy. (This later changed, after the philosopher had retired and Lakatos assumed control over the department: academic politics took priority with him over intellectual matters. He was a genuine, passionate intellectual, but his passion for politicking was obsessive.)

I went to each and every class and seminar and public lecture that the philosopher gave in London between January 1953 and August 1960 and I participated in his intensive seminar in Alpbach, Tyrol, in summer 1954. In 1962 he gave a guest lecture in the University of Hong Kong where I chaired the philosophy department. The lecture was well received, and it was judged captivating by the more knowledgeable members of the audience. Thus, my judgment that it happened to be one of his rare poor lectures may well be an error.[6] For a moment I felt sick during the lecture, as it was a particularly hot and humid evening and the ventilation in the lecture hall happened to be extremely poor; I left for a few minutes to catch a breath of fresh air. He noticed this and told me many years later that it was my protest against a lecture I deemed poor. As if it were possible to lecture habitually and never give a dull performance. If anything, the quality of his lecturing was regularly so high that I remember every exception as extremely puzzling; and the number of these is very small: I have mentioned two and there is one more to mention later on. Yet I have heard well over one hundred very exciting ones, and I have been told of many more. And every lecture of his contained some new nugget. There is no better evidence for the high frequency of his exciting lectures than the tale of the great disappointment that in summer 1984 in Salzburg his audience experienced as he lectured to an international philosophical gathering and they deemed poor his performance then. (Thus, I understand, nobody was ever disappointed to hear a poor lecture by the

---

6 Popper's lecture in Hong Kong comprised a discussion of the transition from the (valuable) theory of *diairesis* to the (confused) theory of definition. This is fascinating. Popper's reading of Plato's situation and texts was terrific. The paper is probably still unpublished. It still puzzles me that he chose for the occasion something so unsuitable. The audience did not object, of course: they were flattered. They misled him in their adherence to the Confucian proscription of all signs of incomprehension.

great Niels Bohr, for example, as he was a notoriously poor speaker.) But if evidence is required, let me mention now that as late as in summer 1988 he was the star lecturer in Brighton to the same international philosophical public. I was absent there too, but on this occasion too I heard comments of sympathizers and critics. I also heard that the public cheered him there like a star and people stood in line for his autograph. And I cannot resist repeating the story that when he lectured to the Royal Society of London on Darwinism people waited in a line blocks long when the usher came embarrassed to report that the lecture hall was full and that a plan to invite him to give a series of lectures was already in the making.

One might suggest that to discuss lecturing is pointless, as it may be something no one can do much about. Not so. Admittedly, it is no use telling people to deliver interesting lectures or not to lecture at all. Still, one can learn to improve one's lecture style.[7] Also, arguments have to be given against those who think that the dull and boring lecture is preferable to the exciting and stimulating one. And, indeed, I did hear people censure the philosopher because his lectures were exciting and brilliant and because they usually included irrepressible asides, including comments on aesthetics and on art and on topical political matters and on education and on current university affairs and on and on and on. This censorious attitude is perverse and effective criticism of it may improve matters much. To my surprise, the reason boredom is preferred over excitement is the (important but false) theory of induction of the great Sir Francis Bacon: small boring facts, he said, are assured of being objective and they slowly but surely deliver the goods, knowledge and power. But this was only an ideology: we must not blame Bacon for the popularity of his advice, especially since it usually reflects a desire as remote as possible from the one he extolled. He advocated boredom as a defense against dogma, yet boredom is nowadays practiced as means with which to bolster dogma against exciting critical assaults: the dogmatic love to demonstrate their open mind without risking their dogma and they do so by engaging in lengthy, steady, detailed, sedate, careful, well-thought-out details: these may very well be pointless, but they have their use, as they do keep the bores too tediously busy to be able to see the exciting forest for some dull trees on which they are studiously determined to fix their gaze.[8]

The manner in which the philosopher lectured[9] to his students or to

---

[7] Benjamin Franklin found it important to report in his autobiography, despite its brevity, on the preacher whose sermons were well received until they were exposed as unoriginal. He rightly dissented, saying, the unoriginal but interesting is preferable to the original but dull.

[8] In *Science*, 238, 30 October 1987, a paper by four leading cognitive psychologists on teaching reasoning, seems broad-minded. Even science students often reason with highly brittle algorithms, it admits. It presents many alternative modes of reasoning but ignores the most famous and best one, namely, trial and error — perhaps due to its famous refusal to adhere to any algorithm. The authors were searching their penny under their lamppost.

[9] John Watkins describes hearing Popper lectures thus:

the professional audiences he so often addressed is not the model I follow, however. I think he always had an axe to grind. I like to compare him to Michael Faraday. Both were crudely ignored by their peers, both saw this as an unforgivable, annoying, frustrating crassness, and both used the lecture podium as their weapon of retaliation: there they scarcely paid attention to those who were assiduously ignoring them and developed simplemindedly their own ideas as beautifully and systematically as they knew how. No, of course I never heard Faraday speak: he died 60 years before I was born. But he gave two famous lecture series for children that he published,[10] as are almost all of his lectures to the Royal Society of London. Apart from the interesting published versions of lectures of his, we have the testimony of people who wrote glowingly about the experience of listening to his lectures [Agassi, 1971, 8-11]. Also, like Faraday, Popper did mention now and then in his lectures the official views, and then, again like Faraday, he presented these views better than anyone around. Of course, the chief difference between them was in the subject matter of the lectures: Popper's lecture courses always deal with logic and scientific method, a two-year course, as it was in the terms of his appointment.[11] His appointment was a readership deriving from the rebellious origins of the London School of Economics. Among its founders were the famous socialist couple, Sidney and Beatrice Webb, who had been greatly influenced by John Stuart Mill's *A System of Logic*[12] which declared that scientific theory (inductively) evolved from facts and so achieved the status of utter and unquestionable and final validity — once and for all. At the time in which Faraday was forging his field concepts and beginning to upset Newton's philosophy and his theory of gravitation, Mill declared anyone not committed to that theory simply prejudiced. Mill was the only philosopher ever to have described in detail the

---

"I was riveted. [He] ... had no notes or other paraphernalia ... almost hypnotic persuasiveness. There were happy improvisations and touches of humor."

10 Faraday's *The Chemical History of a Candle* and *The Forces of Matter* are both still in print.

11 Popper's method of lecturing in class reflects his educational philosophy: dogmatic teaching cushioned by encouragement to criticize. He usually began a lecture with an invitation to comment on the previous one followed by a straight lecture interrupted by questions, usually for the sake of clarification. When I replaced him, which I often did, I had the contents of the lecture decided beforehand, but could experiment with delivery. My aim was not to limit the question period and yet to cover the ground of the lecture as intended. This was easy to accomplish because the lecture course was transparent and superbly structured. Popper must have noticed this: he often entered the lecture hall in the middle of a lecture. I would then ask him to take over. He would refuse and ask me to continue as if he were absent. After one or two sentences of mine he would interrupt me and take over. After a while I would sit down. This became standard ritual and amused the class.

12 The Webbs were eager to present their work as scientific [Radice, 1984, 148-58]. They did that by following John Stuart Mill. His *A System of Logic* is still popular. Its popularity is not due to its logic, much less due to its methodology, but due to his claim of scientific character for his social philosophy as described there. During the stormy years of the nineteen-sixties students' revolt Lewis Feuer spoke to the American Sociological Association, suggesting to them that social scientists would do better to appeal to the authority of Mill rather than to that of Marx. See [Mill, 1987, Feuer's Introduction].

technique of forcing facts to yield a theory, or of squeezing theory out of fact (like wine out of grapes, to use the lovely metaphor of the great inductive philosopher Sir Francis Bacon).[13] Mill's description is called the four inductive canons; they were still taught in the sixties in many respected philosophy departments around the world, let me report an observed fact, and they may be still taught in some; in Toronto they are now taught in ambitious courses in some high schools, let me also report. Adolf Grünbaum, a leading American philosopher, has attacked Popper — as a mythmaker,[14] no less — and declared his allegiance to Mill [Agassi, 1988, 247, 250, and 256]. I think he did not mean it. Mill's canons were popular as they promised success.[15] Yet the only people ever to have used them to churn out science were Sidney and Beatrice Webb: each time they found a new rule by which to reshuffle the cards on which they had sketched empirical information, they wrote a new book.[16] And they wanted this technique taught to all students in their newly founded college and so they took the unusual step and instituted there a department of logic and scientific method. The philosopher was always extremely disparaging about their establishmentarian views of science as an unquestionable authority: "science with a capital S", he called it. He was similarly scathing about the idea that the mere churning of a handle can secure scientific progress: he called it "a science-making sausage-machine". At

---

13 Bacon's claim that theory emerges from observations is as popular as the readiness to deny it when under fire. Hence, it is a myth proper, surviving doubt as quasi-belief. Bacon's view of the mechanism of this emergence is *passé*: he said that observation is the prompter of proper intuition which is true learning. He compared facts to lenses: a distorted vision is due to too small or too big a database. He was clearly uncomfortable about it all: having presented diverse data in tables of comparisons and contrasts, he gave "permission to the intellect" to make a small hypothesis, admitting that this is a substitute pro tem for the intuitive insight. (Using medieval terminology, he described true learning as the placing of Nature and the Intellect on a par; *Novum Organum*, ii, Aph. 19.) The editor of Bacon's collected works and his chief and still unequalled commentator, the admirable Robert Leslie Ellis, mistranslated this remark and erroneously noted that there is no description of induction anywhere in Bacon's writings; he also rightly declared Bacon's permission to make a hypothesis an admission of bankruptcy. Generally, a theory that survives obvious refutation is on its way to becoming a myth.
14 Adolf Grünbaum and Hilary Putnam dismiss Popper's claim that science has no use for induction, calling it a myth. For Grünbaum's inept insult see also Chapter Eight and [Agassi, 1988, Ch. 24]; for Putnam's see Appendix to Chapter Ten.
15 The promise that all inductive ventures would be graced with success helped Mill's methodology to oust Whewell's (*Dictionary of National Biography*, Art. Whewell; [Wettersten and Agassi, 1991]; [Wettersten, 2005]). Bacon had conditioned this guarantee on purity of heart. (He borrowed this idea from the Cabbala, which is not far fetched, as his view of science was somberly utopian.) Refutations of errors were then used as proof of the impurity of their advocates' hearts. Lavoisier was not exempted; Newton was. Consequently, Bacon's demand promptly dropped out of sight. (In the 1818 original Dr. Frankenstein's defect is the usual sin of arrogance. In B horror movies he and his heirs are deranged.) The refutation of Newton's optics (1818) should have refuted the claim for guarantee, but only the refutation of his astronomy did (1919). These days the philosophy of science of inductivists who do not expect a guarantee is idle metaphysics. Wittgenstein rejected induction [Wittgenstein, 1922, 6.31-32], replacing it with conventionalism (6.341) completely overlooking Einstein and his robust realism. (This is surprising: from the start Russell knew of Einstein and appreciated him.)
16 The Webbs owed Mill more than methodology. They owed him (and thus also the sadly ignored, astute George Jacob Holyoake) the very idea of cooperative socialism. Mill's methodology reached them through the mediation of Charles Booth [Beatrice Webb, 1926]; [Radice, 1984]; [Simey, 1960].

the beginning of each course on the subject he mock-solemnly proclaimed himself the occupant of the only established post on the subject in the entire British Commonwealth and thus the proper authority to judge the subject itself non-existent.[17] To put it sharply, scientific method is an algorithm that should generate science. Hence, it is a computer program. Nobel laureate Herbert Simon and his associates have published in *Science*, the official organ of the American Association for the Advancement of Science, precisely such a program, which they aptly called BACON.[18] The response to them, as to Mill and Grünbaum, is, *Hic Rhodus, hic saltus!* Do use your algorithm and offer us one little lovely new law of nature [Agassi, 1992]. As the philosopher used to say in his lecture course and published much later [Schilpp, 1974, 1031], if such a science-making sausage-machine and researchers have access to it, why did Planck and Einstein labor during the last decades of their lives with no success?

Perhaps because these arguments against inductivism are forceful, the philosopher did not see the need to defend that doctrine in his classes before attacking it; he was convinced that his students should not be apologetic and admit with no hesitation that his criticism is devastating.[19]

I have no information[20] about the efficacy of his years of forceful

---

17 The popular ordinary-language (mock)-critical response to Popper was this. Although there is no paramount method, there are many small ones. For successful results such methods are obviously neither necessary nor sufficient. Since inductivism is a promise for success, this response is a red herring.
18 [Langley, Bradshaw, and Simon, 1983]; [Langley, Simon, Bradshaw, and Zytkow, 1987]; [Langley and Simon, 1995, 55-64]; [Langley, 1996]; [Agassi, 1992].

The persistence of faith in the promise of successful results is an impressive instance of Bacon's astute observations on the power of illusion. The computer gave this illusion new wind. *Statistical Package for the Social Sciences* User's Guide warns beginners not to try all possible correlations, since (by Bernoulli's law) in the long run some will succeed for sure, but, unless they are repeatable, they will have no scientific value. Much fuss was made over the obviously a priori guaranteed success of a computer program for the rediscovery of Kepler's first law by placing this law among the functions for the computer to inspect as possible a correlation between Brahe's data within prescribed limits. This way one can find hidden messages in Scriptures: write a set of possible hidden messages, a list of rules for decrypting and scan the Bible for any message. By Bernoulli's law, the likelihood of success of this exercise increases with its length. The search for a fit between data and formulas follows the same logic.
19 Admirable as Popper's wish to train his students to admit error is, he had no right to demand it of them as he did, especially if he gave the impression that he did not provide a suitable role model for this [Munz, 2004, 14, 30, 18]. (Whether the impression is right or wrong does not matter here.) More important, it is incumbent upon teachers to inform their students of their rules at the earliest stage and acquire their consent before imposing these rules on them. This is essential though hardly sufficient, as students hardly ever take it seriously. (They take it as empty admonition instead. The more this is reasonable, the more it is difficult to transcend and find a way to communicate with students.)
20 Perhaps it is no accident that the literature in the philosophy of economics pays more attention to Popper than does any other field in the philosophy of a specific social science; this may be due to his having taught in the London School of Economics and the friendly attitude to him that economists displayed — peers as well as young colleagues. But this may also be due to the greater readiness of economists than of other social scientists to attend to serious problems. For, economist or not, you will naturally ignore Popper — and Watkins and Bunge — if you prefer game theory and similar scholastic exercises to genuine problems. (Jarvie finds it intriguing that Popper read Von Neumann and Morgenstern on the boat from New Zealand but never mentioned game theory in print or in lectures.)

teaching in London. Did his influence on students grow and did it contribute to the spread of his new attitude to science? I do not know. For my part, Popper regularly thrilled me with his positive attitude to science that he regularly exhibited as a matter of course, coupling it with a lighthearted dismissal of the scientific establishment[21] and their authoritarianism. I could never have enough of it. Yet I was, of course, not typical of his audiences, undergraduates or professional.

The sophisticated among his professional listeners endorsed the professional opinion that he was flogging a dead horse. Some of these thought he had killed the horse but would not do him the courtesy of saying so out loud. Lakatos, the expert in the art of exasperation, until his last word (that appeared in print posthumously) repeated the claim that no one endorses Baconian induction these days, except for some antediluvian cave-dwellers.[22] Max Born[23] and Konrad Lorenz[24] are among these antediluvian cave dwellers, as well as the whole school of artificial intelligence researchers, not to mention Adolf Grünbaum. After Langley, Si-

---

21 Wonderful as Popper's lighthearted dismissal of any establishment is, he knew that there is no substitute for the analysis of its positive functions and the critical assessments of its views, which requires sympathetic understanding, of course. This is no complaint: he did more than his share and he left lacunae that invite studies that, performed in accord with his lights, should be quite useful and quite intriguing.

22 The Sixth International Conference on Logic and Methodology of the International Sociological Association Research Committee on Logic and Methodology, Amsterdam, August 2004, seems to me characteristic. Its greatest part was devoted to data collection, processing, and mining. The lead invited speaker there was Lynn Richards of the Australian business firm "QSR International". She reported that many researchers collect data, waiting for theory to emerge from them. She recommends the use of means for coaxing it to emerge (Baconian "aids to the intellect"). She also suggests the "rigorous testing of a hunch" (Bacon's "permission to the intellect").

23 The inductivism of Max Born is out of character. A close friend of Einstein, he had to be familiar with his anti-inductivism. Also, his forceful response to Sir Edmund Whittaker's denial of the revolutionary character of the special theory of relativity [Born, 1969, 103, 105] is inherently quite anti-inductivist. His major philosophical work [Born, 1949] is exciting despite its inductivism. His strongest argument there for induction is the derivation (129-33) of Newton's mechanics from Kepler's laws. This is precisely what Whewell had refuted a century earlier [Whewell, 1860, Appendix H]; [Sambursky, 1971]; [Shae, 1982]. On this Popper criticized Born [Popper, 1972, 198n]. Before publishing this criticism he wrote about it to Born, who responded angrily.
Langley and his team report that their computer emulation of the inference from Kepler to Newton rests on Clark Glymour's version [Ford, Glymour, and Hayes, 1996]. The discrepancy between the different variants of this obviously invalid derivation is minimal and insignificant as the inductivist canon does not grade the culpability of jumping to conclusions. Thus, Laplace's reliance on his own intuition to decide this question (*The System of the World*) now seems odd.

24 The learning theory that Konrad Lorenz advocated is anti-inductivist, as he replaces the inductivist *tabula rasa* with the internal release mechanisms (IRM) that release certain choices and the scope of options available [Agassi, 1877, 139 (note 51 to Chapter 2)]. His Nobel Lecture ("Analogies as a Source of Knowledge") follows the traditional inductivist view of the extraction of theory out of data as deduction. He told Popper that they were in agreement on scientific method. When Popper told me this I regrettably burst into rude laughter. Probably Lorenz was alluding then to his own famous remark that it is healthy to refute a couple of hypotheses before breakfast. And presumably he deemed Popper's equation of learning from experience with refutations a merely exaggerated but just emphasis on the value of refutations, the way he deemed Freud's reduction of all motives to the sex drive a merely exaggerated but just emphasis on the significance of sex. Tradition always condoned all distortions meant to conceal disagreement as positive acts of good will. Historians of science often misjudge such efforts, viewing them as deep insights instead of rectifying distortions while reporting them, as they should.

mon and their co-workers won great acclaim, influential Carl G. Hempel said, now is the time to rethink the whole matter.[25]

So much for the sophisticated in the philosopher's audience; others sincerely thought he was flogging a dead horse, as they held the conviction that he was not criticizing their opinion. This conviction follows from their conviction that their opinion was right and that his criticism was valid. This dismissal of the philosopher's critical efforts would be harder to sustain were he to follow his own canons of critical presentation that require that one explain a problem clearly, defend previous solutions to it and then criticize them and only then try to offer an alternative. These are often recognized as good rules, yet they are seldom in evidence. It is a strange fact. Collingwood, for example, was a terrific champion of these rules yet he almost never followed them — perhaps because he confused criticism (especially of living opponents) with contempt for what opponents stood for (and regrettably even for their persons, as his wonderful autobiography sadly indicates). A problem well presented is half-solved, they say, yet they seldom present a problem at all, and it never occurred to them to devote half-a-book to the presentation of any problem. The problem of induction to which Popper devoted his first masterpiece of 1935, as well as his lecture-course on scientific method, has a literature devoted to it, in which it — the problem — is not stated;[26] in Popper's first masterpiece the presentation of the problem of induction takes a few lines. (The reason for that presentation may very well be his new and rather elegant wording of the problem: inductive inferences are from particular statements to universal statements; how can such inferences be valid?) He had hardly reason for aggravation, then, when he heard people say, as they still do, that his attack is on the wrong problem: they are so confused or they want reassurance and nothing else. His patience was boundless, yet he seldom had enough patience to explain problems — at least not as efficiently and forcefully as he could and as he occasionally did — unless people asked him expressly to do so. And he assumed (rightly) that his interlocutors were familiar with the problem, and so he took for granted (erroneously) that they were open to examine solutions to it. They usually are not, and this is quite an aggravation.

This aggravation came from the sophisticated, professional audience; it had little place in his scientific method classes: there his dismissal of Baconian induction seemed no impediment as most students there did not know the problem. They took for granted the views that he was

---

25 Carl G. Hempel rejected the idea of inductive inference, assuming that no artificial intelligence could discover natural laws ("Recent Problems of Induction" [Colodny, 1966], reprinted in [Fetzer, 2001]; see also there 420). The computer program BACON made him show readiness to rethink.
26 The standard modern version of the problem of induction occurs first in Russell's classical *The Problems of Philosophy*. It turns up also elsewhere, especially his *The History of Western Philosophy*.

attacking, and as very important. And so they engaged him in lively debates. He enjoyed the challenge enormously, as the students were naïve and put their arguments as fresh even when quite unoriginal and well-worn; and so he was extremely gracious and admirably patient with them. But the cost of avoiding the kind of aggravation caused by his peers was another kind of aggravation — in the form of a certain stubbornness on the part of some students. And every year the class had at least one of those. He knew this, of course, and to combat it he was exacting: he demanded that they either admit error or come back with some reasonably good response. This would hardly be necessary were the ground better prepared. Things could have been done in a less exacting manner and with little effort. For this the general problem-situation had to be better discussed and the different parties to the dispute better introduced — by way of a fair debate about the question at hand (what, if anything, is the method of science?) and the received answer to it (the method of science is that of deriving theories out of large deposits of information). Had he started by elaborating on the problem as much as was necessary in order to see the received opinion about it as an answer to it, had he been ready to defend it better, then he would also win some sympathy with his attack on it. Students would then become aware of their earlier endorsement of the views that he was criticizing and that he expected his criticism to make them give up.

Now, of course, the philosopher assumed that educators should be exacting and that everyone's opinions always should be at issue. To a large extent this is a matter of temperament and even of circumstance: there is no denying that the philosopher's own place in the commonwealth of learning largely determined his very classroom techniques. It really mattered to him — as a self-respecting thinker and as a self-respecting educator — to combat the official theory that science was inductive. Such an attitude may easily make those students feel tense who happen to consider betrayal what he was demanding of them. It does not matter what they thought he wanted them to betray. It could be the scientific attitude and it could be their own heritage and it could be merely their other professors who were playing a bigger role in their curriculum. (Remember: Popper's was a mere service course.) He put pressure on them. This surprised them. They resisted. Their resistance led to tension. The situation invited a projection, a search for a cause that they felt they could not betray. (The same urge to vanquish induction, you will soon see, created much worse tensions with his former students, with his disciples, whom he had expected to enlist in the same war yet who took it to have been won.)

The demand of the philosopher for a quick decision — for an admission of error or a presentation of a reasonably good rejoinder — is clearly outrageously excessive: why is it not possible to ask for time out? Yes, time out. We all do this anyhow (when we admit defeat by not returning but running away in silence, which is understandable even if also regrettable and useless). Moreover, not all students know that they may ask for time out: they then have to be repeatedly told it and reminded of it. Otherwise they experience pressure buildup: they may feel unconvinced without knowing why and they may feel the fear that it is not authentic to give up a conviction which is still resting secure in their breasts and that accepting the new conviction which the philosopher was offering was more than they had bargained for.[27] This brings forward the need to think things out slowly and quietly. But the pressure to respond and the bewilderment bring about mounting stress; and stress debilitates.[28] And then the tendency to wriggle out starts to manifest itself.

I have to qualify the point of the previous paragraph, or at least to put it in a proper context. There is a general agreement that the task of teachers is to include the ironing out of all unsuitable, deep-seated prejudices. And the philosopher was doing no more than that. Yet there is here a world of difference between the authoritarian and the liberal systems of education. The one requires submission; the other requires cooperation and builds exclusively on active participation and even initiative. The success of the authoritarian system, of forcing pupils to submit, is usually the success of the tasks for which teachers have the backing of the community as a whole. Teachers who cannot rely on community support must enlist the cooperation of their students or impress them in a very great way. Indeed, all those who stayed willingly in the philosopher's class were individuals who decided on their own that the class they were attending was special: they did not have to take it and there was strong incentive from some quarters to the contrary.[29]

The tendency, which even some of the best students exhibit from time to time, to wriggle out of a debate (I know I did so regularly, though

---

27 A longitudinal study [Douglas *et al.*, 2004] claims that out-of-school influences are the decisive causes for educational gaps between social classes.
   Young people go to university to acquire credentials and, if they must, some knowledge and skills too, not education. Classroom discussion of the best way to achieve their aims is becoming increasingly vital. The establishment of the new on-line universities may facilitate this.
28 Lorenz has observed that pressure reduces the ability of animals to think. This may be more pronounced in humans as pressure may cause defensiveness. Otherwise, pressure may improve thinking. Even the mentally ill may think well under pressure, although only temporarily [Fried and Agassi, 1976]. Sociologists of work have shown repeatedly that unclear assignments debilitate, especially when crucial decisions must take place in haste. A very famous yet hardly studied example for this is students failing in exams due to over-preparedness. Most students have no idea about how examiners grade. Thinking about this already causes immediate relief, reduced pressure, and improved performance. (In the opening scene of Satyajit Ray's 1975 movie, *The Middle Man*, the fate of its hero is decided by a minor accident that determines a poor grade for his finals. This should have drawn more attention.)
29 Watkins told me early in the day that the sociology department in the London School of Economics had demanded that their leading obligatory classes be set to coincide with Popper's service classes. When requested to put this in writing, he added, they complied. I never checked this information.

less out of defensiveness and more out of bad habit and poor training, I dare say) regularly drove the philosopher into a quiet rage: he quite rightly demanded that debate proceed honestly, that, particularly, the avoidance of shifting one's ground surreptitiously. No doubt, the rule the philosopher was applying is correct: this is the thesis of Plato's *Euthyphro*, of course. It is a pity that even trained thinkers do not know the rules: we cannot then expect students to do much better. And it is really not so hard to master them: the early dialogues of Plato center on one methodological rule each and illustrate it by some interlocutors' violation of it and, for amusement, Plato makes the culprits regularly find their own faults in Socrates. It is not obvious to the untutored eye that Plato's humor serves to keep the reader relaxed; it was Kepler who confessed he consciously used humor to that end [Baumgardt, 1952, 83]: "humor", he said, "is by far the best seasoning of any debate." The philosopher had a tremendous sense of humor and he could use it most effectively, but when he worked hard at what seemed to him to justify his existence, when he illustrated his achievements as justifying his existence, and when he found his results then as poorer than he had expected, he then got understandably impatient[30] and his having made very great efforts rendered his falling below his expectations from himself patently unjust, and this injustice made him uncontrollably angry, and then his anger at the world could easily land on a poor unsuspecting student too busy trying to wriggle out of a tough spot to comprehend their situation.

(Sadly, too many people are reluctant to acknowledge even the fact that a critic has raised a new point that invites deliberations. Obviously, this impedes progress. Still worse, the editors of *Science*, who pride themselves on open-mindedness and fairness and professional decorum, violate etiquette all the way thus adding insult to injury. For instance, they reject many powerful letters against papers they publish, such as that of Simon and his collaborators on the BACON project. Thus, they silently suppressed some controversy. This way they impart the false impres-

---

30 Popper always responded angrily to small talk and to poor talk. This was his way of blocking it. This is not the best way, but the blocking itself was valuable, more in his seminar than in class. Wittgenstein fans now forget the frivolous conversation style popular at the time in English philosophy seminars under the sway of his mannerisms: nowadays days they seek deep ideas in his frivolous output. This is some improvement, although reading ideas into texts is inappropriate [Goldstein, 1999, Introduction]. Goldstein, for example, portrays Wittgenstein as an advocate of some sort of trial-and-error ethics, in disregard for Wittgenstein's pride in his rigid ethics as well as his demand for certitude. Advocates of such ethics can hardly be too fond of his "I follow a rule blindly" [Wittgenstein, 1953, I, §219]. This dreadful sentence comes in the end of a paragraph that describes his having no choice. This is often true of obsessions: he was obsessive and, in the customary fusion of obsession with sincerity, he justified his ramblings by appeal to his sincerity. So at the time of my story the literature of philosophy was full of trivia, perhaps in efforts to emulate him. A person whom Popper asked to leave his seminar (Brian O'Shaughnessy by name) soon became a small luminary in the Wittgenstein galaxy on account of his good emulation of the master. (What emulation is good, incidentally, was a matter for the Establishment to adjudicate: they never specified rules [Munz, 2004, 56-7].) Fashion altered, and deliberate rambles left center stage: the analytic style changed. (O'Shaughnessy changed his style accordingly.)

sion that the fairly mild debates that they publish are all there is, and that just those papers in their journal that call for sharp criticism raise no controversy at all.)

The most embarrassing part of the philosopher's whole lecture course was this: he took expressions like "what I meant to say" and "you have misunderstood me" as almost always signaling surreptitious changes of opinion. It is hard to judge, but my impression is that he was right on this point. Yet it was wrong to take it amiss, as he invariably did: as students could scarcely be expected to know that, he would have done better to explain it to them than to blow his top. He would invariably lose his temper whenever students used these expressions in class. And then he would be in no position to explain anything; after some pressure-buildup on both sides, a beautiful workshop carefully and most lovingly and intelligently developed for weeks would be over in a flash. The class would continue, usually with a few less students, but without the previous enthusiasm and excitement.

This was a recurrent and famous phenomenon, the talk of the School. Yes, he was aware of it. He once asked me what he was doing wrong. I could not answer: it is very hard to analyze such situations: to do so I needed a long time out, and when I had an answer he was already not there to hear it — or rather I was gone and in my absence my analyses were no longer requested.

I once had a beginning of an analysis of what went wrong. Once a guest who was follower of Wittgenstein introduced in the philosopher's seminar in a rather aggressive manner the kind of prattle that then went for philosophy almost everywhere in England. The philosopher openly blocked him, and made him leave. Afterwards we briefly discussed the matter. Since that blocking was deliberate, the philosopher had not lost his temper and so the matter was not pressing. People who work very hard discuss only pressing matters, of course. I only managed to mention the humor Plato makes Socrates exhibit all the way, only to be told that my image of Plato's Socrates is erroneous: in the *Protagoras* he declares he has no time for long irrelevant speeches. Interesting, even though not quite relevant: Plato describes in great detail the way the encounter took place: Protagoras was holding court and Socrates was almost dragged there. He would never say he was too busy when he was speaking to a guest of his, or to a fellow guest, or even to a participant in a chance encounter. Moreover, even here Socrates is very gentle and blames his poor memory rather than appeal to etiquette. This is my image of Socrates, of course, but at least there is no case that I remember in which Plato allowed Socrates to embarrass anyone (except for the *Ion*, possibly). This is perhaps one of the themes of his *Symposium*, where (for fun) his company cleverly brings into conflict his reluctance to embarrass and his refusal to make a speech. He escapes embarrassment, incidentally, by making a marvelous speech leading to a story about a debate, a debate that has fascinated many a commentator. But then we do not know, since Plato admits frankly that his portrait of Socrates is idealized.[31]

Embarrassments hardly matter anyway: a little adult attitude and a little humor dispel it. Ian Jarvie, now one of the best advocates of Popper's ideas as well as an important scholar and researcher and editor in his own right, when he was first present in the philosopher's class on scientific method, expressed frank incredulity upon hearing misgivings about the theory of science as inductively based on facts: in the department of anthropology where he was registered they constantly pumped this theory into students' heads. "Surely, you do not mean to say that ...", he earnestly said, his first syllable highly intoned and his head cocked sideways most charmingly. The philosopher responded to this expression of incredulity of the kid — Jarvie was a teenager then and he always looked much younger than his years — in a naughty imitation of his style, cocking his head and intoning the first syllable too: "Surely I do", he said, and proceeded to elaborate. Lovely and to the point as the response was, I feared the sneer would drive the kid away. The next week the story was repeated; this time the philosopher lost his temper: "surely" was an improper word; Jarvie was forgiven only once.

It is not hard to see the philosopher's point: it is really an expression of disregard, however unintentional, to reject offhand the statement of a speaker, even if this is done in an inquiring mood. But he would not have lost his temper at this expression of disregard, were he more aware of its being unintentional, were he sensitive to his students' ignorance and anxieties and habitual defensiveness. Indeed, when you observe that some expression or another just used in class is rather unbecoming, the response is, but the expression was used with a pure intention. The philosopher developed in class a very interesting theory of communication[32] as objective, so as to place the intention of speakers in their choice of words and not in the objective (or social or inter-subjective) meanings of the chosen words. But he could not apply his theory all the way, first, be-

---

31 Plato, *Seventh Letter*. But then, its authenticity is contested [Boas, 1948]. Popper has suggested to me once that possibly Plato's *Letters* comprise the earliest known epistolary novel.
32 Popper's theory of communication owes much to Külpe and to Selz, as it touches on psychology. As a theory of comprehension, it is semantic (unlike Shannon's information theory). Recognizing grades of comprehension, it makes simple and congenial sense of Kant's strange observation that he comprehended Plato better than Plato did (*Critique*, B370; cf. B862). It also offers the most striking refutation of sensationalism by drawing attention to the uses of approximations in perceptions. We perceive a reasonable fit to some default option first; we then notice the deviations from the default case, and even these are usually standard in the first instance. A striking example for this is intonation. Another is present in the diagrams of Béla Julesz [Julesz, 1960]; [Julesz, 1994, 41-3, 155-63] regarding binocular vision [Agassi, 1987c]. There, the comparability between the approximate information and the more exact one comprises the ability to estimate depth. This refutes the sensationalist view that we are aware of our data. This is why pedantry is necessarily frustrating.

cause it is too sophisticated and so it requires an open discussion and explicit agreement to abide by it, and second, because it was not clear even to the philosopher that his theory invites the conclusion that speakers' beliefs are private[33] and so not parts of communications unless they comprise the specific objects of the communication. His demand for assent was personal and so he would have done better had he not made it public, or, alternatively, had he made it a part of the educational process. Be it as it may, the demand is quite excessive to comply with a new and sophisticated theory of communication.

The picture I am trying to convey of the philosopher's lecturing techniques is riddled with tensions. On the one hand, as he used it, he demanded self-discipline and critical thinking to a very high degree; on the other hand, he claimed that one should stay in the world and not fly to remote spheres. The philosopher himself complied with ease with both demands, thereby performing a most enjoyable act while demanding a level of performance from his students that was intolerably high. He never suspected that his lectures were very special because they were both simple and sophisticated, as he required of himself to work very hard and to remain humble. Yet he craved praise all the same.

A point about his regular references in his lectures to current affairs. He boasted regularly that he had no access to the mass media. He had neither a radio nor a television set, and he read no newspapers, he repeatedly declared. In his opening address to the Salzburg Mozart Festival of 1978, which I happened to hear on the radio, he expressed contempt for the mass media.[34] His familiarity with current affairs came from such things as reading headlines of newspapers displayed in the streets, especially when he purchased a paper for his wife, and his references to them included explanations and interesting comments.

An example: as a standard comment on power politics he observed that the proper use of power is a necessary evil. He would patiently explain that, and then add two comments, on the necessity and the evil of

---

33 The prevalent theory of meaning as reference directed the study of language to expressions that refer, to parts of speech that can serve as terms proper. In 1879 Frege's refutation of that theory opened discussion of other parts of speech. He viewed the meaning of a noun as a dual reference to objects — in space-time and in Platonic Heaven. Russell rejected this Platonism (calling it metaphysics). In 1905 he offered a partial substitute to it for noun phrases. In 1926 (preface to the German translation of his *Problems*) he reverted to behaviorism and made the slogan that remained the battle cry of the analytic movement for decades: "meaning is use". That slogan had all the defects of behaviorism, including merger of actual with potential communication, such as a symphony, a specific interpretation of it, and a specific performance conforming to it. In 1940 Russell gave up his project of a non-Platonist theory of meaning. Quine considered numbers real like sticks and stones [Tennant, 1994, 345-51, 348]. He told me he never expected to be properly understood, as no critic of his work was ready to assume that he meant this literally and quite seriously. It is not only his metaphysics that they could not take seriously; it is the objectivist theory of meaning that it is a part of.

34 Popper's lecture to the opening of the 1978 Salzburg festival appeared the next day in a Frankfurt newspaper (on a full page with his photo), later in a volume in German, and then in English translation [Popper, 1994b, 223-32].

power. As to necessity, he said, it is not based on any observation of fact. Suppose the anarchists were right on the point of fact: suppose that people are naturally friendly. Suppose then, that were there no governmental power, as there would be no need for it. Nevertheless, he added, some might find this condition deficient: they may feel that without proper protection, the weak would be unable to enjoy their right to security without owing a debt of gratitude to the strong for curbing their strength. As to the evil of the necessary power, he said, it behooves governments to admit that when they use force they are performing evil acts — even though with full right, since regrettably what they do is necessary. This seems to me important and especially pertinent in the England of the fifties when a judge could add to a death-sentence a nasty, self-righteous[35] sermon to the person just condemned to death. In this respect the political situation these days is much improved, though not sufficiently, since veracity is still not required of politicians. The philosopher had a lot to say in his lectures about veracity too, but I shall not discuss it here.[36] I wish to return to my experiences, as a student and apprentice of the philosopher as an educator.

May I echo two points: I was already allergic to apologetic conduct before my apprenticeship, and yet I was then far from being free of it. I remember having engaged in it in the philosopher's presence, eliciting a wince and later even a sudden burst of ill temper. I hope he had some success reducing this bad habit in me, if not totally eliminating it, and I am deeply grateful to him for this success, achieved by stern disciplining and by lengthy discussions.

Popper's theory of science is best seen as the view that science abhors apologetics. This wording is very advantageous. It makes science ideal as ignoring friction[37] — in Galileo's sense of idealization (and sim-

---

35 The argument from doubt is from fallibility. It is decisive against capital punishment. During my early days in England the case of Timothy Evans was in the public eye: he was hanged in 1950 and fresh evidence turned up in 1953 that disproved the guilty verdict. The Home Secretary admitted error [Times, 27 Feb. 1956]; [Woffinden, 1987]; [Independent, 24 May 2001]. The abolition of capital punishment there took place *de facto* in 1965 and *de jure* in 1999.

36 In 1981 the Austrian College in Alpbach honored me by an invitation to open its summer session. I chose to speak of rationality and truthfulness. I offered a trivial thesis: truthfulness ins necessary for those who wish to invite criticism and thus to have some rational assessment of their views. To my surprise, this provoked immediately immense hostility: the Austrian minister of science attacked me instead of reading the usual ceremonial greeting address. I could not get the paper published and was not re-invited. Things may change now, as mechanism design theory won recognition (Nobel prize, 2007), and as this theory describes the economic usefulness of truth-telling.

37 The word "friction" denotes marginal effects. Writers who rightly recommend ignoring them often sadly forget to qualify their recommendation to the initial stages of research: they forget that their judgment is context-dependent. Most of the empirical evidence against Popper's theory of science relates to friction. It obviously counts to the same extent that the criticism of Galileo's theory of gravity from friction does. Both invite decision as to the conditions under which they tolerate deviations. What theory is science-with-friction and what is non-science? The distinction is not sharp. Sharpening it requires the formalization of scientific theories that may be gained only after much work. (Lakatos rightly spoke against some early formalization, even in mathematics.) Even demands for increased clar-

plifying matters), in Spinoza's sense of reaching the ideal (of attaining peace),[38] in Kant's sense (of regulative idea)[39], and in Max Weber's sense (of ideal type, the Platonic Idea or the ideal image of the scientist).[40] This leads directly and elegantly to a number of suggestive avenues of developing Popper's critical philosophy. This is also advantageous for his critics, however: the severest and best known criticism of his philosophy is that abhorring apologetics is an (established) ideal, not a reality. It is a lamentable fact that Popper himself rejected this criticism out of hand, hesitating between claiming that it is utterly false and that it is in no conflict with his view, as his theory was meant to be a proposal, not the true description of the facts of the matter.

All this said, the application of the critical theory to education is important and should be aired. I had a few occasions to employ it very successfully; I will report one later on.

I did not finish my story of the initiation of Jarvie. He was not the easily deterred type, and he was genuinely interested: he persisted. (Perhaps rudeness has the function of deterring people from painful casual relations; possibly rudeness humiliates as the price exacted as proof of eagerness, of true desire for friendship. This is the explanation often given

---

ity may be excessive, not to mention the demand for an absolutely clear language for science that was the battle cry of the "Vienna Circle". Popper's alternative suggestion is terrific: the study of the content of a theory should lead to decisions on the limit of the degree of deviation that it permits. This makes the content (and so the meaning) of a theory depend in part upon a decision that should precede tests. To the extent that empirical evidence helps determine this decision, Popper's critics are right in asserting that such evidence may comprise harmless refutations; they are in error when they declare this a refutation of his view, since he explicitly allowed for it (with no *ad hoc* need for any assumptions).

38 The literatures on peace of mind, East and West, ancient and modern, overlook the contribution of criticism to it. This is masked by the common view of criticism as negative. Donald Hebb notes [Hebb, 1958] that his graduate students were upset when they refuted their own hypotheses. In response to this he told them not to take things too seriously. Eccles was upset the same way. Popper helped him by telling him that refutation is progress. He commented thus [Bunge, 1964, 275]:
   "Thanks to my tutelage by Popper, I was able to accept joyfully this 'death' of the 'brain child' that I had nurtured for almost two decades, so that I was able to contribute immediately, both theoretically and experimentally, to the successful rival ... hypotheses."

39 Kant introduced regulative ideas or ideals quite *ad hoc* — to allow for the extension of his theory that he hoped to implement (*Critique*, A204; B249). His proof for their validity is that he needed them. He deemed the need for an idea a transcendental proof of its validity; this renders all his ideals valid quite contrary to his intent. This is no accident: a classical rationalist, Kant polarized assertions to the arbitrary and the scientifically valid, dismissing the arbitrary as counterfeit. This excluded ideals, as they are heuristic devices and thus mere suggestions [Agassi, 1977]. Whewell destroyed the idea that ideals are proven, a priori or otherwise, by allowing different thinkers to follow different suggestions. This does not hold for the ideal of the absolute truth, however. As this cannot hold for Kant's ideals, he invented a Pickwickian concept of truth [Agassi, 1988, 60]. Whewell restored it, and postulated proof as both *a priori* and *a posteriori*. This is excess. Mach went the opposite way: he said, we have only one intellectual need: the need for economy. Ignoring the Kantian roots of Mach's philosophy, Planck and Born trivialized it and made it look silly: economy is easiest to serve, they said, by simply not thinking.

40 With all the extant applications of Weber's theory of the ideal type to diverse cases, ideal scientists are conspicuous in their absence [Agassi, 1981, 73-4]; [Jarvie, 1994, 50]. The nearest to this is Robert Merton's list of scientific values. Was Merton a Weberian? He could not decide, not even when he reported the hostility in the early days to his ideas as quasi-Weberian.

for the rudeness Beethoven regularly exhibited.)[41] I was delighted to observe that Jarvie reappeared and became a regular. He did suffer clashes with the philosopher, but to my relief there was soon little or no fear of hurt: the two became friends for life, with all the usual ups and downs that such lengthy relationships usually include. I do envy Jarvie, of course.

The logic lecture course was a real masterpiece lost on most of the participants who were beginner students in the social sciences with no interest in the material they were hearing, much less in its revolutionary character. It is foolhardy to recapitulate a pioneering idea for a whole course in one paragraph, but let me try, if only in an effort to convey the wonderful flavor: the torture of the dull, incomprehensible run-of-the-mill logic course contrasts sharply with this flavor, I hope. The philosopher offered a short history of logic as the theory of both rational debate and science — this is in line with Aristotle — via stoic logic and Boolean algebra, leading to the logical paradoxes. I never found anywhere a discussion of them remotely as satisfactory as what I heard in those lectures, barring a brief remark of Fraenkel [Fraenkel, (1953) 1968, 71] and the more detailed discussion in Fraenkel and Bar-Hillel [Fraenkel and Bar-Hillel, 1973, 1-14, 275]. He then explained why the paradoxes require an artificial measure to overcome them. This, he added, refutes the naturalistic view of language, the idea that the rules of grammar and of inference (or transformation) must suffice because they are fixed (as inborn or as ideal). He then defended conventionalism and discussed in this spirit Tarski's idea as the idea of artificial languages (where there is a sharp divide between formal and descriptive words). Artificiality is now much easier to comprehend because of the popularity of computers and the artificial character of their programs. (He presented the theory with his own important modification which can be found in his publications on logic and which is still scarcely commented upon.) He then gave a terribly simple and lovely proof of Gentzen's completeness theorem: defining a

---

41 See the note on friendship in Chapter Five. Popper viewed Beethoven's famous *Heiligenstädter Testament* as a self-portrait. He saw it as full of despair overcome with great effort and invincible optimism [*Autobiography*, §13]. His criticism of Beethoven's expressionism notwithstanding, he identified with his person: hard working, suffering endless physical pain due to chronic illness, embittered, and still hoping that his work will bring him posthumous recognition. Popper suggested that Bach's objectivism fit Beethoven's character better than his own expressionism:
    "Bach forgets himself in his work; he is a servant of his work. Of course, he cannot fail to impress his personality on it; this is unavoidable. But he is not, as Beethoven is, at times, conscious of expressing himself."
Not so: to be conscious of expressing oneself goes very well with expressionism but not necessarily so. To my regret Popper refused to discuss this with me. He found it not sufficiently important. Perhaps it was too important for him to expose it to criticism, and so not in the best critical spirit. Popper mentioned to me Bach and Beethoven as hard working in response to my objection to his attitude to hard work. He was surprised to hear me agree with him about the facts but again he stopped the discussion. Much later, when he wrote his autobiography, he told me he thought it would not appeal to me because of the sentiment he expressed there regarding the obligation to work as hard as possible.

theorem as a conclusion that makes any inference to it valid and a tautology as any statement true in all substitutions, he most elegantly and easily proved their equivalence in any formal language. This, naturally, disposes of the Aristotelian theory of science (as it permits only tautologies to be provable) and what remains then of logic is the theory of formal proof (by the transmission of truth from premises to conclusions) and of critical debate (by the retransmission of falsity from conclusions to some premise). The calculus of composite statements comes next, and the proofs of the equivalence of the axiomatic method, the truth-table method and the (forcefully simplified) method of definition of the logical operators by their role in natural deduction — always as exemplifying the idea that the role of logic is to detect error by the repeated and central use of the idea that a valid inference is an inference that does not allow valid deduction of falsehood from truth. Boolean algebra, including (greatly rectified)[42] Aristotelian logic, then follows the definition of Boolean inclusion (as, given that definition, the transitivity of the conditional guarantees the transitivity of Boolean inclusion). And all this in a one-year freshman service course!

It was a terrific course. The philosopher once gave permission to Paul Feyerabend and me to write it up jointly and publish it. He took this back very quickly: he was never satisfied with his own work, let alone with the work of others[43], especially two unruly novices. I have emulated the course in different places, with a bit more accent on the disagreements between the leading thinkers as to the nature of logic and of language. In my last summer at Boston University I taught the course there on logic and language designed for students too limited to do what counts as real introductory logic (like the terribly boring and pointless proofs of composite tautologies). These allegedly limited students ended by working out the sketch of Gödel's theorem in Nagel and Newman [Nagel and Newman, 1958]. I know: it is not "in" [Putnam, 1961].

It pays to treat students as intelligent beings and to explain one's intentions and reasons for presenting to them whatever one does. The trouble is, many teachers are too ignorant of the reasons for the traditional curriculum and present it as mere good form. My model teachers were mainly the mathematicians Abraham Halevi Fraenkel and Menahem Max Schiffer in Jerusalem and, of course, the philosopher. I learned from them not to follow the trodden path as a teacher but to present to my students even the most common ideas as responses to some questions that have whetted their appetite. The courses in applied mathematics with Schiffer

---

42 The concept of judgment drops out with all formalization. Rectifying Aristotelian logic by formalizing it is thus not unique. Compare [Lukasiewicz, 1951] and [Bar-Am, 2003].
43 Popper was dissatisfied with at least three translations of his *magnum opus* — by Joseph Woodger, by John O. Wisdom, and the one by Lan and Julius Freed that he used as the draft for it [*LScD*, 6].

were real masterpieces and a proof that the knowledge of the rationale of an exercise greatly facilitates its comprehension. I was stunned to realize that a part of Schiffer's slow and easy course covered with pleasure the material that was covered in another full course that was tough going and required much drudgery.

The trouble is that for some reason or another, students often refuse to have their appetites whetted, and the question is, what is a teacher to do then? In the London School of Economics this was unthinkable. In most other places I have taught this is usual and I do not know what to do about it. I know what not to do. I refuse to be a bore. Wherever I teach, I find work with concerned students a great pleasure and teaching indifferent ones painful: I am lucky tenured professors do not depend on grades.

I was graded once, come to think of it: in Boston University's students' evaluation publication (a great idea, this sort of thing, provided its misuse is in check). Though my grade then was very low, I felt complimented: believe Professor Agassi, the report began: he means exactly what he says. This is a gentle way of indicting our profession for bad faith, and who can deny the indictment? After all, the bad faith is usually so obvious that it is covered up by the fig leaf of motivation, as if students have to be given some concocted reasons for doing what is in their own best interest. The reason I was graded low, incidentally, was admirably also provided: students found it wrong that usually they had to make me talk if they wanted to hear me, and first say specifically what they wanted and why. And this explains much of the situation. The bad faith is there because professors are pressured by students to demand less and by the administration to demand more. The classroom is thus a site for concealed haggling between students and administration with professors in the middle.[44] The assumption, then, is that learning is not naturally desired by students but imposed on them by the university administration. A famous president of Boston University said this out loud and emphatically and in many insulting ways [Agassi, 1190c].

Assuming that students do want to learn, or that they know that learning is in their own best interest, why, then, do they behave so differently in the classroom from grade-school to college and beyond? It is a matter of traditional perversion. Sexual appetite, after all, is much more

---

44 The logic of the situation of the teaching faculty is obvious. Refusing to see the obvious, they can do nothing to rectify their inbuilt conflict of interest, as they ignore the observation (of Abraham Flexner) that all grading is of both teacher and student, so that it is wrong to appoint one person as both teacher and grader. The wishes of students for cheap, high grades clashes with the wish of the administration for a reputation for excellence. Both parties put pressure on the teaching faculty. Administrators demand grading on curves in efforts to prevent cheap grades, thus implying that all classes in all schools are equal in talent and in investment of effort. Consequently, every year some hopeful high-school graduates armed with local top grades arrive in top universities and run into trouble at once. They drop out silently. The system thus causes them unjustifiable damage and misery. The on-line universities may force the administrations of traditional universities to be less callous.

natural and much stronger than the desire to improve one's mind, and yet throughout the ages our culture has managed to make many people find sexual intercourse enormously repulsive. Nor does one need be a psychoanalyst or endorse this view[45], as is evident from Tolstoy's perceptive detailed description of this hate in his novelette, Kreutzer Sonata. As we can spoil the fun of sex, surely we can spoil the fun of learning too; and the hatred of science or of the arts is notoriously popular among art or science students. And who can deny that this hatred is not inborn but bred? My having made my students demand more or not get it, and their finding themselves having demanded more education and more answers to questions that interested them, simply brought to the surface their ambivalence about it, or, in their own words, it confused them. It is no surprise that the said president of Boston University disliked me and wanted me to leave on the pretext that I am a controversial teacher. Truly I am.

Why, then, do teachers who treat students as adults so often fail to have their students demand the education they deserve? The answer was given by my late younger brother who was a superb teacher in a relatively liberal high-school and tried to teach in a truly liberal way and failed too: when teachers put much pressure on students, then the likelihood is that those who put much pressure gain all the attention and so no attention is left for those who put less pressure — unless students are adult enough and clear enough and determined enough and knowledgeable enough to make an effort to prevent this process. When they do, the result is exhilarating; when not, it is a sad failure that I, for one, find very hard to take in my stride despite repeated experience this way.

I debated education with the philosopher a few times. His opinion always was that children are authoritarian by nature and that teachers have to charm and educate them in an authoritarian manner — in order to have them grow out of their authoritarianism, need one say. I do not agree: A major argument in his *The Open Society and Its Enemies* is, after all, that we do not know what human nature is [though we may refute some views about it and avoid immunizing them against refutation defensively[46]]. His view is refuted by democratic schools where authority is democratically controlled and pupils learn no worse than in authoritarian schools. (The fate of graduates of democratic schools, as far as longitudinal studies can show, is not different from that of others in similar situa-

---

45 Freud's *The Ego and the Id* presents his hostility to sex as aversion to the loss of semen [Freud, 1953-74, 19, 3-66]. Did he approve of Tantric sex? He could have, as some Western authors of his day discussed it (and suggested that educated Easterners are free of sexual inhibition).

46 Ideas about the natures (or essences) of things are usually seen as irrefutable definitions. Perceived wickedness then does not refute the view of humans as essentially good, as it is allegedly unnatural. The same goes for perceived goodness. This is only one option. Some people admit valiantly that the extremist views of human nature are refuted by certain experiences, such as the horrors and the heroism exhibited in horrendous conditions such as the Shoah [Agassi, 1977].

tions: some fail some do not; those who do not fail constitute the refutation of the tradition or at least the hope for improvement.[47]) Worst of all, rationing love as a means of pressure is the unfailing source of ambivalence and, as such, more harmful than rationing food. Popper's support for the romantic element in education [*OS*, ii, 275-7] amounts to his making allowances for manipulating pupils for their own good.[48]

Popper's Tuesday Afternoon Seminar was famous all over the London academic community and beyond. I brought there guests regularly, including Imre Lakatos, who took it over when the philosopher retired — in 1969 I suppose, since my talk there in 1969 was chaired by the philosopher and my next talk there was chaired by Lakatos. I understand the flavor of the seminar stayed until the philosopher retired. (And then, I noticed, Lakatos made a mockery of it: after my talk there, he said the talk was not worth discussing and proposed to discuss someone else's paper instead. Charming.) My last talk in the presence of the philosopher was short enough — it is published in *The Philosophy of Karl Popper* — and when I finished it there was incredulity in the air and someone said, no one finishes a paper in Popper's seminar! My success was due to the paper's brevity and to my not having dodged comments. He really was easy to please, yet somehow too many members of the seminar did not know how to do it. It is hard to answer the charge that this way he was somewhat manipulative.

Example: just before I left London a young Australian lecturer in economics was obliged to read a paper to meet the conditions of his grant and since he was a member of the seminar he wanted to read a paper there but was extremely reluctant to put himself in the position of extreme frustration, of not being allowed to state his thesis — an event he saw happen to many of his predecessors there, since the philosopher would interrupt everyone more or less on the spot and see to it that they were as frustrated as possible unless they did things the right way (whatever this was). That economist was advised to consult me. He did. We worked on his paper for some time. I could not help him beyond advising him strongly to declare his aim and state his thesis in the very first sentence; and having expressed my fear that postponing this task to the second

---

47 Longitudinal studies of education are very scant and there is almost no discussion of the variety of methods for its assessment. Perhaps American community colleges lead the way [Heaney, 1990].
48 In progressive schools teachers often ration love as a potent means for breeding ambition. Popper expressed strong opposition to them on psychological grounds: this conduct is bound to create severe neuroses. (Rationing food does not create the emotional conflict that rationing love does.) These schools are nevertheless liberal in some sense, but not democratic. In democratic school the hypothesis is that children are curious; this is true but it is no solace to children whose natural curiosity was destroyed in "normal" schools and who are in need of rehabilitation and much patience. Most alternative schools work on the supposition that the removal of the pressure will do that all by itself. (A. S. Neill almost says so; Russell was more realist but he left his school after a very short time.) When it does, it is a great boon. But this is not always the case.

would be too late, I felt the choice was his. Yet it pained me to see him struggle for well over an hour to say the second sentence of his paper: the philosopher wanted him first to give up the first — the theory of science as the body of solid knowledge based inductively on facts, wouldn't you know — a minor matter to settle fast before reaching the agenda. Naturally.

The best seminar I attended was in the year 1952-3, which I attended only in 1953. It had a theme: "the debunking of Man". The philosopher always responded to whatever paper was read there, from any field, most intelligently and knowledgeably — though, of course, hardly ever fairly. He was exciting. He combated positivism when practically everyone around was a positivist of one color or another, presenting all sorts of very clever arguments against it from commonsense, from science, from metaphysics, taking commonsense realism for granted and arguing that positivism devalues humanity one way or another. Unless pain is real, he said, perhaps its prevention is not as urgent as it seems. It was the most exciting seminar I have attended, and the only thing more exciting comprised the private meetings with the philosopher in his home.

This is no endorsement of the argument just presented, as I do not know what he meant by "real".[49] (In traditional philosophy any substance, any immutable thing is real; by this standard both pain and pleasure are not real, of course. But as today most philosophers do not discuss substance, the word "real" cannot retain its traditional meaning.)[50] But this is another story. Let me only say briefly here that I see in this mode of arguing for realism and against positivism a remnant of the philosopher's own early somewhat positivist, rather anti-metaphysical stance: this way he tried to solve too many problems which traditionally belonged to metaphysics, seemingly with no recourse to any metaphysics, or taking recourse to commonsense, namely, to commonsense metaphysics. For my part, I prefer the mode he adopted later in his career, that of contemplating how different metaphysical views offer different avenues for solutions of given problems, or, more generally, how different metaphysical systems offer different criteria for the adequacy of solutions, and so on;

---

49 The traditional philosophical term "realism" denotes the view that matter exists as a substance (since its existence in the common sense of the word is never denied). Transporting it carelessly to modern philosophical contexts that ignore substance is bound to confuse. Analytic philosophy traditionally identifies all senses of "realism", the traditional philosophical, the commonsense, the modern philosophical, and the rejection of the Oriental idealist slogan, "the world is a dream" (not to mention the terrific idealism that Lewis Carroll invented jocularly: we are a part of the dream of Red King, destined to disappear upon his awakening). This is bound to confuse.

50 The vague commonsensical meaning of realism and of materialism survives the doctrine of substance. Russell, a staunch realist and a confessed materialist, said [Russell, 1927, 254],
"the notion of 'substance', at any rate in any sense invoking permanence, must be shut out from our thoughts if we are to achieve a philosophy in any way adequate either to modern physics or to modern psychology.".
What is here (putatively) permanent is (putative) natural law [Agassi, 1975, 233-9].

and then debating the pros and cons of the various metaphysical systems by examining such solutions, of course. Attempts to solve metaphysical problems with no metaphysical assumptions are barren.

Perhaps I should present here my reading of Popper's attitude to metaphysics and its change through the years.

Popper's earliest publications, known only to some specialists, are charmingly anti-metaphysical in the (Machian) style of the day. In his first vintage, his *Logik der Forschung* of 1935, he demanded that every theory should be rendered empirically testable and severely tested prior to its endorsement. This demand expressed an austere attitude towards the endorsement of opinions. By contrast, the respect for religion that he expressed in his *The Open Society and Its Enemies* of 1945 is permissive rather than austere. He reconciled these two attitudes in two distinct ways. First, already in 1935 he declared the endorsement of metaphysical opinions private. In line with this, in 1945 he declared religion a private affair, despite its being an institution and thus, still in his view, given to rational reform. (He even misread the Sermon on the Mount which is a famous conservative manifesto as a plea for reform; and this in a book in which much more defensible misreading — of Plato's texts — was sternly exposed. [51] When I said this in his seminar he responded with a denunciation of "the" Jews of Vienna of his youth (which has since found its way into his autobiography) and promised to make amends in a later edition of his book if I would refrain from publishing my criticism. I postponed publication of it for over a quarter of a century.) Second, he changed his opinion and declared criticism a broader category than tests, thus admitting the possibility of criticizing moral (and historical) judgments [*OS*, Chapters 11 and 24]. He then declared metaphysics open to criticism though not to empirical refutation, especially since already in 1935 he had recognized the ability of a metaphysical theory to become scientific through the increase of its contents [*LScD*, §85]. (My doctoral dissertation is a mere elaboration of this, rather traditional recognition.) Also, he adopted the view that metaphysical theories are points of view so that they can be pitched against their contraries (for example, we can write political history as the history of freedom or as the history of slavery, without having to endorse the metaphysics that humans are free or slaves [*OS*, Chapter 25]). This is Kant's advocacy of "the principle of neutrality in all disputes" in matters metaphysical (*Critique*, A756, B784) as echoed by Ernst Mach, who was the leading philosopher in Vienna early in the twentieth century. Both of Popper's moves are permissible, but they are

---

51 Popper presented religion as a private affair [*OS*, i, 65-6, 235, ii, 23-5], institutions as hypotheses [*OS*, i, 126, ii, 260], the Sermon on the Mount as reformist [*OS*, i, 65], "the" ideas (philosophical or religious?) of "the" Jews as (essentially?) tribalist and (thus?) historicist [*OS*, i, 19, 21, 24-5, 55, 75, 78, 84] and their conduct as arrogant [*Autobiography*, §21]. We may dismiss all this as blemishes.

rather unpleasant, as they make some ideas mere second-class citizens. Popper's admiration of the intellectual value of modern science is impeccable and so is his critique of the "Vienna Circle" denial of legitimacy to all non-empirical assertions. His low view of all non-science, however, his grading of even the greatest and noblest non-scientific one below any scientific idea, is somewhat distasteful, and presumably unintended.[52] In any case, science can do nicely without degrading non-science. Indeed, Popper's later writings are more appreciative of some non-scientific ideas and even of their contribution to science[53] — in great variance to his earlier writings. Had he discussed this change more clearly and openly than he did, he would have noted the excess of his critique of some pseudo-science[54], such as Freud's views. But here his moral condemnation of psychoanalysis enters the picture. I will return to moral condemnation later.

Let me return to the conduct of intellectual discussion, private or public. It is always reasonable to keep the ego less involved than the love of learning. My Jewish education was in the company of all sorts of Jewish scholars, from the most naïve to the most sophisticated, from the most temperamental to the most placid, and all of them would be distressed to see how much ego-involvement interferes with the intellectual love of God in the secular commonwealth of learning. And this is so from the very beginning (as documented in my doctoral dissertation that sketches the rise and growth of the modern scholarly traditions). The philosopher, a most avid devotee of debates, was extremely sensitive and habitually read his interlocutors' objections off from their facial expressions (espe-

---

[52] Putting down any idea is unpleasant, but at times it is admittedly necessary. Einstein dismissed such things as UFOs (that by "logical" positivist criteria are empirical), and he did so light-heartedly, never addressing the philosophical issues involved, as he deemed them rather trivial. This way he also dismissed off-hand the view that certain knowledge is possible. Yet he said a harsh word against the view of knowledge as synthetic *a priori* valid. For, he assumed that it leads to conventionalism, and he viewed its attraction as still strong and pernicious. Only on this point of methodology did he take Popper's writings as challenging rather than as obviously true. By contrast, Popper began his methodological studies by condemning all non-science, and his views slowly mellowed. Maimonides went even further than Popper ever did, viewing his religion as scientific (and opening his *Codex* with a Book of Science) and blaming astrology for the destruction of the Second Temple, namely for all of our woes. (This judgment is naturalistic when understood sociologically and magical otherwise.)

[53] Popper has observed [*LScD*, final section] that the ability to render a metaphysical theory (like atomism) scientific refutes the "Vienna Circle" view of them as inherently ungrammatical. He nevertheless rejected the suggestion to use metaphysics as a heuristic (*loc. cit.*). On this he valiantly changed his mind [Popper, 1982, final chapter].

[54] When popular-science writers present metaphysical theories that serve as programs, they usually imply that their implementation is completed. This renders them pseudo-scientific. This is an understandable pitfall, especially in view of the need of these writers to yield to the aggressive mode of the system of public-relations of science [Agassi, 2003, 2.6]. Freudians and Darwinians often err this way, except that only the Darwinian program has helped develop very powerful researches [Fried and Agassi, 1976 and 1983]; [Agassi, 1996b]. (As geological and other evidence piles up, Darwin's program is for ever unfinished. As to Freud's program, his evidence is unrepeatable, it is unbegun.) Popper tried hard to clarify this matter; he seems to have yielded to pressure from the Darwinian Establishment to declare Darwin's theory scientific rather than a program for scientific research.

cially since his deafness worsened with time). For lovers of wisdom this is as God-given as dreams come true, of course; not so for those who must hear their own voices and feel that they contribute to debates with giants. And when they feel defensive, they can declare Popper's intuitions faulty. Alas, the result was that he always suspected that all questioning of his intuitions of foul play. Paul Feyerabend[55] told me that in Vienna after the war the philosopher's reputation was still that of one who would not let you get a word in edgewise. Do not consider this the standard of the then Viennese philosophical community; this standard, common as it is, is too high to ascribe to them: they did not think much then — or later for that matter — of their having driven out of Vienna the whole of the "Vienna Circle", Jew and Gentile alike. Their conduct towards the philosopher and towards other Jews

Yes, of course; he should have been intuitive enough to notice his interlocutors' impatience; you are quite right. He was. He repeatedly discussed this matter with me. He said he could not help it: it was incumbent on him to take his opponents seriously. I do not agree: when others request one to indulge their failings rather than pretend that they are serious, then, respect for them requires indulgence, not pretense. Moreover, the pretense is useless: it brings about bad reputation instead of the intellectual love of God. Furthermore, it may be suspect as having to do less with respect for others and more with the need for their recognition, with ego involvement. One is always suspect when one does not take others at their word — even out of seriousness. And the philosopher often used shock tactics; most effectively, I admit, but this is no excuse — a redeeming quality at most. With honest interlocutors, shock tactics could work miracles and elevate them from uttering poor views to new heights. But usually

Yes, I agree: we all use shock tactics, often unknowingly, with this or that frequency, with this or that excuse; I am speaking here only of the (objectionable) conscious (or careless) use of it. So let me conclude. The few years I spent in close contact with the philosopher were the most seminal in my whole life. Under his tutelage I concluded my years as a student — which began with failure and misery and, with his help and guid-

---

55 Excessive sensitivity was characteristic of Feyerabend too: he was extremely sensitive to the possibility of causing discomfort to anyone around; the very thought of it visibly filled him with anguish. This made him excessively shy. In his young days he used to hide in any possible manner. For example, he used dark eyeglasses even at night. I dissuaded him and helped him come out a bit more. I do not think that this was to the good. For, when the anguish was excessive he would all of a sudden become insensitive and as coarsely assertive as his tremendous intelligence enabled him to be. He then enjoyed embarrassing his company in front of their friends, at times by repeating tactlessly things they had said thoughtlessly or in confidence and at times by aggressive public debates with big shots. He once intimidated Max Black, which is no small feat. I stopped him and it angered him tremendously that I did. Some of his written expressions are extremely aggressive too. I repeatedly drew his attention to all this. In response he was always contrite, but he did not change his behavior pattern. I presume he could not.

ance, ended with excitement and progress: I learned from him how to write and how not to, how to argue and how not to, what signifies and what not, and how to try to do one's job as best one can. It was my period of apprenticeship with the philosopher, no doubt, that was intellectually[56] the most exciting time of my life. The years I spent as his colleague were no less exciting, but in different ways: they were intellectually not as profitable.

---

56 Life around Popper was exciting both emotionally and intellectually, of course. My greater emphasis here on intellectual aspects is due to the context.

CHAPTER FOUR: AT THE FEET OF THE GREAT THINKER

> Here faith is not demanded ... One repays a teacher badly if one remains only a pupil ... Now I bid you lose me and find yourselves; and only when you have all denied me will I return to you.
> (Nietzsche)[1]

My period of apprenticeship lasted seven years all told. I met the philosopher first in his seminar in the London School of Economics in the very beginning of the year 1953, when I was in my mid-twenties and he was twice my age. Less than one year later, I was to become his research assistant and later on his colleague at the School until August, 1960. We were extremely close for a while, when I worked with him on all of his manuscripts. He then announced that we were tied by a sort of family relation. He kindly expressed great appreciation and high hopes for me at that time and he generously offered me a lectureship in his department. I was determined to leave: his personality was too strong to resist in proximity and I could not let him succeed in his effort to mold my character in the light of ideas that I had refused. These ideas may be correct, of course, but I would not allow him to force them on me without debate. Of course, he had no time for lengthy debates, as he was so overworked. I could understand that. Yet one of his ideas that I refused was that we should have much work to do and struggle in a losing battle to do it; another was his view that ethical issues are uninteresting and debating them is a waste of precious time. The worst of it was, he simply could not discuss ethics with me: for one thing, he was too busy for it, as he had to work very hard; for another, even while solemnly expressing willingness to pay attention to my agenda for my sake and so showing readiness to discuss with me what interested me, even then he sabotaged discussion on ethics. So I had a simple choice: I could leave and I could stay and quarrel and remind him repeatedly that I was an apprentice no longer.

When my apprenticeship was over, however, I was not quite let go. I left as soon as I could. We parted company amicably after almost seven years of intense cooperation, in August, 1960, when I left London for a post in Hong Kong that I would not have been offered without his kind help. (I would have left a year earlier, as I was offered a Rockefeller fellowship that I gave up for a junior post in the University of Chicago that I gladly accepted, but I was barred from entering the United States as a Communist of sorts. I can hardly express my gratitude to the philosopher and to Sir Walter Adams, the then Director of the London School of Eco-

---

1 [Nietzsche, (1889) 1969, 1, 22, Foreword; *(1892) 1969, 1, 22*, "On the Gift-Giving Virtue"].

nomics, for their extremely generous last minute extension of my position for a year.) When we parted, he observed with satisfaction that although I had come to him in very poor shape, I left ready for a position of leadership in the profession: he thought I was ready to take over.[2] It was extremely flattering and yet I was not impressed — much to his surprise and incomprehension, even some hurt. But I was not moved: I knew what he had demanded of me and I knew he still hoped I would follow his example — personally and professionally — and as I knew full well I was determined not to follow him in ever so many ways — ways that I deem comparatively marginal but that he took extremely seriously: for example, his opinion that ethics is a dull subject for debate. I cannot abide by that opinion, even on the doubtful supposition that it is compatible with taking one's duties very seriously. Incidentally, interestingly, he took the Socrates that emerges from Plato's early dialogues for a personal model for a philosopher (in all matters except for the technical error that presumably Socrates made when he endorsed Pythagorean essentialism), yet he ignored the fact that Socrates had lost interest in physics.[3]

Let me make do with one small example of the philosopher's dedication to work[4] that I found excessive. He read proofs with a passion. He took it for granted that misprints are unavoidable yet he tried his best to weed them out, and invested endless efforts in these attempts. His poor eyesight, even his detached retina — he nearly lost his eyesight once — did not stop him, all pleas to the contrary notwithstanding. While I worked with him, I did the same, of course, but with no intent to make it a character trait of mine. I told him so. In response he said, this way I would bar myself from having my ideas published: editors and publishers

---

[2] As I was leaving London, Popper told me of his assessment of my capacity as a possible future leader. He thus came closest to admitting that he was one. He should have kept to this consistently. Sadly, he could not. Another time he admitted his being a leader was much less agreeable. In 1962, he took exception to my public response to a review of his *The Logic of Scientific Discovery*. Although it was a response to an attack on his book, he said scornfully it did nothing for him. This troubled me: my intent was to respond to a reasonable challenge, not to serve anyone.

[3] Plato, *Phaed.* 99d. Cp. *Apol.* 19. Aristotle reported that Socratic essentialism is Pythagorean. G. C. Field [Field, 1925, 13] disagrees: "... there is next to no evidence to connect Socrates with the Pythagoreans; in fact, most of the evidence, such as it is, is rather the other way."

[4] There is no need to praise the joy of work as one of the greatest gifts in life. Paul Lafargue noted this, *How to be Idle*, as did Jerome K. Jerome, *Idle Thoughts of an Idle Fellow*, Shaw, Preface to his *Parents and Children* (*Misalliance*), Russell, *In Praise of Idleness* and *The Conquest of Happiness*, Chapter 14, and William Somerset Maugham, *The Summing Up*. It even appears as the apex of the autobiography of movie star Bette Davis. The perseverance in work is even a greater gift, though less often praised. I owe it to the training that Popper gave me. He told me how important he always considered challenge and how intolerably boring he found routine work. On this we disagree: routine work too can challenge, as Jack London describes in his *Martin Eden*. Challenge is personal and variable, anyway, unlike its absence that is always the cause of textbook-style depressions. It is obvious at the entrance into and exit from school and workplace. (Teachers often mistake it for laziness and leave it unattended.) The educational reformers Maria Montessori, Homer Lane, and Janusz Korczak are admirable fighters against boredom, as are reformers of the quality of working-life (QWL), Einar Thorsrud, Fred Emery, David Herbst, Eric Trist, Louis Davis, Bertil Gardel and their associates.

do not allow my attitude. I shrugged my shoulder: then I will not have my output published. This greatly annoyed him. Rather than further debate the morality of my choice, however, he tried to train me. I repeatedly asked him not to talk to me of misprints, and I suppose I was childish taking his utter disregard for my wish as a personal insult, but I did take it so. I knew I had to leave London and strike out on my own elsewhere. Once, much later, we discussed my refusal to stay. I explained and gave this same example; he then started talking to me about misprints. I reminded him of a particular misprint of which he had spoken with me many times despite repeated pained pleas that he avoid doing so; this he remembered; so he went to the bookshelf and pulled out the book in question and showed me that same misprint yet again. It was too much for me.

(In 1975 I published my *Science in Flux*, a voluminous collection of essays that I dedicated to the philosopher — "in gratitude, admiration and dissent", the dedication ran. His response reached my ears in no time. It was negative: I should not have done so without seeking his permission — which he would have withheld, of course. (There is no rule on such matters.) One review of that book was by another former assistant and apprentice to the philosopher, David Miller, the editor of the exceptional *A Pocket Popper* and, I understand, the philosopher's heir, now that Bartley is dead. (I gather this from the interview published in the superb [Koelbl, 1989].) I had important things to say, Miller noted in his review [Miller, 1978]; but rather than tell his reader what, he filled the body of the review with a list of some of the numerous misprints in the volume and some of the humorous entries to its index. Was he obliged to act in such a manner? I wonder. I do not know: we met subsequently and on quite friendly terms, but I did not raise this matter with him.[5])

I never returned to London for more than a couple of days, a week at most, and our relations deteriorated almost at once. In 1962, to repeat, he generously visited me in Hong Kong and we did not hit it off well. In the same year I visited him in England for a most dramatic encounter, of which more later. In 1964, when I was teaching in the United States, Imre Lakatos managed to convince him that I was scheming to undermine his influence, such as it was, and take over whatever there was to take over. He created enough misunderstanding to be able to maneuver the philosopher to a decision to sever relations with me — and then with Bartley, and he would have gone on and on but he then isolated the philosopher himself: it was the way Lakatos built his reputation in the world. (His model was not Iago but Stalin.[6]) Yet I would not blame him for my rift with the

---

5 Later Miller and I did discuss his review of my book. He rejected my surmise then, saying, he had never discussed the matter with Popper, and he is a pedant in his own right. So I withdrew my surmise. He also took exception then to my description of his review. On this we differ.
6 Stalin used to pacify his victims. So did Lakatos, usually with no more than an apology in private. After

philosopher: he could not sway the philosopher without appealing to his famous immense sense of persecution [Munz, 2004, 50-1] that I had occasions to witness repeatedly while I was his assistant and close associate. (In particular, in agony about his deafness he quarreled with the telephone operators in the School and with the phone company at his residence.)

Things got complicated: as close associates for quite some time, we still had a lot in common even after he broke off relations with me; so often we simply had to rely on mutual friends to transmit hidden messages between us, like children cross with each other, and usually the messages were from him to me, usually reprimands. At times we would communicate, usually at my incentive. After each exchange he declared strong reluctance to continue. On and off we met a few times, but most of the time he was not on speaking terms with me; he could still call on me any time he liked with the assurance that I would do anything he wanted (other than agree to a thought I deem erroneous or to the suppression of my thoughts and my criticisms, whatever these happen to be). Though he sharply and clearly denied that he had ever expressed the wish that I refrain for a while from criticizing him in public, I equally clearly affirm this: he said so to me and he said so to other of his associates and visitors. Obviously, at stake here are not the observed facts — we were both reporting matters *bona fide* (he was at times of stress prone to ask, are you calling me a liar? The answer, as far as I could ever make it, is always clearly in the negative: no one has ever called him that, of course[7]); so the disagreement was always a matter of interpretation. He constantly reaffirmed his readiness to accept criticism and he constantly considered the publication by his associates of criticisms of his opinions a slight on his character, an expression of the refusal to accept his promise to admit the criticism and publish it with acknowledgement.

It is possible to take this as his refusal to accept criticism. It is also possible to take this as a refusal to accept his technique of dealing with criticism of his ideas. And it is possible to take this as recognition of the fact that he made promises and forgot them all too often. One way or another there is a clear case of disagreement and a need to discuss the background for it. The disagreement was clear: he wanted us to come to him in private and offer him our criticism, and he promised to accept it in

---

many such occasions I refused to accept his apology until he made it as public as he had made his insult. He shrugged his shoulder and never apologized to me again. The most revealing moment in my experience of his attitudes was early in the day, after we viewed together Sergei Eisenstein's *Ivan the Terrible Part 2* (released in 1958), the film that is still the best portrait of Stalin available. Of all of Eisenstein's works, that film impressed me (as well as the Odessa Steps scene). The scene that captivated Lakatos most was Ivan's switch from a sense of horror at his own cruelty to the opposite, as he muttered, "not enough": Lakatos repeated the Russian words appreciatively and expressed empathy. I shuddered.

7 Alas, this is no longer true. In our last meeting Watkins told me that he considered Popper a liar, and he said this in obvious distress. He was utterly mistaken, of course, and hopefully in a fleeting aberration.

good spirit and do it justice and more than justice. And we crassly refused this offer. I may be in gross error and I may be extremely unfair and unjust, but for what it is worth I still declare that he forgot his promises to take account of specific critical comments that he had admitted, and, anyway, no one can judge oneself open to criticism and no one can judge adequate one's acceptance of a criticism and one's subsequent response to it. (I have already mentioned an example of this, but it seems to me to be a point that needs no evidence, one that we should all admit on general (democratic) principles. They are ones that Popper advocated most emphatically. This very criticism he rejected on *a priori* grounds: it is a criticism of his view of his own critics among his associates and former students[8] that regrettably[9] he did not open to criticism.)[10]

What it all amounts to, on any interpretation, is his understandable

---

8 Munz declared Popper hostile to criticism and disagreement. He offers some examples and an explanation [Munz, 2004, 18]. I disagree. In private, Popper did welcome criticism and showed sympathy with the critic, although, he admitted to me (in 1974 as we had our last long conversation; it was in his home), he found criticism upsetting as he thought everyone should, as incentive for relentless self-criticism. Now liking or disliking criticism is psychological, and training for it is educational. The observed fact on common attitude to it, however, that the common dislike for it rests on the common view of it as rejection and strife. This, in turn, and still as a matter of empirical observation, rests on the equally common view that error and strife are avoidable and that hostility to dissent is inborn and thus permanent. (This possibility of error avoidance, let me hasten to add, is not a metaphysical doctrine that only credulity admits; it is the erroneous but commonsense view that by-and-large it is possible to avoid clashes with the law and even with custom. At most, this is true in some pockets of some highly civilized societies. Were it always true, tragedy would be inconceivable.) Popper, of course, would have none of this. His view of criticism, to repeat, was educational, and rested on his moral and psychological views. It is the only case in which he could benefit from heeding Plato, who repeatedly declared that criticism is beneficial and never hostile and that its target should always consider it friendly. This is strange, as Popper profoundly admired Plato the dialectician. Plato's social view of criticism is better and pertinent, since he viewed it as significantly social. Socially, criticism is utterly impersonal, so that it does not matter whether the source of an idea and of its criticism are the same or not, and it does not matter what motivates either.
9 The educational question remains in midair: what values should we instill in our young to make our society increasingly open? For this, truthfulness matters most, of course, and so also disinterestedness, and this runs contrary to Popper's demand for unease about being a target of valid criticism. Popper was in error even psychologically; what upset him was that a critic saw something that he had missed. Whenever presented with a new idea, he tried to add something to it (or at least to its presentation), at times most brilliantly, even in discussions remote from his studies, but at times embarrassingly so. When the other party acknowledged his being one up, he shined with pleasure, but only very briefly.
10 The workshop mentality that Jarvie and I recommend seems to me vastly superior to Popper's invitation to self-flagellation. It is an interesting empirical observation that in workshops properly erected the normal hostility to criticism is absent. This includes even some of Popper's public performances in which he exuded bonhomie for whatever reason.

The claim that the hostility to criticism is inherent is thus empirically refuted by an easily repeatable observation. It is open to better refutation by the institution of a system of rewards for targets of criticism from illustrious sources. (A learned society can extend an invitation to join it to any target of public criticism by any of its members.) The benefit from this to society would be in its institutionalization of the view of the discovery of a new effect as criticism; the source of the criticized idea would share the glory with the discoverer of the new effect. Thus, Kepler should be credited as a party to the discovery of planetary irregularities, Lavoisier should be credited as a party to Davy's discovery of halogens as elements that act as oxidizers, and Niels Bohr should be credited as a party to the discovery of the neutron. Thus, rendering Popper's methodology public should have enormous benefits.

but not easily acknowledged anger with me as well as his rejection of my autonomy that was quite unintended — counter-intended, to be precise. Now that I am considerably older than he was when we parted company, I hope I have reconciled myself to his lasting rejection and can approach things with some sense of proportion.

Very early after we had first met, already early in 1953, the philosopher kindly invited me to his home, and in no time we were in close interaction — he teaching and myself sitting at his feet — soon it was together with Feyerabend. The two of us had met in his seminar. I befriended him at the philosopher's advice or instruction: he informed me that we — Feyerabend and I — had extremely similar interests (mainly in the philosophy of physics) and we could visit him together to everyone's advantage; he assured me then that Feyerabend had much regretted his past in the German forces.[11] We met at the philosopher's home a few times; we would catch a morning train to the nearby township, consume a simple hearty early lunch and proceed to the isolated country house by bus and on foot and get there as soon as he had consumed his. We spent intense hours until it was time to catch the last ten o'clock train back to town, debating physics and the philosophy of science together if we still had some energy left in us and returning home exhausted; we often fell asleep in the train.[12] In summer 1953 Feyerabend left for a holiday in Vienna — to come back in the fall. I accompanied him to Victoria railway station, though it was to be a brief absence: we were good friends by then. He never came back to London except much later for brief visits, and then he usually stayed in our place and we have great time together.

The disappearance of Feyerabend surprised me.[13] He had met the philosopher in the Austrian College, in Alpbach, Tyrol, on an earlier occasion, and then received a British Council grant with his help to come and sit at his feet for a year. This, at least, is how I remember his story as he had told it to me at the time. He was much closer to the philosopher than I was — which is not surprising, as they were both Viennese and had

---

11 Feyerabend flatly denied this in his reply to my review of his autobiography.
12 In his autobiography Feyerabend mention these visits fondly [Feyerabend, 1995, 96].
13 [Feyerabend, 1987, 312]:
"Popper asked me to become his assistant; I said no despite the fact that I had no money and had to be fed now by the one, now by the other of my more penurious friends."
Not true. He accepted the invitation and then stayed silent for a year while serving as assistant to Arthur Pap. He was thus very inconsiderate. Popper offered me the job after months of futile and vexing waiting. Incidentally, in the summer after that Popper and Pap ran a seminar together at the Austrian College Alpbach, Tyrol, and had us as their respective assistants. Soon after, on Popper's advice, Feyerabend applied for a job in Bristol. He was successful, partly due to the support of Popper and of Schrödinger [Feyerabend, 1978, 116]. His explanation [Feyerabend, 1995, 97, 99] that his refusal to be Popper's assistant was in the interest of his intellectual independence from Popper does not explain his silence. Moreover, he never achieved this independence [Motterlini, 1999, 192-6].

quite a few mutual acquaintances[14] and liked each other tremendously. When the philosopher's first assistant, George Zollschan, departed, he naturally chose Feyerabend as a replacement. I could not understand why he failed to show up — or even to say he was not coming. Later he explained it to me: he did not come because he was the assistant of Arthur Pap, who was a visiting professor in Vienna then, and who bullied him and forced him to stay. (Pap denied this to me.) Anyway, as I learned later, Feyerabend was always susceptible to pressure and he was always very reluctant to travel or suffer any change in his environment.[15]

Feyerabend told me all this when we met again a year later, in summer 1954, in the Austrian College, in Alpbach, Tyrol, actively participating in a seminar run by Popper and Pap: we were their respective assistants. We then spent three weeks in very close contact, arguing and horsing around and generally having good time. The seminar was amusing: Pap had a hard time trying to break Popper's monopoly; he was outclassed in every way, fair and not. He was the most dogmatic positivist I ever met in my whole philosophical career: he could not entertain the idea that criticism of naïve realism was deserving of attention, nor could he imagine any disagreement between the philosopher and his "Vienna Circle" peers. Feyerabend and I remained friends for decades. While he taught in Bristol he regularly visited us and stayed with us; one semester while I was new in Boston he stayed there, frequenting my philosophy of biology seminar and spending much time in my house; he befriended both my children — he could charm any child. Our relations deteriorated later: we went in very different directions, and his public statements were plainly false and overtly insulting — mainly concerning the philosopher but also involving myself. He kept telling me not to take it personally. I am afraid I did: I do not wish to have friends who behave this way: perhaps he is right to call me a prig.[16]

---

14 This is a gross understatement: Feyerabend's doctoral dissertation adviser was Victor Kraft, the author of the once very influential *The Vienna Circle: The Origin of Neo-Positivism, a Chapter in the History of Recent Philosophy* (1950) who was a closet Popper fan (see Index to Chapter Ten).
15 Always on strong painkillers, Feyerabend showed remarkable courage and perseverance. He suffered from a severe case of agoraphobia yet he held a dual appointment and traveled often besides. Every act of writing a paper precipitated a crisis for him, yet he kept writing and publishing relentlessly.
16 [Feyerabend, 1978, 139-40, 183] describes me as "not too bright and badly informed", "incompetent professional" who has helped standardize "professional incompetence". He also called (Gellner and) me illiterate. In a letter to Duerr [Duerr, 1995, 153] he said of me, "he can dish it out but he can't take it". Admittedly, I cannot take willful distortions from a friend; but then I cannot dish them out either. During his last years we had no contact — at my instigation and much to my grief. He felt the same way, and argued against my refusal to keep in touch with him. He viewed as a weakness the inability of Popperians to keep friendship apart from philosophy and from politics [Duerr, 1995, 113]. He had a strong sense of loyalty to two early friends, one of whom was a Trotskyite lecturer in chemistry in Vienna University and the other a Wittgenstein scholar of repute, G. E. M. Anscombe, who had stayed for a time in his Vienna apartment. This made him stick to Trotsky and to Wittgenstein [Motterlini, 1999, 151, 272]. He also felt partial loyalty to me, but not to Popper — I suppose because of the rude interference in his personal affairs and the indifference to his ambition that Popper displayed. He re-

110   Chapter Four

I do not pretend to understand Feyerabend any more: the very memory of his ever having sat at anybody's feet angered him so, that he utterly denied the facts, but only in print; not in private encounters, nor in correspondence: his very last letter to me still expresses his preference that public statements (such as his calling me the capitalists' hired pen) should generally be disregarded on the personal level. He then wrote his memoirs. This is intriguing: are memoirs private and so truthful, so that he could tell there how exciting he found the philosopher in those young days, or are they public and so a mere added occasion to insult the philosopher and tell the world he always found him a pompous, shallow preacher (advocating rationality)? I do not know. The complaint about preaching (and about meddling with private affairs) is valid, though. Feyerabend did suffer from this.[17]

In 1954, some time after Feyerabend failed to show up as the newly appointed research assistant to the philosopher, the philosopher invited me to replace him. I was flattered but not surprised, as the philosopher had already helped me get some financial aid and started consulting me about his manuscripts. It was really incredible luck that this took place, and I should interrupt my narrative again to tell how it happened, since the honor was unwarranted.

It began with the 25th anniversary celebration for the Hebrew University of Jerusalem in 1950. For the occasion a guest speaker was invited, the celebrated Morris Ginsberg of the London School of Economics. We met accidentally, as he was visiting with Judith's grandfather, perhaps because of professional connections, as they were both sociology professors (the Hebrew University still forbade then the teaching of com-

---

sented this very much [Duerr, 1995, 179]; [Motterlini, 1999, 192, 194, 196]. Despite his strong sense of loyalty, he was thus mean and disloyal to Popper, the one person who in hours of need was kind and friendly and understanding and helpful to him more than anyone else on the face of this dreary, cheerless planet.

17 Munz viewed not too kindly Popper's aggressive interference in other people's personal affairs [Munz, 2004, 49]. Despite my having had my traumatic share of it, I consider it as rooted in his obsessive generosity. This, of course, is no excuse. Noticing the evil that good intentions might easily cause, Popper recommended limiting love to friends [OS, ii, 236-7, 240, 244]. This is a mistake: we may freely love or hate whomsoever we choose, but help only those who ask for it or at least explicitly permit it. But let me pause: as everything may have its misuses, proscription should be limited to misuse and subject to common sense. Here alas Popper lost his. Banning a drug, such as alcohol, because of its misuses, may easily misfire. Even limiting the use of anti-biotic drugs by making them prescription-drugs has misfired. Popper supported such measures for paternalist reasons. See his letter to Thomas S. Szasz of August 10, 1984 and his late publications on the topic.

Popper did owe Feyerabend a grudge for interference, but this is no excuse for his retaliations. In retaliation Feyerabend redirected against Popper the criticism that Popper devised against induction, as if it was his own invention, and as if Popper was an inductivist. This amounts to the (tacit) identification of rationalism with inductivism. And so he attacked all rationalism. Later he softened this, saying [Feyerabend, 1991b, 123], limiting his hostility to the rationalism of Kant and of Popper, to the exclusion of that of Lessing and of Heine. Later he withdrew this too [Feyerabend, 2001] but never explicitly. He never withdrew anything explicitly. This habit is usually due to cowardice; not so in the case of Feyerabend, who was very brave. His downfall was his addiction to fame.

parative religion, which was Buber's truest vocation next to philology, for which there still is little demand), and perhaps because he was the initiator of the very idea of a Hebrew University. As I was a student then, Ginsberg took the opportunity and consulted me about his intended lecture since it was to be delivered to the students. My comment was negative: the lecture rested on the premise that Marxism is refuted yet it was meant to engage the students there who were then remarkably unanimously of the opposite premise (and I, as usual, in the smallest minority). My response was not appreciated; it later earned me a few ruthless efforts of the said gentleman to sabotage my career. Only after he had retired and I had some measure of success in the School did he befriend me.

Yet the experience had its reward. In 1953, when we ran out of money and in a profound sense of disappointment prepared to return home and face our failure to further our careers without the aid of our Alma Mater, Judith's grandfather visited London for a couple of lectures and exactly then he was engaged writing a paper on some sensitive Israeli issue; the Israeli weekend press published some criticism of his sympathy with Christ ("my older brother" [Berry, 1977]) and the Christ syndrome ("a recurrent phenomenon in Jewish history")[18]. While in London, he drafted a reply. He consulted me on it, and that impressed the philosopher sufficiently to do likewise, although this was only half a year after we first met and he had barely a reason to suppose that my input could be useful. He also enlisted Judith's grandfather's help in securing for me a small grant from a philanthropic banker. I was remarkably lucky; without unusual luck I would not be today anywhere near where I am.

I do not remember how long it took before I started working really closely with the philosopher. I do know that before long we moved to live in the country, close to his home, where for some time I spent 12-hour days regularly — soon it became seven days a week, and, when Hennie — Lady Popper — complained that this was too hard on Judith, it was reduced to six or five. At the peak of our interactive period, I would get up late in the morning, except if he was ringing our landlady and calling me to a long discussion on the phone concerning exciting material he had worked on overnight that he had no patience to postpone to the afternoon. After a late breakfast I would cycle a few miles to arrive at his place as exactly at midday as possible, but never before midday, so as not to interfere with his lunch, and stay till midnight. It is hard to describe the intensity of the activity. It was unbelievable, yet he complained of slackening due to old age. I hope I am forgiven if I report that it was so hard to disengage that I often relieved my bladder only at the very last moment. Naturally, he was by then in the same condition. I would rush to the toilet at

---

18 Messianism and the Christ syndrome are indeed recurrent in Jewish history; so is the view of Jesus as a Jew [Herschel, 1998].

the entrance and he to the one upstairs. And then, rushing downstairs, he would report to me some new idea he had in the meanwhile.

All the nourishment we had during one 12-hours visit was an afternoon coffee with half-a-dozen biscuits with some cheese or tomato on them and an evening instant coffee. It took months before I dared say I could barely eat tomatoes unsalted. I would arrive home more hungry than tired and would eat and fall asleep. I do not know how long I could sustain this kind of existence, and I know it did not last for long — since in July 1956 I received my Ph. D. and to that end I took some breaks from his service. We then traveled to the United States together for one year — he was invited to the Stanford Think Tank and requested that I be invited as his research associate — and there the intensity of our relations slightly wore off for a while and later it was radically altered — first it was increased at his initiative and then it was drastically reduced at my initiative and despite his effort to alter my decision. He never forgave me my abrupt break then.

Before I received my Ph. D., I was relieved now and then of my duties to work on my dissertation. During these brief periods another of his graduate students, Cesław Lejewski, took over as his research assistant. Lejewski was terrific; a perfect gentleman, kind, well organized, and very quiet. But as he was unusually slow, he easily exasperated the philosopher. Also, he was dedicated to the reconstruction of the contributions to logic made by the Polish school of logic before the war, that were lost with the brutal murder by the Nazis of hundreds of Polish logicians and the loss of ever so many important manuscripts. The philosopher always had time for his students, but Lejewski worked on his own and scarcely consulted anyone.[19] His comments on Polish logic I am not qualified to judge. His expression of appreciation of Popper's contribution to logic (in *The Philosophy of Karl Popper*) is a classic that will one day find the recognition that it amply deserves if logic will ever return to its old robust expansion; in it he compares the deductive systems of Popper and of Alfred Tarski. (Popper's response to Lejewski's essay there is most intriguing, but his comment on his criticism is disappointing.)

The very brief periods of my absence made me feel upon my returns something like coming home — inasmuch as the austere atmosphere in that household permitted such feelings. At times I was also made to feel a bit more at home there: I was invited to assist in minor household

---

19 As Lejewski told me, he had read a work of Stanisław Leśniewski in manuscript and, an undergraduate reading the classics, he could not comprehend it. It was lost and he could barely remember it. He devoted his research career to its reconstruction. I found his devotion admirable but could not judge the value of his project. Few logicians take Leśniewski seriously, none of them mainstream. Popper admired him from afar. Arthur Prior and Mario Bunge are his most serious followers, each in his own way, but neither as a disciple in the sense that Lejewski was. Popper respected Lejewski's devotion and never tried to influence him.

chores, that I loved, and to accompany the philosopher, whether locally — I particularly remember a few visits to second-hand car dealers — or to town. The visits to town were almost always either to the School or to a physician: he suffered from chronic illness and at least a touch of hypochondria.[20] (I am told that this changed. This stands to reason: Lady Popper's long terminal illness was terrible and he was dedicated to serving her to the last.)

The philosopher wanted me to spend more time with him on his work than on mine. This did not disturb me. On the contrary, it delighted me and I would have been happy if I could only bring him to agree that possibly I did have already in 1955 enough material to submit for my doctorate: he insisted that I finish the whole work as I had originally planned it. I never did. Finally there was so much of it, so hurriedly submitted, that most of it was not submitted nor even typed. I used it only years later. What I did submit was never published: the philosopher had no time to read it and he could clearly see that it was a rough draft that was far from living up to his stringent standard of presentation that he expected me to embrace. I much regret that consequently I started my career a year later than was possible and I kept a petty grudge that went away when the philosopher explicitly admitted error and apologized, to reappear late on. The manuscript work on which he was spending endless efforts while I was his assistant, his famous Postscript: After Twenty Years, was in galleys well over twenty years and then it got to the printer for resetting. In the published version, he says that he had asked me to study the case of Faraday. He was referring to the last part of my dissertation that I never submitted and that was not even typed, but appeared all the same in 1971 [Agassi, 1971]. This seems to me to be a withdrawal of the apology (especially since at the time he allegedly asked me to study Faraday all he knew about him was the little he had heard from me in one brief session): he took credit for having forced me to write on Faraday as if there was a risk that I would not.

It is hard for me to report on my dissertation: it was a bizarre experience. The philosopher had no idea about how to help students write dissertations as he did not know how to write a book; he even boasted to me that the books he wrote, other than his collections of essays, grew out of essays all by themselves. Indeed, he was working on his *Postscript* then and it blew up under my eyes from appendices for a book to three vo-

---

20 Clinically, the symptoms of hypochondria are more than what Popper displayed [Fried and Agassi, 1976, 171]. He experimented with medications and looked for an ideal physician. Lady Popper said she was tired of his playing a five-guinea pig. (Harley Street physicians charged five guineas a visit.) He resented their self-confidence. Once I spoke with a specialist over the phone, arranging an appointment and advising him not to hide from Popper his doubts. He concurred and won immediate great rapport. It did not last: he followed my advice only because he mistook me for Popper's physician. Realizing this, he returned to the usual style.

lumes.[21] He was as undisciplined regarding the framework, in his insatiable craving for hard work and for pressure to work even harder, as he was disciplined inside the framework with a fine feel for rigorous logic and for marvelous clarity. Whenever I criticized him for his colossal inefficiency he would say, but I try so hard! This looked even to my untutored eye like an excuse, and a cheap one at that. Once, when he was in a relatively relaxed mood, I told him he was escaping from planning work to unplanned hard work. He stopped, thought for a while, and then solemnly spoke. Joske (he insisted on calling me by my nickname; it is the one close friends used to call me since my undergrad days, and as they and Judith use it to this day it stuck), he said, you are a clever psychologist, much cleverer than I am; but I really have no time to discuss all that. No, I am not taken in by the compliment: of course, he was much the cleverer psychologist between the two of us and he knew it: he only meant to appease me. I know: most of his compliments are highly goal-directed; normally they were also accurate; not this one, however, which is somewhat understandable as at times I was called to reconcile the two members of the department, Popper and Wisdom, whose idiosyncrasies usually caused friction.[22]

During my studies for my dissertation, every time I told him about something interesting that I had discovered in the literature, he would say, put it into your dissertation. Fortunately I seldom obeyed; rather, I tried to put some order into things. But without a frame and with no training I moved from contemporary physics, which greatly bothered me, to discuss the philosophical error (positivism, what else?) popular (then even more than now) among physicists and then looked for a scientific example of an old, by-now settled controversy. I chose the case of electrostatics, where action-at-a-distance and fields competed. I thus came to make a study of Michael Faraday. It turned out to be a most happy choice. My attachment to it was personal: I had learned electrostatics not in school but while listening to my elder sister doing homework with a friend; only

---

21 All of Popper's reports on the dates of his work (on his *Postscript* as on other projects) are inaccurate.
22 Popper suffered an almost constant irritation. He did enjoy his work, and even very much so, but as a workaholic and a perfectionist he always managed to turn the fun into irritation.
  Perfectionism is obsessive and any obsession can appear as perfectionism. As excesses, each reduces the sense of discrimination and of proportion; together they destroy these valuable qualities, being a mix of craving and aversion. Excesses rest on the confusion of the maximum with the optimum. Perfectionists (con)fuse self-criticism, self-censure, and self-dislike. They ignore a common observation: performing a whole job perfunctorily first and then rectifying some of its most obvious defects, is much more efficient and enjoyable than doing it perfectly piecemeal. This way also helps boosting self-reliance and avoiding waste as the law of diminishing returns sets in: there are always better tasks than upgrading sufficiently well executed jobs [Agassi, 1999a]. It also causes irresponsibility: perfectionism is a strong method for fending off criticism. Popper, who repeatedly used perfectionist excuses, escaped the irresponsibility, as he worked in almost total isolation. Consider, by contrast, the best advice that many scientific advisers have famously offered politicians: conscientiously but in foolish disregard for its unattainability.

years later did I understand the significance of it. My physics teachers had told me that metaphysics is the physics of the yesteryear and I tried to show that metaphysics may be — just may be — the physics of the morrow. As was the case of Faraday. Faraday's electrostatics was at the time equivalent to classical electricity (the electrostatic field energy equals the energy due to the Coulomb central forces), so that at the time the choice between the two looked as undecided as the choice in quantum mechanics between the two standard representations of quanta.[23] So I delved eagerly into the works of Faraday. And, not to weary you with details, I moved to Faraday's background and so on. Fortunately, Sir Francis Bacon and Robert Boyle were both the natural stopping point — the beginning of the modern era and the establishment of the scientific standards — and there was neither need nor possibility for me to go further back (especially since I have no Latin at all and the works of Bacon's major predecessors were not available in English). The dissertation concerned standards of scientific discourse as created by these two unknown seventeenth-century giants (they are slightly better known these days)[24] and comments on some of their successors: some 800 typed pages. My material from the eighteenth-century went to serve as illustrations in my *Towards an Historiography of Science*, 1963, that is still the chief basis of my reputation such as it is and that I wrote (at the kind invitation of Bartley, who then became an associate editor of a new journal) in Hong Kong as soon as I arrived there; the rest went to my *Faraday as a Natural Philosopher*, 1971.

A year or so after I went to Hong Kong the first message came: I publish too much; I am too facile. I did not like the intervention: I was a big boy by then; yet, big boy or not, the only compliments the philosopher ever granted my work concern these two publications. And I think he meant them. And I am flattered. An aside: Once I bumped in a most unlikely place into Menahem Max Schiffer, the applied mathematician whose teaching style I admired despite his positivism, as he skillfully used audience participation to cover much ground slow and easy. He remembered me and knew my book on Faraday and praised it. I was elated. On the whole, I had to learn to make do with few compliments and little encouragement.

The dissertation. At least I received my Ph. D.: the University of

---

23 Einstein's reverence for Faraday and Maxwell and gratitude to them encouraged my choice of their major contributions for my first independent study. Later on I admired Koyré for his courage in his earlier choice of Newton for a similar end. He came to pay Popper a visit and his company was always most enjoyable.
24 The literature on Bacon and more so on Boyle is growing, mainly due to publication pressure, I daresay. It concerns mostly marginal matters. I may be in error, though, since I am not free of bias: this literature assiduously ignores my views on their immense contributions to the growth of scientific etiquette.

London is not as particular as, say, Cambridge. Others had harder times. I remember particularly John Summersbee, who was in the middle of his research towards his doctoral dissertation when I arrived. The philosopher spent long hours helping him by giving him long uninterrupted private lectures. (The philosopher was always unbelievably generous with his time when it came to instruction, and he gave many private lectures to many students and colleagues.) But, again, the discipline was only internal and, intelligent and able and devoted as Summersbee was, he knew nothing about the frame: how to organize a book. In desperation Summersbee did what no dissertation adviser should allow: he looked for an original idea: he tried to solve a famous problem in logic. He risked finding that it was beyond his reach, or that someone else might reach it first. After many years of hard work, he changed his dissertation adviser; finally, an external examiner failed him. I identified with him: we were in the same boat. Afterwards I tried to help others by discussing structures with them. I have a standard structure borrowed from Popper himself: it works miracles. It is simple: take a question that interests you and that is under controversy in the literature, on which you can report intelligently with ease: the question, the competing answers, the arguments offered in the field for and against the competing answers; and do not choose a question, even if it is under controversy, if you think one side clearly is in the right. You may agree with one side, on the condition that you can report the debate respectfully. Do not report on a debate on a disagreement that is not respectful. And write the whole first draft fast even if sloppily so as to see if the structure can sustain a more serious second draft. (More on this one can find in the exchange I had with my former student John R. Wettersten on the educational philosophy of Selz, Popper and myself. You can imagine how embarrassed I was to be in such an august company; yet the philosopher liked Wettersten's essay! He refused the offer of the editor to comment on the essay but he did praise it.)[25]

What I know about writing techniques I learned working with the philosopher. Most of my techniques he simply taught me, some I learned by observing him; the rest I learned by avoiding what I did not enjoy about his way of doing things, namely his perfectionism, or, if you wish, his way of spending lots of energy for the sake of hard work. For example, I developed my feeling that one should leave embellishments to the last by seeing him waste time on unneeded embellishments. He once wrote a reply to Bar-Hillel's defense of Carnap; we worked together hours and hours on his preface, discussing it, comparing his variant wordings, formulations and what not. When the paper was ready it was clear that the preface had to be scrapped. (When its proofs came it he scrapped

---

25 See [Wettersten, 1987], [Agassi, 1987], and [Wettersten, 1987b]. See also [Agassi, 1999a] and [Segre, 2005, Chapter 11].

it *in toto* and a new version went to the printer.) Consequently, I never cross the t's and dot the i's of a paper before it is accepted. I then have a better idea as to what I should polish and how. Moreover, it gives the editors a good excuse to reject papers they do not wish to publish anyhow.

Perhaps I am a bit unfair: the philosopher was very excited when he found an occasion to corner Rudolf Carnap Himself — the then leading philosopher from old Vienna, who once had mentioned the philosopher in print, when he was utterly unknown, but then rudely ignored him (in print: he wrote him long private letters on his ideas; they remained unread)[26]. The exercise worked: Bar-Hillel came to the rescue. When the second barrage came, Carnap himself chipped in, but only with a brief introduction to lend Bar-Hillel moral support. The philosopher was very excited indeed. Myself, I felt it was all so devoid of interest, and so eerie that two Israelis follow the footsteps of two refugees from a vanished Vienna[27] who still lived in its ruins, not seeing the destruction of the old and barely noticing the new. I told the philosopher from the start that I found the debate with Carnap uninteresting. He asked me for my reason. In response, I mentioned a very minor idea that renders the whole debate superfluous.[28] The philosopher added it to a passage in his essay on Carnap and named it after me (no; it is not my idea; it led Nelson Goodman, one of the best known members of the profession, to a funny priority dispute; later I both refuted his priority claim and showed that it belongs to Russell [Agassi, 88, 61] — if we may ascribe priority at all to trite ideas). I am still unimpressed, but I am now not insensitive to the excitement and have less trouble understating the undisciplined, inefficient investment of so much work on the project: it did mean a lot.

I know. You wonder why I am writing in such a free-association manner rather than follow my own advice and get a bit disciplined and put some proper structure into this melancholy account. It is my pleasure: if you want to see how well I can structure, if you want to enjoy a well-structured work of mine, take, for example, my two books mentioned

---

26 I think Popper scarcely glanced at Carnap's detailed letters that were full of neatly typewritten formulae: he did not know them (or he forgot them) when he saw them in print. He noticed then that that they are vulnerable to criticism present already in his *Logik der Forschung*.

27 Carnap and Reichenbach represented the "Vienna Circle" then; texts of other members were not translated and scarcely noticed. Later apologists of the "Vienna Circle" shifted the leadership of the "Circle" to Neurath [Agassi, 1998]. Carnap's immense popularity rested on his method of combating metaphysics, now dismissed offhand [Friedman, 2000, page x]. Reichenbach was a terrific popular-science writer. His works on probability and on physics, much respected then, are now mercifully forgotten. Current efforts to explain his popularity are pathetic even by today's pathetic standards (see Appendix to Chapter Ten). Russell and Einstein wrote respectfully about him in their respective Schilpp volumes, only hinting that they considered him lightweight [Russell, Replies, 683, 686]; [Einstein, Replies, 676]. Reichenbach's works on probability won then public praise. The disregard for Popper's obviously superior alternatives to that is a colossal unfairness that regrettably embittered him.

28 Consider any characteristic that all past observations share. By the theory of confirmation, whatever it is, most probably it holds of all things. Examples: all events are observable; all events were observed; all events are in the past.

above. This account is written to a different structure, strong enough to carry the material, but loose enough to permit liberties to achieve some inner balance, some coherence, in a highly complex matter.

The rudest message I received indirectly from the philosopher was a composite one, worked out in great detail with great care, I fancy, by himself and his chosen messenger for the occasion, one of my later successors as the philosopher's assistant, Jeremy Shearmur, who looked like he was very well prepared to deliver it. He said to me almost as soon as we first met (I recognized the intensity) that he found my work literally crazy:[29] a stream of consciousness, he explained. (A funny view of the literally crazy, though.) I do not know what he expected as a response. My response was — you should not be surprised — a request that he give up reading my stuff. He said, he could not promise that. Responding to my query, he explained: he found my stuff irresistibly interesting. I melted with pleasure. This infuriated him; he explained that, as I understood, he meant it as no compliment: I think he tried to accuse me of being an intellectual teaser, a provocative-and-love-it sort of a writer.

He offered an example of a terribly interesting item in my writings that I had not bothered to document. As it happens, it is indeed a lovely and important idea. (Later on Wettersten and I wrote a paper on it [Fisch and Schaffer, 1991, 345-69].) It was the discovery that Newton's celestial mechanics is inconsistent with Kepler's theory of the planetary orbits. This discovery belongs to the great Dr. William Whewell. It is almost two centuries old, yet it still is a source of constant annoyance to most philosophers of science, including Whewell himself. Surprisingly, the same holds for Karl Popper, who makes a big thing out of it when he argues that science is not based on factual information (but rather contradicts it), yet pooh-poohs it when Feyerabend and I argue that it refutes just as well his own theory of science as hypothetico-deductive explanation (namely, as deductive, since the premises of a valid inference cannot contradict its conclusion) [Wettersten, 2004, 273-4]. To return to the story, Shearmur was understandably interested and thought me sloppy and negligent for

---

29 The label "literally crazy" is offensive; "crazy" is not. A letter of Feyerabend [Duerr, 1995, 153] and Raphael Sassower's essay in my honor [Sassower, 1995] call me crazy. Bernard Corey's report on life in the LSE [Tribe, 1997, 188] refers to me as a "very nice, mad guy".

Lakatos declared me literally crazy in our last conversation, shortly before he died. It sounded to me then like a mere exaggeration. I later learned that he had publicized this declaration for years. Its source, it seems, is Popper. He wrote this in a letter to Watkins that he probably did not send and something similar in one to Albert that he did send [Morgenstern and Zimmer, 2005, 255]. He alleges there that I wrongly ascribe to him a pessimist epistemology and explains this allegation of his by ascribing to me megalomania. As it happens, this allegation is false. (Russell judged Popper's epistemology pessimist because he denied that induction is possible, as Russell could not let go of his identification of induction with learning from experience.) On this, by the way, I feel closer to Borges than to Popper: since out of the depth of despair we have to judge every goodness, life itself, as a gift, it is epistemologically easier to take despair as the default position, although morally we must act on optimist assumptions, so as to grant the Lord the benefit of the doubt, as it were. Perhaps this is truly crazy, who knows.

having ignored the reference. Apparently, he meant I was such a scatterbrain that I did not know how important the idea was. I suppose the absence of scholarly reference somewhere in my writing to some important information discussed there looked to him as forceful evidence that I had no sense of significance — that I was literally crazy, perhaps. He learned there and then that I knew the important reference by heart — as sheer by-product, I assure you, since I had written repeatedly about it. He was baffled. Perhaps I should dedicate this section to him.

Sorry to sound so flippant. I have a point to convey. I was apprenticed to the philosopher who was thus my master — in the very old-fashioned sense of the word. He also was an educator. As my master he educated me; as his apprentice I worked for him. It was a fair exchange, and I was very satisfied with things as they were: I did not ask for more; I did not request promises and I did not offer any. I was in charge of my fate in that I was there by choice. There were *a priori* things I did not like. There were equally *a priori* more things about me that the master did not like. He was short-tempered, you may remember, and I had the choice: leave or let him behave naturally. On top of it we were also friends — as much friends as a great master and an unruly apprentice can be, yet two individuals cannot undergo intensive interaction for many hours daily and not become friends of sorts. But I never promised to be a sequel to my master; I never planned to enter his shoes on his retirement. I know: in the old days the apprentice was expected to do so and marry the master's daughter too. Not in this case, though: the master was childless and I had a wife and a daughter; and he and his wife did very kindly befriend us all. Busy in the extreme as they were, unbelievably tidy as they were, they invited the three of us to their home from time to time. They once took us to Whipsnade Wild Animal Park and to a picnic lunch there. And later, when Judith gave birth to our son, they visited her in hospital. I deeply appreciate all that.

This is not the sum-total of the situation, but never mind all the details: we can take them together. He had his quirks and I have mine. Naturally, some of his quirks I have adopted — consciously or by symbiosis, it matters not. But there is a limit to that. Beyond that limit, if one wants to go further, there is room for rational debate between civilized individuals. Whatever is not settled this way or that way has not yet been settled in the courts of reason. The philosopher, I think, violated this point, as he deemed unsettled points settled; so did all his emissaries if and when they acted as more than mere messengers. (They usually did.) The moral superiority of the pedant, in particular, may be questioned and is questioned; by me; now: he and I had a legitimate disagreement and I will not be treated like a moral inferior, much less like a mental case. I did not want our friendship severed; I wanted my apprenticeship severed. I got the reverse of this: year in year out someone of the philosopher's coterie found a way to help him treat me like an apprentice — from afar. Also, every time a paper of mine got into the hands of a referee who happened to be a member of the coterie, senior or junior, I received an extremely carefully reasoned referee's report appended to the rejection slip, naming all my small faults and tendering me grandfatherly advice. I do not mind the rejections, nor the grounds for them: I have nearly five hundred publications in the learned press so-called (*i.e.*, in refereed journals) and thousands of rejections (not counting rejections of book manuscripts), and most of the rejections rest on unbelievably incompetent referees' reports [Agassi, 1981, 156-63]; [Agassi, 1990]. What I minded in referees' reports from the philosopher's coterie was their paternalism: I did not want it. I wanted to be treated like a stranger; in the normal friendly way that becomes a stranger.

I may be unfair to Shearmur: he is not responsible for my grievances against the inner circle. The case of Lakatos, for one, is much more obvious. He not only warned me repeatedly that he would make me work really hard; he constantly publicized that I was a charlatan and certifiable, literally (where does all this literalness come from?) certifiable; he invested inordinate effort in harming me in any way he could devise, justifying it by feigned paternalism — while systematically and craftily intriguing between the philosopher and myself. Yet the philosopher it was who legitimated such conduct that, as illegitimate, is readily kept under control and easily tolerated.

I do not wish to blame the philosopher either. He had his quirks and in his isolation and devotion to hard work he had little interaction and less experience in group dynamics. And we all found in him some treasure and we selfishly invaded his loneliness and also absorbed his quirks and thus amplified them somewhat. So things got a little out of focus. There is no other way to explain the gross exaggeration involved in the anecdotes circulating about the philosopher and his devotees. (I like Hans Peter Duerr's locution: I as a precursor of myself.) I know: some prefer gossip to arguments and so they would dwell endlessly on the unsavory details mentioned here. But there is room for both gossip and arguments — though in some proportion and without malice. Some gossip matters because some ideas matter — the ideas that the gossip relates to — ; never the other way around.

The philosopher asked me once, why did he have so much trouble with his former students? He quickly appended to his question the reply he had received from Sir Ernst Gombrich, the leading art historian, his close friend for decades, mentioned in the philosopher's autobiography as a close friend in endearing terms. Gombrich had said that this is due to the philosopher's dedication and understanding. I do not wish to involve

Gombrich, whose opinion I have second-hand. But I do agree with the idea ascribed to him — partly and perhaps even in full. It is hard to say without some further checking

Perhaps this is the place to say something of the philosopher's quirks — those that unfortunately we amplified (yes, myself included: I did think that I was more philosophical, yet I do regretfully share the fault). He was, first and foremost, powerfully, tremendously intensively seriously thinking — and, in company, while speaking. He was highly intuitive. When I told him of an utterly forgotten, interesting old book that I had found under old dust somewhere — this is a Talmudic expertise I had acquired in my adolescence — he would correct my reports. At first I was uneasy, but he was often right. The trouble was, when he was in error I could not tell him so, as he suspected I was cheating. In conversation, to repeat, he would learn more from the expression of the face of his interlocutor than from words,[30] especially later in life, since he was increasingly deaf — he said he was literally deaf and reported that he found even hearing aids quite useless. At times he would be eagerly engaged with an interesting interlocutor and enjoy the debate. The interlocutor would possibly be enormously pleased to interact with the philosopher on equal footing. In the excitement the interlocutor would feel increasingly free and off guard and then take some liberty or make a wrong move. The philosopher would then be reluctant to stop an animated discussion and gladly exhibit towards faulty moves as much patience as he could. This could encourage the insensitive interlocutor to continue in the same manner. Sooner or later, the philosopher would lose his temper. The result would be a profound sense of surprise, shock, and hurt. The very magnitude of the outburst, the very sharpness of the switch, made the damage usually irreversible. Even when some titanic force — an intervention from the usually very withdrawn, always sweetly reasonable Lady Popper — would stop it and knock some sense of proportion into the philosopher, the quick rescue operation, a clear, explicit expression of regret, was out-of-the-question: normally he could not offer a clear, explicit apology.

It was odd. The philosopher demanded apologies to be explicit and clean. He always accepted these gracefully — at times inviting explanations, at times not. But the very expression, "I am sorry, but", could send him into rage, like the expressions "what I meant to say was" and "you did not understand me" — and "surely", surely you remember.

He once owed Judith a very small apology — I do not remember why. He entered our dilapidated apartment humbly and looked around for a prop. There were a few books around, and a couple of toys; nothing more. He picked up a book. It was, of all things, the latest by Ludwig

---

30 I learned to conceal corrections that would annoy him. He noticed this and complained strongly.

Wittgenstein, the only person with whom he had exchanged total dislike on all fronts, the only person with whom he had an exchange that was more than unpleasant, an open row it was, during a public performance. He resigned himself even to this humiliation. He read the German text to her (it is the mother tongue of both of them, and I then could scarcely understand it) and discussed its style (as he rightly refused to see in any late Wittgenstein text anything remotely resembling philosophy)[31], being immensely charming. Judith, bless her soul, hardly responded; all she waited for was something practical: an explicit small apology and a brief discussion of the incident would have sufficed amply. Nothing like it took place. After a short while the philosopher addressed me and showed me that in the book's preface Wittgenstein refers to thoughts, not to investigations.[32] This sufficed for him to dismiss the book. His oblique apology to Judith remained unresolved. This was usual with him. For my part, I suffer from an allergy to oblique apologies — since my father often used them — and every time the philosopher lost his temper with me, regardless of the circumstances, I would cringe in the prospect of the next rapprochement.

I am reaching my point; do not rush me, please. In 1956 the philosopher was invited to America: twice; once to a brief three-week seminar and once for a year of research in the Center for Advanced Studies in the Behavioral Sciences, in Stanford, California (the Think Tank, so-called). It was out of the blue, and to me most surprising in view of the claims he always made about his being so utterly unrecognized. It brought me to America on his coattails, as the American expression goes, and this brought me enough recognition to be offered a Rockefeller fellowship and a position soon afterwards in Chicago (that I preferred but could not take, you may remember). But the greatest boon for me was that, as he expected to die in a plane accident, he demanded that I submit my dissertation pronto. I did. Three typewriters hammered simultaneously day and night. The external examiner received daily portions by special delivery

---

[31] Popper's refusal to ascribe ideas to the later Wittgenstein agrees with Wittgenstein's insistence that he had none. That all philosophers have nothing to say he considered a fact and he only wished to illustrate it. Today leading Wittgenstein scholars profess admiration for him while trying to ascribe some philosophical ideas to him with alarming disregard for his wishes.

Nevertheless, by the motto of *Logik der Forschung* that Popper took from Kant, he [Popper] is in error: even Wittgenstein had a real philosophical problem. The only place I could find even an allusion to this option is Jane Heal's [Heal, 1995] brave effort to present his posthumous debris as solutionns, on the true supposition that philosophy is "in the business of trying to get answers to fundamental questions ..." [Heal, 1995, 73], while timorously concealing his resolute insstence that he had none to offer. She views his texts as dialogues with the names of the interlocutors omitted. But at least she admits (penultimate paragraph) that her technique "has its pitfalls" and is "intrusive and impertinent", appealing to Wittgenstein's "sincerity and intellectual integrity". She finds in his output a critique of the traditional theory of meaning (ignoring Frege's explicit and sufficient version), and a replacement for it, which regrettably she does not present [Heal, 1995, 78]. Sad.

[32] Wittgenstein is not to blame for the choice of a name for the book: *Philosophical Investigations*.

so as to meet the deadline. The pressure was unbelievable.
In America he raised the pressure on me again. In the Californian sun members of the Think Tank and their assistants and secretaries mingled and played a lot. I loved it. The philosopher saw me once from his office window play volleyball in the yard. He did not quite censure me, but his comment sufficed to make me stop playing. Pressure mounted. I was a smoker then (not in his presence, of course) and I increased my daily dose. He noticed this somehow; he demanded that I stop smoking. He had one explanation for his demand, and I had another. He was extremely pressured himself, as he had galley proofs of his *Postscript* and he feared adverse public response (yes, he told me so himself, though much later; the work went to the press almost two decades later), and so he worked ever harder on it; probably he did not notice that he was raising the pressure on me beyond endurance. I owed him a few dollars, that he gracefully refused to be paid back at once, and he took this as more license to meddle with my budget. I was very poor, but he had been poorer at my age. He regularly gave me some (very much unwanted) pieces of financial advice.

Once he was talking on the phone with me and he heard some music in the background: we had a tiny garden shack for an apartment and it was a Saturday morning: my daughter was watching TV. He thought I was listening to music and not to him (knowing nothing about TV, much less about children's programs, of course). He stopped talking. I waited patiently. It was silence all the way. Finally I asked him to continue. He took this as evidence that engrossed in the music I had not noticed how very long he was silent. He slammed down the receiver. Later he explained to me this line of reasoning, expecting me to humbly corroborate his surmise and to apologize nicely. He was ready to forgive me wholeheartedly. I found his Christian forgiveness[33] worse than his oblique apologies. Yes, I know; it is my childish quirk.

He was extremely surprised when soon afterwards I announced to him the termination of my apprenticeship. He reacted most maturely: he was always extremely mature when surprised. He admitted, if I may abbreviate, that he had been bullying me; he would from now on treat me like a colleague, he promised, and he requested that in exchange for this promise I should take back my decision. He expressed clearly and quite unequivocally and unconditionally the readiness to discuss with me whatever I wanted. Even though he fancies himself egalitarian with his former students, I knew what an effort it was for him to discuss with me my own interests in preference to his. It was an offer I could not resist. I invited him to a discussion on ethics: I said he had given me much moral instruc-

---
33 See my last correspondence with Popper reproduced below.

tion but never any debate on it. He graciously gave me the green light. The discussion was a disaster. I said that my opinion was (it still is) a modified hedonism. He interrupted me at once: hedonism is not worth discussing, he said. (This is what he means by taking his opponents seriously: telling it to them like it is, never mind their feelings.) Hedonism identifies morality with the moral sentiment and is thus the reduction of ethics to psychology. Hedonism thus explains away morality, leaving no room for genuinely moral sense. I found this lecture an insult: he should have known that I knew it by heart. He should have asked me, what was the modification to hedonism that I was proposing? I would have said, it is the recognition of the moral duty to make the most of the little we have; in appreciation of our fortune we ought to enjoy every bit of it.

It is not that the philosopher did not know this (hardly new[34]) idea; it is that he had promised to discuss it with me and did not.[35] If further evidence is needed, let me mention that he did indicate to me on other occasions, earlier as well as later, that he knew where I was at. He said, for example, that he rather enjoyed my constant cheerfulness (alluding perhaps to Dr. Johnson's famous anecdote about a friend who was prevented from becoming a philosopher by his cheerfulness). He also said, Joske, you protest for the sake of protest: you loved working hard with me. I confess I did. Very much so: I always try to be true to my modified hedonism. And at times I miss the old days, of course.[36] When the philosopher and I met in 1969 in the Boston area, as he had a semester in Brandeis University soon after his retirement, I slid momentarily into our old stance. He protested and I apologized: he said he was too old. He meant too old for the strain involved: I had given him a lot of strain without noticing it, as he bore it heroically without giving s clue. And nevertheless, these old days were glorious. At other times he, too, fell momentarily into some old stance. On the whole things went well during his stay in Bran-

---

[34] The hedonistic imperative is conspicuous in William James. Depressive, he splendidly decided to embrace a positive and optimistic attitude [Myers, 1985, Chapter 13, The Ethics of Optimism]. He went too far and endorsed the therapeutic techniques of the Church of Christian Science despite his realization that its doctrine is superstitious. A more critical attitude would have informed him that this attitude might boomerang: phony optimism leads to painful disappointment and to needless crises [Laor and Agassi, 1990, 71, 179-80].

[35] Popper often forgot his promises. At the time he wanted me to compose an index to his *Poverty of Historicism* and to abbreviate a part of his *Postscript* so as to render it a paper ("Probability Magic"). I refused on the ground that he would then criticize what I would do and I wanted no criticism from him on such matters since he offered them highhandedly — while refusing to discuss them with me. He solemnly promised to spare me his comments. He then spent hours on end criticizing my output, again blocking all response. His publisher told me that he had complained that my index of his *The Logic of Scientific Discovery* is incomplete, obviously due to my refusal to recognize some great ideas of his.

[36] In an unsent letter (File 266.24, n. d.) Popper reminds me of the good old days:
"(... until quite recently I cherished the idea that you would write, when I am dead, about those splendid years — or so they appear to me — when we encouraged and helped each other. But I wish nothing more than that you should never write another word about me — and least of all in praise of me.)"

deis University. On the whole. Those were our last bright days. A lecture he gave then in Brandeis University, where he was then teaching, was not very fortunate. He spoke then of the myth of the framework.[37] (He called the tenet a "myth" pejoratively, in the Enlightenment sense of "popular error" or "vulgar error"; this is not as good as his descriptive use of the word in his superb "Back to the Presocratics", where he says, myths can evolve into science.) The myth in question is that of necessity some principles are shared by all parties to any dispute. In Medieval Latin this reads, "*contra negantem principia non est disputandum*": there is no arguing on principles. In contemporary American slang it reads, "you can't argue with a commie"[38]. Inasmuch as this is a testable tenet it is amply refuted: otherwise, I would have remained to the last a practicing Jew or a Marxist. Of course, in defense of this tenet one may easily deny that I have ever been a real Jew or a real Marxist: one may deny that my case is relevant as I never was committed and so I never was serious. This would not alter the biographical fact, that I did accept criticism and that it did lead me to alter even the framework of my views. In any case, refutable or not, in any reading this tenet is highly objectionable, since it makes rationality secondary to some principles: it is fideism, the demand to settle principles first by commitment, by some irrational decision process, and leave only secondary matters to rational debate. Still worse, any opinion that is vulnerable to rational criticism, or even one already beaten by it, may be rescued and fully buttressed by promotion to the rank of a principle.

The philosopher said all this in his lecture and more; he beautifully elaborated on it. Yet by his own standards the lecture was poor: it included no problem, particularly no problem that the tenet ("myth") under attack might be a solution to,[39] and it included nothing else that presents it as valuable; yet unless the attacked tenet is valuable, the attack on it is not valuable either, of course. This shortcoming was somewhat met by the (true) comment, placed in the beginning of the lecture, that the tenet under attack is extremely popular. I do not think this is a sufficient reason for an attack: after all, astrology, for a conspicuous example, is more popular, yet having no intellectual merit it merits no serious criticism.

---

37 "The Myth of the Framework" appeared much later [Popper, 1976], also in the eponymous book [Popper, 1994a].
38 Popper's "The Myth of the Framework" displays his originality in his broad application of fallibilism. It contains many new significant ideas. Its central message, however, is not new. It is that holding on to a framework at all cost is dogmatic. Shaw and Orwell had described it. F. Scott Fitzgerald wrote to his daughter about communists (March 15, 1940),
"the important thing is that you should not argue with them. ... whatever you say they have ways of twisting it into shapes which put you in some lower category of mankind – 'Liberal,' 'Trotskyite' – and disparage you both intellectually and personally in the process."
39 Silly as the myth of the framework is, it is a possible solution to the problem of rationality, of course. Indeed, Popper attacked it not because it is silly but because it is a solution. Pity he did not say so.

This shortcoming too he somewhat met, by the added (doubtful and irrelevant) comment, also placed in the beginning of the lecture, that serious people sanctioned this tenet, including everybody in the audience, including even those who declared it false. (I was in that audience, and I do not see that this fact was taken into account.) A sweeping remark about the audience[40] being in error no matter what they say raises audience hostility, of course; it is therefore advisable that a speaker be aware of this and expect it. The speaker's intent may vary. Thus, it may be to raise hostility and leave the audience hostile to the last; and it may be to end up with some tour de force that will turn the audience around towards the end of the lecture. The case at hand was quite different: the audience was hostile and showed it during the question-and-answer period; the display of hostility stunned the philosopher and he was profoundly disheartened.

After the lecture we were given a lift home in the same car. He was depressed. He asked for my comments. I had no solace for him and he depressively demanded that I say my piece. Finally, I did. We gloomily enlarged the topic. I was drawn in. At a certain appropriate juncture, I forget which, I very foolishly mentioned to him the details narrated above about our truant discussion on ethics. He said, we all make mistakes. I burst into a loud, rude laughter: I know that much, I responded. In the darkness of the car rushing us to our respective homes gloomy silence prevailed.

Commentary. The philosopher did not wish to inform me that we are all fallible: he was admitting error, though obliquely. I, however, did not want that oblique admission, nor an apology, however direct or explicit. I wanted from him the question I felt he owed me: what modification of hedonism? he might have asked. What modification of hedonism? But I was in error: he could not ask. He knew the answer and he could not follow it with the expected debate: it was too late for that. Maybe in our next metamorphosis.

Corollary: I was in error. I should not have laughed — aloud or quietly: I should have gracefully accepted his oblique apology as direct and explicit enough and as the best he could do then. Upon reflection I concluded that should I ever meet him again, it would be my turn to tender an apology. And I promised myself that if we ever did, I would.

By now the gates are closed.

---

40 The printed version of that paper is a little less hostile: "I fear that the majority of my present readers may also believe in the myth, either consciously or unconsciously" [Popper, 1994a, 34]. The popularity of the myth deserves praise and an explanation.

CHAPTER FIVE: THE PHILOSOPHER AND HIS FRIENDS AND PEERS

> (Democrats who do not see the difference between a friendly and a hostile criticism of democracy are themselves imbued with the totalitarian spirit.)
> (Karl Popper)[1]

In the fall of 1957, my position as apprentice radically altered for the better and stayed improved until I left London, in the summer of 1960. The improvement was both in external and in internal relations. Externally, I ceased to be an assistant and became a colleague. Internally, after I had told the philosopher I was to be his pupil no longer, he reduced his pressure on me. It was wonderful. Though what I had meant to say was that I had terminated my apprenticeship, and though he respected my wish, I still was his apprentice. Nevertheless, intellectually it was the peak of my career.[2]

He helped me then in my own independent work. He discussed with me my papers and even imposed his style on me, as I have already reported. I remember with amusement that he vetoed a word in a paper of mine as non-existent in the English language: it was not in his pocket dictionary; for his own work he often took recourse to the two-volume Shorter Oxford English Dictionary, but he thought that I should first learn to speak as plainly as possible. When I left for Hong Kong I told him I had the ambition to develop an English style of my own; he responded saying this was immoral: like George Orwell, he considered style best when invisible and unnoticed.[3] His sole concentration was on the clarity of the content, and he usually worked and reworked on it in tremendous detail. When I prepared for a conference he would help me with the paper and even with the discussions of the diverse papers on the program: we would conjecture what each speaker would say and composed a succinct response to each in the form of an essay three to four sentence long. (This is a terrific exercise, not unlike Rossini's transcriptions of famous soprano lines and attempts to fill the accompaniments as best he could before comparing his result with the original accompaniments [Michotte, 1968]. He said he owed much to these exercises. Of course, the speakers in the conference were no great masters; but they echoed some, and this was the point of the exercise.) The result in the conference was something Watkins found most incredible: I participated in each debate there, in an impressively well-informed manner, and yet (therefore, rather) managed to arouse tremendous hostility (akin to the well-known hostility that nor-

---
1 [*OS*, i, 189].
2 This was the peak of my career, not of Popper's. That came very soon after [Watkins, 1977a].
3 [Orwell, 1968]: "Good prose is like a windowpane."

mal students exhibit towards over-prepared classmates). When I applied for a position he sat with me and together we prepared a whole lecture course on the supposition that the exercise would raise my likelihood to get the position. This did not help me get a position, but it did greatly help me prepare my lecture courses, for which I am eternally grateful to him. He taught me that a well-prepared plan for a lecture course is very useful and helps set up each lecture: a course with a good overall scheme is more likely to be a success than a course with a poor overall scheme — even if in the one case one puts much less effort into each separate lecture than in the other.

Why was he a great expert at planning a lecture-course but utterly unable to design a monograph? I suspect that this expertise is the result of a simple and obvious fact: the clear delineation of the audience for a lecture-course is of a great help to the lecturer. This is a tremendous help. I therefore always design a paper or a monograph with a distinct audience in mind. I find that it also helps students to decide in advance what is their intended audience — as delineated by a level of background knowledge plus an area of interest. Why, then, was there no clearly delineated intended audience to essays and books of the philosopher? Was his secretly intended audience always his colleagues from old Vienna? Perhaps he tried to address everyone in sight, thus making his task impossible. This explains his working hard on his manuscripts. Why could he not write for smaller audiences? He could, as is evident from his technical papers on problems within logic and probability and physics. If his philosophical papers are intended for his old Viennese peers who, he knew, would neither read what he wrote nor comprehend it, then clearly his ordeal was irrevocable: but then he doted on hard work and so he regularly set himself impossible tasks.

He wanted me to stay in England, and even then he had already enough influence to be able to help me find a position; but he had a very low view of his ability to influence people and was greatly inhibited. He did help me seek a position, and it was his influence that had acquired for me the early positions I had, in the School, where I replaced him in his courses to a large extent, in the University of Chicago, which had offered me a position which the American authorities prevented me from taking, and finally in Hong Kong, though he was convinced that the move to Hong Kong was a great error, especially since he had kindly persuaded the School to offer me a tenured position.[4]

This brings me to the external relations. I was then no longer the

---

[4] Stephen Körner, the head of the philosophy department in Bristol, asked Popper for his advice about hiring a lecturer. (They were both Nelson fans and on good terms.) Popper recommended me although I was still a student. I was in doubt and pleased that Feyerabend showed interest. He took the job. Later he invited me to lecture in Bristol. David Bohm chaired the meeting and entertained me afterwards.

philosopher's research assistant, but his colleague, yet my job description included assistance to him, chiefly in administrative matters, since for the first time he had administrative duties proper, with the establishment of a degree course for philosophy and economics, with Alan Musgrave among its very first students. I tutored the students and substituted for the philosopher when he was unable to attend a class.

My position in the School was extraordinary.[5] I made friends with a number of distinguished scholars, especially the leading economist Lionel Robbins. I made friends with young exciting scholars, especially Kurt Klappholz and Percy Cohen. Klappholz invited me to write a paper with him [Klappholz and Agassi, 1959] which is still cited surprisingly often and which is my earliest entry to an anthology [Kamerschen, 1967]; Percy Cohen quoted me in his sociology textbook, no less [Cohen, 1968, 94]. Klappholz introduced me to a staff seminar in economic methods. Neil de Marchi edited a book on the legacy of Popper in economics. In the central essay in that book he describes that seminar and its significance. And he gives me there a prominence I do not deserve.[6] He is right, though, when he notes that I was invited to report on Popper's methodology. The same holds for a less known but more interesting anthropology group of undergraduate and graduate students, among whom were Jarvie and Cohen and others who are better known now than then. The group was very exciting and I was very grateful to them for their having helped me with anthropology more than I had helped them with methodology. Unfortunately, some of the brightest members of the group failed their final exams and the others took the hint: the group disbanded. Whereas I learned with the help of the economics seminar mainstream economic theory and prob-

---

5 My fortune was unbelievable: in the London School of Economics I had the occasion to receive the best training in the diverse social sciences. My friendship with the economists opened many doors for me. So did sociologist Ernest Gellner and anthropologist Maurice Freedman. My contact with political science was through Judith, who received her doctorate in the department of government. She also taught me the important field of the sociology of work that no one in the School could teach me. The School had no experimental or clinical psychology, only social psychology that at the time did not interest me. So when I wrote a critical review of a hot volume in psychiatry, I knew no one competent to comment on it and no possibility to get it published. With a very narrow and very inadequate education, I acquired there a more rounded education that allowed me to pursue diverse interests. This taught me the value of a stimulating environment [Agassi, 1981, Chapter 28, What Makes for a Scientific Golden Age?] I taught courses in introductory physics and the history of medicine in Boston, economics in Frankfurt and in Tel-Aviv, political science in Tel-Aviv and in Rome, and criminology in Bar-Ilan; I was a researcher in the school of medicine in Jerusalem and in the school of education in Calgary. I gave lectures and authored and coauthored papers in diverse fields. I owe all this to the exciting London School of Economics and to the critical philosophy as practiced in Popper's seminar. I cannot fully express my gratitude to all my friends there, but especially to Popper and his associates.

6 [De Marchi, 1988b, 148, 161]:
"Agassi was keen to proselytize. He became mentor to the economists and, as Lipsey recalls it, over about half a year they 'learned and came to accept, much of Popper's views on methodology'". "They were feeling their way toward empirical testing when Agassi showed them Popper."
I hope the assertion about my keenness to proselytize is false.

lems and history, George Morton taught me the by-ways of economics, including linear programming, storage and queuing theory and such; he also helped me study operational research and similar material and he invited me to conduct meetings in workshops for business managers that he was organizing. But for his help I could not write on technology as later on I did, when I attempted to argue against the theory of rational belief. (I say, belief is private and theories are never obligatory to endorse in any sense; reliability is required not in science but only in technology legally so and as a precondition for the permission, not any obligation, to use novel ideas.[7]) At the time I was also invited to a debate with Gellner in the School's sociological society, with the result that I published in the prestigious *British Journal of Sociology* an essay on methodological individualism, and a decade and a half later in the same journal a follow-up on it on institutional individualism (merged in [Agassi and Jarvie, 1987, 119-50]). I was also invited to speak in various seminars, including one on the sociology of education and the awesome seminar on statistics, where I presented Popper's new axioms of probability and explained their significance and revolutionary characteristics. (No, this is still not public knowledge, though they are becoming so through Hughes Leblanc's studies on probability Popper-style.[8]) Maurice Freedman then began *The Jewish Journal of Sociology* and invited me to write a couple of reviews for him. Just then I had my first serious publications, reviews that appeared in The *British Journal for the Philosophy of Science* and were instant successes; my very first review, on Duhem and Galileo, is cited in the *Encyclopedia of Philosophy* [Alexander, 1967]. Though a foreigner and a beginner, to my surprise I was even invited by some non-academic circles to speak — about Popper's ideas, of course. In a way it was the very peak of my own intellectual career; except that I was the mouthpiece of a thinker from whose views I had begun to diverge. I was a colleague to the great philosopher, and he was treating me as gently as he knew how, but I had to exercise my independence, and he was too supportive

---

7 The application of technology is seldom obligatory — as in an officially required responsibility, such as the duty of guardians to administer medications to minors in their care in accord with expert opinion. Liberal legal principles recognize faith only in the sense of good faith; otherwise, faith appears only socially: the paradigm is English law: it allows physicians to apply magic only where it is common practice.

8 [Leblanc, 1960], [Fraassen, 1976], [Leblanc and Fraassen, 1979], [Spohn, 1986], [Roeper and Leblanc, 1999] discuss Popper's system of probability and record its value, at times quite reluctantly.

I met Jaakko Hintikka at the London School of Economics in the late fifties at a party in his honor. He told me then that he had just encountered Popper's ideas and so he planned to revise his philosophy. In the 1964 annual meeting of the Association for Symbolic Logic in Chicago, he read a paper offering a compromise between Carnap and Popper on confirmation. I asked him once for a reference for the published version, and he said he has too many publications to remember. Leblanc showed more courage and made the switch, with a notable contribution [Leblanc, 1981; 1989]. He wanted to report this to Popper. He failed. I tried to help and failed too. This surprised me, until I read the bitter remark on Leblanc's early work in a late edition of Popper's *Logik der Forschung* (*Anhang* *V, *Anmerkung* 6).

and too demanding for that.

One last item: Lakatos. He was a Hungarian refugee with an impressive past, including top administrative positions and a four-year jail term, including one year in solitary confinement. During his days of power he had access to some forbidden literature, including *The Open Society and Its Enemies*. He left Hungary after the Hungarian Revolt of 1956 and registered in Cambridge because in that year the philosopher was in the United States. I met him in fall 1957, in the British Society for the Philosophy of Science Oxford meeting. He was eager and like a newborn baby. He was surprised at the unfairness of a chairperson (A. C. Crombie) towards me. He was puzzled at his own level of competence, since he was a newcomer and the crowd was top professionals. He cracked excellent jokes (now that Wittgenstein's enemies are publishing all his output, how can anyone still take him seriously?) and he told me of his doctoral research project (about his heuristic division of all mathematics to the pre-formal, the formal and the post-formal, as open to different kinds of criticism). We discussed his work and I invited him to Popper's famous Tuesday Afternoon Seminar where he read an excellent paper, the *Urtext* of both his doctoral dissertation and his epoch-making *Proofs and Refutations*. He was most impressive in his ability to make non-mathematicians deliberate mathematically and to use their deliberations as means for the development of his mathematical presentation. This technique deeply influenced me. (I wrote in his praise on this matter in the volume in his memory (now in [Agassi, 1988].) He was unbelievably uninhibited and he could always make up for his faults or for a very bad move by inventing terrific jokes or telling fantastic anecdotes. We became close friends almost at once, despite his terribly crude attitude to women and the streak of cruelty he showed from the start. When he moved to London we helped him and he stayed in our apartment for a few weeks. Yet when I left and the philosopher consulted me about offering him or someone else the position I had refused to take, I advised against him: the two of them were very restless and Lakatos was a master schemer of intrigue. Enough.

Though I spent much less time with the philosopher as a colleague than as an assistant, I could watch him better, and observe his interaction with his friends and colleagues better and on different dimensions, the personal and intellectual (and in between the professional). Personally, it was best epitomized in my impression from his repeated report, early in the day, that he had no more than four friends[9], most of them not in England and none of them frequently met. This is puzzling — since I soon met in his company many more than four of his friends — that it requires a

---

9 This contradicts Popper's autobiography, §23, where he lists nine "lifelong friends" from New Zealand. I met only three or four of them.

context.[10] This should include the claim — not much exaggerated — that the philosopher used to work day and night for at least 360 days a year and when he went for a rare vacation he packed the typewriter so that Lady Popper could type his output at once. (She typed every version of his work and he always wrote endless variants of each fully-fledged paper before sending it to the printer — and often even afterwards.[11] On the average he wrote his *The Open Society and Its Enemies* thirty times [Watkins, 1997a]; he had five complete versions of it and he worked on the final version and even rewrote Chapter 23 of it entirely after it went to the printer. He took pride in his habit of regularly messing with proofs so much that they had to be reset — at his own expense, of course. Gilbert Ryle, his only friend in the Oxford philosophy school (to the extent that he had any),[12] was even more extravagant in rewriting proofs than he ever dared to be, and he took this as a legitimation of sorts, or so it seemed to me when he told me all this. During the time we were in close contact he did take vacations, usually to visit his sister in Switzerland, but they were, indeed, both rare and very brief. Otherwise he worked seven days a week, often from dawn to midnight almost nonstop; and, as he reports in his autobiography, he worked on problems of logic and on other calculations when he was sick in bed too ill to go on with regular work;[13] and he got headaches when problems he was not working on were brought to his attention for longer than he was ready to be distracted — which was usually a few minutes; fortunately, hardly any important problem failed to interest him and he was willing to discuss almost all of them. (Some exceptions come to mind, but none nearly as important as ethics, which he deemed not controversial.)

Spending time with friends, unless on work, then, was just intolerable, and working with friends he deemed more the burden of work than the pleasure of friendly company. I once met one of the aforementioned four friends, the one who lived in England. He was a charming artisan

---

10 The paucity of friends may be due to limiting friendships to close ones. My study of Faraday drew my attention to this. By received views it is due to vulnerability [Oldham and Morris, 1995, 189], to the fear of betrayal — as detective Kinsey Millhone says in the early pages of Sue Grafton's thriller *N Is for Noose*. Kant and Freud feared intimacy [Scharfstein, 1980, 229]. Beethoven feared desertion [Solomon, 1977, 83, 115, 119-21, 129, 258-60; 1998, 63-4, 80, 145, 151-3]. Nietzsche feared emotional dependence. Popper confessed to me as well as to Bartley that he feared desertion. He was always suspicious, but he consciously and valiantly ignored his own vulnerability to the possible betrayal of his students and did nothing to prevent it although he viewed us all as destined to betray him.
11 Hacohen refers to drafts of Popper's autobiography.
12 Maybe not. Arthur Prior was a friend of sorts; he had inherited Popper's position in New Zealand and after his arrival in Oxford he occasionally dropped into the LSE for brief visits. Fritz Waismann was in Oxford too. Popper disapproved of his output and of his person; his published expressions to the contrary, I am afraid, seem to me questionable. He was present in a lecture of Waismann in London. Lakatos and I participated in the discussion. Our comments were regrettably unfriendly. Popper approved. He greatly disapproved of the way Wittgenstein had treated Waismann, though.
13 Lady Popper's response to Popper's report that he worked on probability when he was too sick to work on more challenging material is renowned: "now I know why I hate probability" she said.

from Austria. He told me he had no intellectual pretensions of any sort. The Poppers treated him like one of the family. His visit was very brief. When we were introduced I did not catch his name, and he left before I had the chance to ask him what it was.

I never met his artistic friends, such as Rudolf Serkin, but I surmise when they were together the philosopher engaged him in intense conversation about music, so that the encounters could count as work. And he did engage in such conversations as I heard from one of his closest friends and admirers, Julius Kraft. He was somewhat older than the philosopher, a serious thinker of the Leonard Nelson school (few know of its very existence), an exceedingly brave and honest person whose open attack on Martin Heidegger [Kraft, 1932], the most influential among the philosophy professors in Germany, made it difficult for him to return to Germany after the war (from the United States, where he had taken refuge) to get a position in philosophy there. For, in post-war philosophy departments Heidegger was still most influential, particularly in Germany (I would like to say, despite his frank Nazi convictions, but the dreadful facts about Germany after the war were quite different[14]). Kraft had been trained as a lawyer and his work was mainly in philosophy; finally he received a position in Frankfurt as a sociologist. He had some influence on the young Popper, as he had introduced him to the thought of Leonard Nelson, who early in the century tried to use the Socratic Method to mix a rationalist philosophy and a socialist politics [Nelson, 1965]. Kraft stayed for a time in London before he went to Frankfurt to take his chair, and he kindly spent some time with me — no doubt due to the philosopher's kind word about me. He told me how they first met: he was playing the piano and was flabbergasted by uninvited, intense, very serious comments on and critique of his performance from an unknown, rather odd-looking adolescent, whom he nevertheless started to befriend and learned to admire. It is thus hard to imagine that the philosopher would not speak with his artistic friends about art. His encounters, then, would count as work.

---

14 Heidegger was reluctant to grant an interview to the press while declaring "that his conscience was clear" [Sluga, 1995, 234]. Sluga comments (244-5):
"In spite of all the philosophical books and articles published in Germany since the end of the Second World War, there remained in the end only one long, ominous silence on questions that should have mattered more than all others. These concerned philosophy itself and its relation to the body politic. For the German philosophers who were left or were growing up in those years proved themselves unable to look at the Nazi system and ask themselves what it might reveal about the nature and role of philosophy, about its relation to politics, and the possibilities and limits of philosophical politics."
"... An entire society had devoted itself to the task of forgetting, and the philosophers were only too willing to participate in the communal act of erasure."
"... Missing a unique chance to learn from their own mistakes, they found themselves ... forced into ever new, ever more elaborate, and ever more predictable evasions and denials."
This situation is changing now, as the interest in the Shoah is growing — with the sense that by now it is past history as there is no one around any more to carry the dreadful burden of guilt.

Kraft was a lovely gentle person, described as a friend in the philosopher's obituary on him [Popper, 1962].[15] (Another obituary for a friend is that on Joseph Woodger [Popper, 1981]. It is hard for me to speak of that obituary, as it was written too long after I had ceased to communicate with him; my impression is that the obituary portrays him, too, as a friend, and my impression of their meetings that I witnessed is the same.) And I have seen him with quite a few celebrated friends in his office and at home, most of whom I forget, as I was too ignorant.[16] All this renders a puzzlement that he should have told me, as he did many times, that he had but four friends. Perhaps he thought only childhood friends count as real friends — and friends from adolescence as mere borderline cases. He once mentioned to me, in a particularly benign moment, Whitehead's report that Russell always remained for him the adolescent he had first encountered in Cambridge, though he met him, of course, also when he was well past middle age. The same, said the philosopher, will be his memory of me. As he later saw me as a traitor, I wonder if this remained true,[17] if he really at times still remembered me in the old cordial light. Perhaps I am missing the point; perhaps it is precisely such memories that made him feel so bitter about me.[18] Do you remember the Marquise of O — ? If she had not viewed her man as an angel first, she explained, she would not have viewed him as the devil later. The story of the Marquise is a romantic tale, of course, with literary license for exaggeration, but also with a significant moral, perhaps. It is a great pity I could not reduce the philo-

---

15 Popper says [*Autobiography*, §15], "... Julius Kraft (of Hanover, a distant relation of mine, and a pupil of Leonard Nelson), who later became a teacher of philosophy and sociology at Frankfurt; my friendship with him lasted until his death in 1960."
16 Economist Abba Lerner, political philosopher and historian Hans Kohn, the Plato scholar Raymond Klibansky and a few others visited Popper and impressed me (I was utterly ignorant of their just reputation). I also met there two or three New Zealand friends, including the already mentioned Arthur Prior; not John Eccles, whom I first met later in Hong Kong. I joined Popper as he visited some positivists in the lobbies of their London hotels, especially Philipp Frank and Tadeusz Kotarbinski. For him they were figures from the past; for me they were authors of unimpressive works. He introduced me in the lobby of the London School of Economics to Bryan Magee as the rising star that he was. We met again in diverse Popper meetings, where I learned to appreciate him. He taught me a significant lesson. I used to emulate Popper's habit of beginning to answer commentators in public discussions the moment I understood their objections. Magee drew my attention to the fact that others want to hear the commentator in more detail. I hope he taught me to be more patient.
17 Let me repeat: in an unsent letter (File 266.24, n. d.) Popper reminds me of the good old days:
"(... until quite recently I cherished the idea that you would write, when I am dead, about those splendid years — or so they appear to me — when we encouraged and helped each other. But I wish nothing more than that you should never write another word about me — and least of all in praise of me.)"
18 Disappointment usually made Popper bitter. With most people it raises anger. Spinoza argued that this is the refusal to admit error: the view of false expectations as rooted in some error suffices to extinguish the anger that they trigger. This is easy to observe. Also, it explains well why failed expectations that rest on calculated risks do not anger. The proper response to disappointment is regret, said Spinoza; in particular we should regret that some people are aggressive. Popper's response was bitterness as he felt he deserved better. I always regret having added to this; I comfort myself in the observation that he could always muster his tremendous sense of humor and of proportion. He was no Dr. Jekyll, yet he managed to abide together two polar traits, deep dark bitterness and terrific, bright sense of humor.

sopher's high hopes for me while we were close, no matter how hard I tried. I am really so very sorry that I caused him much grief to the end, when he had enough grief anyway, after his wife and most of his close friends were gone.

Once, in a fleeting moment, in 1969, when I was teaching in Boston, I met an Austrian-American couple quite by accident. I know nothing about them, except what they told me then: the philosopher was coming to Brandeis University for a semester [upon his retirement from the University of London] and he had asked them as old friends to find him suitable accommodation. I suggested to them that they seek a quiet place near that university. When the Poppers arrived they were taken to an apartment rented for them at the other side of the metropolitan area, about one hour's drive from the university, overlooking a truck route. The philosopher could not sleep all night; in the morning he paid the monthly rent and cleared out. Just before that he gave me a ring and told me of his plight. I promised to meet him in an hour in the hotel to which he was moving. He rang me again almost at once, canceling the meeting: his wife, he explained, did not want him to be in my debt. (Since Lady Popper was very favorably disposed to both Judith and myself, and since we never fell out[19], I find it hard to think she could think that way; I suppose she wisely wanted to save him any aggravation regardless of rights and wrongs; but I do not trust my judgment on such matters.) For some days we did not meet at all. I sent him alternately two of my graduate students to help him seek a more suitable place. A few days later the situation got out of hand; he threatened the university with returning to England at once and this distressed the head of the philosophy department there. Finally he noticed that the two alternate helpers were always in the same car and he surmised it was mine. He rang me and informed me that he had decided to permit me to make a slight effort on his behalf: say, something like two phone calls. I rang my estate agent and offered him a handsome bonus for an immediate solution. I then rang the head of the philosophy department (we were on very good terms) and asked him if I should pay the bonus or if he would pass the bill to the university. Of course, he would. The philosopher was grateful too and even accepted my company for a while, beginning with an offer of an invitation to my family and myself to visit their really lovely place and accepting my subsequent invitation to Lady Popper and himself to return a visit. He later on accepted my service of driving him to work, as he had no car then. On these trips, as well as on some other occasions, we had our last pleasant conversations.

The philosopher's works, especially his autobiography, reflect his friendship, his personal and intellectual rapport, with many colleagues —

---

19 A surprising, bitter, unsent letter of Lady Popper to me (File 266.24, n. d.) may very regrettably refute this assertion of mine. I hope not.

especially the two who gave him the status of a philosopher, primarily in his own eyes: Herbert Feigl and Rudolf Carnap, who considered him a philosopher proper before he dared do so. He says so in his contributions to the Feigl *Festschrift* and to *The Philosophy of Rudolf Carnap*, where he mentions a debate on semantics with them during a mountaineering holiday they all had spent together in the Alps — discussing the stone that does-or-does-not think about Vienna, mentioned in the Prologue above. He did admire Carnap and he felt grateful to him for his (Carnap's) publication of his (Popper's) ideas on science as soon as he had heard of them. (Otto Neurath, the then head of the "Vienna Circle", privately censured him for this. I do not think the philosopher knew about this fact, as it was published much later. Paul Feyerabend was quick to agree with Neurath, but the philosopher did not read Feyerabend's output either.[20]) Carnap's early kindness to the philosopher made it all the harder for him to see his later distortion of his criterion of demarcation of science as if it were a criterion of the meaning, as a theory of language (the language of science, the only one Carnap recognized).[21]

This is lamentable: it is a bit of nasty distortion: Popper repeatedly declared that he had never demarcated science as a language but always as sets of theories stated within some given language. To no avail. But the fact is lamentable on the more important level of public awareness of important ideas. I have met a few established philosophers who could not or would not comprehend this difference. I report in bewilderment that, some time back, the celebrated and justly beloved Hempel, the last representative of the "Vienna Circle" on earth, insisted on Carnap's version as if he deemed it correct. He did so speaking in a meeting in Tel-Aviv University that to my profound regret I had consented to chair, and I

---

20 Popper's reading habits were very unusual and they reflect his unusually high degree of excitability. For relaxation he re-read nineteenth-century English novels, everything by Shaw, and some glossy mountaineering literature. He read avidly the prestigious *Journal of Symbolic Logic* and refuted ideas published in it surprisingly frequently. His familiarity with the literature in philosophy was limited to the not-too-irrationalist. He spent more time writing than reading. A paper written not in the manner that he approved of he could not read from beginning to end; much less a book. Rather, he leafed through it, making conjectures about the contents and then testing them. He had terrific insights and kept an open mind. After a time it was very difficult to criticize his reading of a paper and he would be able to correct it only if he was sending something about it to the press. However rare, distortion prevailed, particularly when some members of his entourage — myself included, I admit to my shame — showed him pieces of text that boosted his complaints. He thus learned of nasty things said about him that he did not check. At times these were inaccurate and all too often they sounded worse out of context.

Given all this, it is hard to judge whether he read Feyerabend. In a letter to Albert [Morgenstern and Zimmer, 2005, 207] he says, it is a long time since he read Feyerabend, yet he calls him fascist ("*heute ist er Anarchist, morgen ist er Fascist*"). He would not do that without careful checking. When Feyerabend was his disciple and defended his views with deep conviction, he was greatly flattered. As he told me this I was displeased and he noticed that.

21 Let me repeat: regrettably, I find Carnap's and Hempel's false persistent distortion of Popper's assertions not simple oversight but a great insult that is barely excusable (his own generosity towards them notwithstanding): it would all be permissible had they not put in his mouth assertions that he rejected.

stretched to the limit the chair's prerogative in attempting to correct his textual reading. I stressed that I was not endorsing Popper's ideas that anyway I have publicly criticized to my satisfaction [Agassi, 1991b] but that I was speaking of accurate reporting. I failed: Hempel flatly refused my correction. I find this hard to swallow, and on two different counts. First, if there is no difference between two versions and some authors insist on being ascribed one of them, surely there is no reason to deny them the courtesy of complying with their request. (The technically adept will recognize, perhaps, that intuitionism constitutes such a case: of some pairs of sentences, one the double-negation of the other, they accept only one into the body of mathematics. Though most mathematicians deem a sentence and its double-negation equivalent, no one will dare ascribe to the intuitionists a sentence on the ground of this equivalence.) Second, there is a difference between the two versions and it is very easy to comprehend, even for readers unfamiliar with the technicalities involved. It is this: the negation of a sentence is a sentence but the negation of a scientific theory is traditionally deemed clearly unscientific. (This last point follows easily from two obvious ones: first, scientific theories were traditionally declared true (or at least probable), and second, the negation of the true is the false (and of the probable the improbable).) Hence, the language of science includes the negation of every sentence it includes, but science does not; hence the demarcation of science and of scientific discourse differs from the demarcation of the language in which scientific discourse and scientific theory are embedded. Popper boldly declared that scientific theories are bold and hence improbable; their negations are therefore most probable, untestable, and hence not scientific.

Carnap and Hempel rendered Popper's view of science silly by their reading of it as a theory of the language of science, since that reading renders the negation of a scientific sentence both scientific and unscientific. Hempel subscribed to the erroneous reading to Popper and refutes him this way. It is no surprise, then, that the philosopher took offense. The error will soon be corrected[22], now that the ideas of the "Vienna Circle" are out of fashion and Appellate Court Justice Quine[23] vindicates Popper on this matter; but the damage is done: a generation of students of philosophy was raised on the idea that they could save themselves the trouble of reading Popper's scarcely available *Logik der Forschung*.

---

22 To the extent that current apologetic commentators on the "Vienna Circle" are representative, the game is over: they admit defeat as nearly explicitly as their modest courage allows.
23 As to "the scientist", says Quine, "Sir Karl Popper well depicts him as inventing hypotheses and then making every effort to falsify them by cunningly devised experiments" [Quine, 1987, 8]. For Quine's criticism of Carnap's views, see his celebrated "Two Dogmas of Empiricism" [Quine, 1951]. Carnap's theory of the language of science implies a phenomenological, idealist metaphysics, he observed. This is deadly criticism, as "Vienna Circle" abhorred all metaphysics, idealism in particular.

Though all this is very sad, and also obviously a case of bad faith, I ask you to allow me to repeat a point I made earlier about bad faith. It need not be a personal matter: it may easily be the result of oversight, especially dogmatism, especially that of not noticing a translation between frameworks, especially when one denies that one is operating within a framework, which is the theology of the "Vienna Circle".

Misgivings about the previous two paragraphs have sent me to the library. I do not like to question personal sincerity or intellectual integrity, particularly not those of Hempel, the sweetest individual among the members of the profession. So I looked up *The Encyclopedia of Philosophy* and *The Encyclopedia of Unified Science*, the one as the currently most authoritative and the other as the quasi-official organ of the remnants of the "Vienna Circle" of its day. The former has an essay on Karl Popper, which begins with a long paragraph that explains that his demarcation of science is within language, not of language, be it the alleged language of science or any other. The latter has Popper mentioned twice. The first reference is in the contribution of Thomas S. Kuhn that is indifferent to the theology of the "Vienna Circle" [Agassi, 2002b];[24] it shows no concern for any matter of language, not even in Kuhn's (significant) discussion of metamorphoses of scientific terminology due to changes of scientific frameworks (which frameworks house paradigms). The second reference is fleeting. The index, compiled by leading members of the profession and of the "Vienna Circle", refers to Popper's demarcation of the language of science — in defiance of both the original and the indexed text. I wonder: why did the indexers deem the distortion so important that they had to drag it in?

I am a bit unfair, perhaps. The members of the "Vienna Circle" were popularizers, and from the start they spent tremendous efforts to explain that the discussion of science in the framework of the language of science makes verification and refutation utterly symmetrical, so that if a sentence is either verified or refuted it is verifiable. The verifiability criterion of meaning was not that a sentence must be verified but that the question is decidable whether it is true or false, whether its truth-value, so-called, is truth or falsity. Today's term is not "verifiability" but "deci-

---

24 Carnap wrote a famous fan letter to Kuhn mentioning affinity between them. This is not to say that this way he endorsed Kuhn's view of science as the refusal to admit refutations. They differed in many respects, while sharing the dismissal of all controversy between reasonable people as rooted in no more than misunderstandings or misconceptions that is easily rectifiable. This, however, is an aside: for my present purpose suffice it to note that Kuhn never used the term "the language of science" or any equivalent term. Kuhn followed Duhem here and claimed that every different paradigm comes with its own distinct language. This is Duhem's thesis that Kuhn labeled "incommensurability". The difference is maximal between this idea and the intent of the "Vienna Circle" to construct one language that could house the whole of science unified. Kuhn's idea that his methodology applies only to the natural sciences, not to the social sciences, is opposed to Neurath's idea of unified science expressed in The *International Encyclopedia of Unified Science* that published Kuhn's book nonetheless.

dability",[25]; the field of mathematics known as "algorithmic" is full of instances of partial decidability: some sentences are decidable under some conditions but not under others, and we may never know which conditions obtain: when certain processes are concluded we know that they can be concluded, but if not we do not know [Harel, 2004]. Thus, the debate was concluded with Popper's decided victory. And so, there was no use for Hempel's move, to his suggestion that, with the replacement of verifiability by confirmability, refutability should also be replaced with "disconfirmability"[26] (his own neologism). The term has stuck and philosophers of science in their ignorance still confuse the demarcation of the language of science with the demarcation of science. Hempel was the last remnant of the "Vienna Circle" and at the time the only authority equal to or even above Quine. What I have demanded of Hempel thus amounts to a full concession. It was, however, too early for either of them to concede. Though Quine clearly rejected the identification of the demarcation of science with the demarcation of the language of science [Quine, 1988] for half-a-century, he could but would not say that this was the chief dispute between the philosopher and the "Vienna Circle", particularly Carnap, and that Popper clearly won — even before the battle began. This will be admitted only semi-officially and only in the next generation.[27] Hopefully.

The strongest image the philosopher left on me, other than his tremendous intellectual ability and intensity and sincerity, was his seemingly constant complaint about unfair treatment — especially the neglect — from his colleagues, especially the ones he had known in Vienna, despite his exemplary conduct towards them. Many of his complaints seem quite just (regardless of his conduct); others were unbelievably naïve. By and large he was right: colleagues treated him shamefully when he was young and they still do.[28] He understandably felt gratitude to them, especially to

---

25 Carnap preferred "resolvability" to "decidability" [Carnap, 1931, 94].
26 Hempel saw symmetry between verification and refutation as both are not final. He thus disagreed with Bacon, and more so with Popper, who had discussed this in detail. (Popper deemed finality irrelevant to the asymmetry: it belongs only to the truth of tautologies and to the falsehood of contradictions, yet the asymmetry holds: a theory in conjunction with its refuting evidence is a contradiction but in conjunction with its confirming evidence it is no tautology.) Hempel invented the word "disconfirmation" for the opposite of "confirmation". (The word "infirmation" or "disfirmation" might serve him better. As to Popper, he preferred the expressions, "supporting evidence" and "undermining evidence".) Hempel's system is silly: his theory of confirmation deems evidence true (in agreement with current literature); his theory of disconfirmation deems it tentative. The simultaneous support of one theory and undermining of another that one piece of evidence may offer (in a crucial observation such as Eddington's), renders Hempel's system ludicrous. Its great popularity exposes the dearth of criticism.
27 To repeat, perhaps I am in error here and the analytic Establishment has already made its required semi-official concession to Popper. Even were it so, a clean and open statement that should close the present needless ongoing debate is still sorely missing. The time lag here will be indicative.
28 Although the dramatis personae are all dead, the philosophical Establishment still refuses to set the record straight. This suits those still unable to stomach fallibilism and intellectual autonomy. It would be nice if advocate fallibilism (Juliet Floyd, for example) would take courage and speak up.

Feigl, for their encouragement to him when he had no publications. When I was his assistant I followed him as closely as I knew how. For example, though I was a smoker then, as he was allergic to smoke, in his company I learned to be more sensitive to smoke than he was and to protect him from exposure to smoke as much as I could. I then also learned to find clues for all sorts of unacknowledged debts to him. Of course, this was pointless and unnecessary; yet I did learn from it that the philosophical leadership was not honest on this: only inner circles were informed of some ideas publicly ignored. Perhaps, like in show biz, any public mention, however critical, is an honorable mention; perhaps, however, it is only natural that some lesser ideas are judged to be of interest only to experts. If Popper's ideas were judged of interest only for experts, why then did the leadership misreport them to the public? Would not then utter silence be more appropriate? But I forget: their misrepresentation was not intentional.

Popper was hardly known in the general philosophical public when I came to London. My own case may illustrate this. Though I came to London as a graduate student in search of a teacher, I had not heard of him. I always sought a teacher (and now that I am too old for that I still seek collaboration) but I was particularly desperate for a teacher after I got my dubious M. Sc., which my Alma Mater, the Hebrew University of Jerusalem, judged not good enough for continuing graduate studies there. I did not mind very much the rejection: I minded the privation: I was desperate for a teacher. I looked for a teacher with effort, I consulted friends and experts, I consulted the official literature and the officials at the British Council in Tel-Aviv. And then I went to London to be a student of Herbert Dingle, who was the only recognized professor of the philosophy of science in England. Later on I wrote to Hugo Bergman, my former philosophy professor in Jerusalem, and complained about his not having told me of the very existence of Karl Popper. Bergman kindly answered, observing that he had mentioned him a few times in his book on the theory of knowledge. (It was on a minor point: factual information is never free of all theoretical overtones. This point is in no way characteristic of any one single thinker: it was forcefully made by Galileo, by Kant, by William Whewell and by Pierre Duhem. These days it is generally ascribed to Norwood Russell Hanson, whom I met for the first time when he invited the philosopher to give a talk to the philosophy of science club he was inaugurating in Cambridge. The talk was a famous classic, "Philosophy of Science: A Personal Report", republished in his *Conjectures and Refutations*.) In Bergman's book many writers are more often mentioned who were known then but are now remembered only by some very old people and by some highly specialized students. (I discussed this in my piece [Agassi, 1985b] in the issue of the *Grazer philosophische Studien* dedicated to Bergman's memory.)

The aggravation the philosopher suffered from Carnap, the source of the distortion in question [Carnap, 1936-7], was more from neglect than distortion. He felt he was forgotten. He could hardly forgive the neglect. And unphilosophical though this attitude was, it is understandable: as the philosopher has put it, Carnap wrote a whole book (published in 1950) splitting the concept of probability into two (statistics and generalized deducibility) and then (erroneously) equating the two; he could just as easily have split it into three or four (the above two plus confirmation and/or credibility) and then (erroneously in the same manner) equate them all. The point of course was that he could not honestly do the further split without mentioning Popper, as the latter naïvely assumed. This is why, as I told you before, the philosopher was so excited when he felt he could nail down the errors in Carnap's book on confirmation as probability (= conforming to the probability calculus). And, to repeat, it is all touching but unfortunately trite. Quite incontestably, it is most probable that the next throw of a die will be a number between one and six. Hence, according to Carnap it is already highly confirmed and not given to further confirmation. That it will be the number six or that it will be a number other than six, to take the standard (inadequate) example, is certain even though it still carries no empirical confirmation, much less any interest. That it will be a six (not a six) is improbable (probable) and still not refutable (confirmable). The question remains: what does it matter? That is to say, gambling aside, what does the conduct of a die matter to science? If nothing, then, personal feelings and ambivalent friendship aside, what did the conduct of Carnap matter to the philosopher? The hallmark of his attitude to colleagues was his insistence on the (undeniable) duty of those who use one's contributions to thank one publicly, on his defense of the ownership of new ideas and on historical truths concerning authorship. I find this unfortunate.

There is a myth that (barring accidents) the commonwealth of learning notes and recognizes[29] at once and properly assesses at once any

---

29 The recognition that the commonwealth of learning grants or withdraws is a subtle matter as it cannot have official representation. Semi-official recognition takes place in conferences ever since the first Solvay conference (Bruxelles, 1911). Scientists go to conferences partly to listen there to gossip about such matters. How does the Establishment control the gossip? Not too well, of course, and in very rare cases it issues quasi-official denials. When is its action more adequate or less? This is a theoretically and practically interesting empirical question regrettably missing from the agenda of the sociology of science — perhaps because this question is naturally critical whereas the job of the sociology of science is to revalidate (Gellner). Kuhn has the only empirically corroborated assertion about the matter: the departments with highest standards, he has observed, those of mathematics and of physics, update their standard knowledge fastest. Feyerabend mentions the denunciation of astrology and parapsychology.

The major way to control gossip is to issue it quasi-officially. Thus, the presidential address to the Eastern Division of the American Philosophical Association in 1946 was an explicit recognition of "logical" empiricism as the new paradigm. This recognition came at least two decades too late. The recognition of its demise is more overdue. Does this indicate a decline of standards?

contribution to human knowledge. This is often declared a distinctive characteristic of learning, especially of science proper, as opposed to the arts. The arts are dominated by personal tastes and by matters of style, says the myth, so that the notice and recognition of a contribution of an artist may have to wait till the public absorbs the style, idiom, and taste, which accompany the contribution — and only then can its worth clearly transpire above and beyond the mere invention of style or idiom. Not so in the world of learning, especially in the hard sciences, in logic and in mathematics, where personal style (and purely verbal innovation) is absent or marginal. Hence, a forgotten work of art can be brought out of the attic, dusted and put on display as good as new. Not so a work of science: if the learned press has rejected one's work and thus a competitor has beaten one to print, then one suffers an irretrievable loss of priority. One who suffers this fate has a reasonable grudge against some editor.

This is a myth, the myth of instant recognition of scientific discovery. It is propagated by historians of science who report every delay, however small, as the outrageous outcome of some outside intervention, usually of prejudice. It is also propagated by the historians of philosophy who declare philosophy an art (not by the historians of medicine who declare medicine an art). The moral of the myth is that philosophers have to study ancient texts (unlike physicians, who should learn their craft from up-to-date textbooks and research-reports).[30]

One of the propagators of the myth of instant recognition in science is the celebrated leading sociologist and the acknowledged father of modern sociology of science, Robert K. Merton. I am embarrassed to write about him: though I have often poked fun at him, he wrote me some unbelievably flattering fan letters. I was once a visiting professor of sociology in City University of New York for a couple of days and then he kindly participated in my seminar there and entertained me afterwards in his posh New York club in the nicest manner. He also tried to be of some help. I once asked him about matters: surely he was aware of the existence of a time lag in the recognition of scientific and other discoveries. I could not be specific and refer to the refutations of the myth: he must have been aware of them earlier than I, of course. He was: alluding to it he said, quite nicely, that things are much worse in the world of the arts, where such a successful piece as Beckett's *Waiting for Godot* had to wait for two decades before its first performance. (Notice the Establishment touch: the delay in the recognition is simply unavoidable. Notice also the shift in position: the delay is not zero, but relatively small; things are not

---

30 Attenuating the myth that physicians are up-to-date is the mechanism of helping them to keep learning or even forcing them to do so. The pharmaceutical industry utilizes the defects of the updating mechanism and pushes advertisements in the guise of up-to-date scientific information. Those in charge of the updating mechanism can use these advertisements to improve their services. They sometimes do.

as good in science as we say, but they are still much less aggravating than in the arts.) I resisted asking him about the contributions to philosophy of such very important yet hardly recognized individuals as Karl Popper and Michael Polanyi, especially as he himself had told me of his high esteem for Polanyi and of the fact that Polanyi had somehow negotiated with him a form of mutual public recognition. (Polanyi replaced the myth of instant recognition with the myth of personal knowledge. Scientific leaders, he said, decide when is the right time to recognize an innovation [Polanyi, 1969, Chapter 6].[31]) Rather, I asked Merton if he would publish his notion that things are not so bad in science as in the arts. This would be an admission that his classical descriptions of the situation as optimal is somewhat exaggerated. It would refute his claim that the difference is qualitative, that even a few hours suffice to destroy a claim for priority. Would he discuss this publicly? He nicely declined.

The myth of instant recognition thus remains above challenge simply because only the established may challenge. Others who wish to challenge should wait for two decades or more.[32] For the two philosophers[33] it was more: it took them about three decades to arrive. Should

---

[31] Polanyi mentioned his own case as an example. He did not complain that the theory (in physical chemistry regarding adsorption) that he presented early in the century won recognition only decades later. He explained that his theory was inconsistent with certain basic ideas in physics and only after their demise could it gain notice — as it did. He even thought his *Doktorvater* had been generous, as he could have failed him. He said, as a science professor, he too had the duty to teach received opinions. I disagree. His *Doktorvater* could not ignore his repeatable data. The scientific establishment has no trouble recognizing competing opinions, and it often does so. Young Galileo had to limit his teaching to received doctrines; Polanyi could and probably did teach some heresies. One should not teach a heresy as the received doctrine, though, and even this happens too often. Even teaching the views of one school of thought as if it were unrivalled is wrong, and yet this is standard, and without having Galileo's excuse: today's Establishment recognizes the need to teach some false and dated and heretical theories. There are examples to the contrary, such as the case of the oversight of Einstein's theory of emission and absorption. My physics professors never mentioned it and my acquaintance with it was through reading Whittaker's heretical *History*. After the discovery of optical pumping, it became a part of the curriculum, of course. Even that took some time: a research assistant in Jerusalem worked on a doctoral dissertation on masers (instruments that produce coherent microwaves) while I was working towards my master's degree there, but it was not a part of the curriculum: he explained this to me only because I showed some interest — much too little, as it happens.

[32] In some respect, the delay in recognizing Popper (and Polanyi too, see next note) was longer than usual — although not as much as Spinoza. Popper draws attention to it gently in the opening of his "inaugural lecture" at the London School of Economics [Popper, 1990, 29], noticing that the lecture suffered a "slight delay" of 40 years: he was appointed reader in 1946 and professor in 1949, retired in 1969, first invited to give a public lecture there only in 1989. (He died in 1994.)

[33] Reviews of biographies of Popper [Blaug, 2002] and of Polanyi [Ree, 2007] notice the absence of instant recognition. Both offer independently the same odd *ad hoc* hypothesis to explain away these refutations of the hypothesis of instant recognition: as these philosophers did not express their views well enough, instant recognition had to wait for the proper representations that these biographies provide. Hence, now all is well. The theory that dogma is comforting cannot explain the high degree of pain that this situation hints at. Possibly, the very possibility that the dogmatic will have to overhaul their worldviews terrifies them: this way they might lose their optimism. This, however, is an explanation of their clinging to dogma, not of the delay in according recognition to these two greatest and clearest philosophers of the mid-twentieth-century. The superficial explanation is that the Wittgenstein cult put everyone else on the margin. This raises the question, why did the philosophical public prefer the inferior stuff? The general answer is that opting for the lesser quality is less taxing. This raises the

we proclaim them artists?

(Even much smaller fish than Polanyi can be offered a deal. Even I was offered a few. The most amusing one was made by the leading Italian philosopher of science, Ludovico Geymonat, who explicitly and clearly proposed to me that I quote some statement from Chairman Mao Tse-tung, significant or not, in agreement or not, in exchange for a more conspicuous position on the horizon of Italian philosophy. I could ask, will "let a hundred flowers bloom" do? I did not.)

The impact of Merton's work was immense, and quite unintended — decidedly counter-intended. He translated the prescriptive (early-nineteenth-century) view of William Whewell on the canons of science to the (early-twentieth-century) descriptive language of sociology. He thus deprived it of any force as an evaluation of science as an achievement, as his presentation of it was in utterly descriptive terms — in a value-free manner, to use the sociological terminology. One is prone to be prejudiced in favor of one's hypothesis, Whewell admitted, and some are prejudiced this way. To prevent this, he demanded the adoption of a critical attitude: hypotheses should be examined severely. Merton said something that sounds almost identical to it yet is utterly different: he said that scientists are critical. Daniel Greenberg — founder of the experimental Sudbury Valley School — wrote decades ago a detailed critique of Merton's description of the scientific community (based chiefly on his own experience as a physicist) and could not get his paper published. The results that are published and publicized agree with Merton's ideas. When I challenged these once, in his presence, in a session of the American Sociological Association in New York, he responded instead of the speaker, sincerely informing me that my objections were known to him and to his associates and that they were working on them; believe me, he said, reminding me of my rabbis.

The test of Merton's ideas was poor as it was conducted on the tacit assumption, sanctified by Derek J. de Solla Price and Thomas S. Kuhn, that everyone with a science degree is a scientist. Hence, no scientist need aspire higher than normal practicing scientists in some recognized academic laboratory; perhaps a technological scientific-industrial laboratory should do; perhaps even less. Look for the lowest common denominator!

The techniques of testing of Merton's ideas were those of participant observation sanctified in anthropology and in sociology. The word

---

question, why do mathematicians and physicists make the efforts that philosophers will not? A possible answer is this. The task of philosophers is to extol science in ways that will please the representatives of science on appointments boards and similar power positions, and these are scientists who have ceased to make the required efforts. (This indeed is why they sit on appointment boards and on similar power positions.)

"participant" is a misnomer designating mere presence: the anthropologist is present all the time and observes rather than performs, say, some magical rites; when an American anthropologist once took the word "participant" literally and became an added member of a New Guinea chieftain's harem (using the fact of her being a woman), a minor sensation resulted.[34] This would not have happened had social anthropologists been as critical and as quick to learn as mathematicians and physicists are. Jarvie has worked on the rationale of fieldwork in his doctoral dissertation (supervised by Popper and Watkins). Its published version, his first-rate *The Revolution in Anthropology* [Jarvie, 1964] is a joy to read and it cannot but make anthropologists think anew the matter of fieldwork. They are not very familiar with that book and their debates about fieldwork still abide by the pious style of the academy without benefit of his rich ideas.

Participant observers of laboratory life are particularly willing to take piously the canons of fieldwork and apply them indiscriminately. Their very lack of discrimination, though a folly, gives them the air of objectivity (by the defunct inductive canons): it enables them to show courage and report observations of the debasing of science by some pretentious and ambitious go-getters posing as qualified researchers. The findings, allegedly, are that researchers sing to the gallery: they are willing to stake any claim that is likely to register. The observers reported their finds allegedly the way anthropologists report magic rites, yet with no mention of the possibility (which anthropologists have learned the hard way to check) that the rite was executed improperly or by impostors (often for the benefit of the unsuspecting observer). Their excitement betrayed the gambler's attitude: on the one hand the test was safe, as it simply revalidated established practices (to use Ernest Gellner's apt expression); on the other hand, there was some possible handsome gain: if natural scientists and industrial engineers can be self-indulgent rather than self-critical and get away with it, then, perhaps social scientists can

---

34 I could not find again the reference to the information mentioned in the text here, but, anyway, it is the thought that counts, as it comprises the strong criticism that Jarvie has launched against the sweeping but vague demand that anthropologists should leave behind their home culture and assimilate into the ones in which they allegedly participate during their ritual performance of their fieldwork. See Bernard's discussion of the paucity of discussion in the methodological literature of sex and fieldwork [Bernard, 2005, 375]. The topic has gained interest recently; see [Golde, 1986], [Dynes and Donaldson, 1992], and [Kulick and Willson, 1995].

Full participation blocks the return home, as a few pieces of fiction illustrate. The required level of participation is not fixed. Sir Edward Evans-Pritchard used informants about sorcery. Paul Stoller objected to this and apprenticed himself to a sorcerer. How far does one go, then? Famously, the inability of Lewis Henry Morgan and Bronislaw Malinowski to leave behind their modern-style sexual problems greatly influenced their work. With an official adoption into the tribe of his study, Morgan went further than most field anthropologists did, yet not nearly far enough. Indeed, it only worsened matters, as it had its political significance for both sides of the adoption. Malinowski made a virtue of the necessity of internment in the South Seas, where he developed his extremist ideas about participant observers. As Western inhibitions prevented him from taking a native wife, he had no first-hand experience of primitive family life. His diaries show how painful this was.

too. Do not try to argue with these people!

These people overlook a major difference that they themselves stress: whereas their reports of magic rites and of criminal life-styles are mere descriptions, not prescriptions, their reports of research signify only because of the tacit (and possibly false) assumption that what they describe is science proper. They thus bespeak approval of the suspect activities that they describe. (Of course, the same goes for Merton's value-free descriptions of science: he valued science and method of its value-free description.) Their approval signifies because it concerns clearly cases of corrupt research practices, since they stress their observation that the activities that they observe fall short of received standards. Not that they are discriminating and can judge that the received standards are too high: though they have some ability to identify a crime and a ritual, they cannot identify a scientific experiment; they take the word of their informants that they (the informants) are scientific researchers engaged in scientific research proper and thus performing scientific experiments proper. And they surmise — at times wrongly, but let this ride — that the activities that they record fall short of received standard. Hence, they prefer the word of their informants over the received standards. Why? I propose that they do so not in the mood of objectivity that they feign but in self-interest: they do so in the hope of thereby legitimating their own researches. Hence, they are far from being impartial, and they are not self-critical in the least. They are unscientific even by the (low) standards currently accepted in anthropology, as they prefer not to examine their information. They will not listen to criticism. Hence, talking to them is a waste of time.

(Nevertheless, the logic of the situation remains: the reliance on acceptability as the criterion of correctness and on public acceptance as the mark of acceptability, is a boost for the status quo. This characterizes the philosophies of Michael Polanyi and Thomas S. Kuhn, of John Austin and Rom Harré, of observers of law-courts like Stephen Toulmin and of laboratories like Bruno Latour and Karin Knorr Cetina. They all play the same game: recognizing the public endorsement of their views they vindicate themselves; recognizing the public rejection of their views they admit defeat. Hence, as long as you and I refuse to endorse their proposals, they face a simple choice: concede defeat or pretend that you and I do not exist.)

Physicists are often uncritical in their thinking about their own social status; so they often reach the same conclusion — that professional public opinion is correct — without the benefit of the studies of Merton and his followers, and without having ever had heard of the great philosopher William Whewell either. In the period when the very first round of the debate in the United Stated about the dangers of the development of nuclear weapons was disastrously lost on fake technicalities and fake patriotism, the concept of Research and Development (R and D) became popular as a means of fusing ("pure") science and (scientific) technology. Most physicists are still reluctant to debate the rights and wrongs of this fusion, and when they do consent to do so they are prone to become rather unpleasant. Do not waste words on them, then!

("R and D" differs from "technoscience", incidentally, as the latter concept depicts the image of scientific-technological innovation with no concern for prestige [Sassower, 2004].)

Until Einstein (1917) and even later, the view generally received was that scientific theory (inductively) emerges from factual information, that it rests (inductively) on facts. These days in fashionable circles this story is replaced by another, according to which normal scientists conform to a given paradigm. (No: though they all claim to know exactly what are the rules of induction and what is science, no philosopher and no scientist claims to be able to present a satisfactory, coherent view of the matter, much less how science conforms to the rules: with the exception of followers of Polanyi, philosophers of science are trying repeatedly to do so and scientists are usually not interested in detailed presentations of such matters. As to the followers of Polanyi, they pretend to know what a paradigm is and how it is given, even though they admit that they cannot explain all this. They claim that they know a paradigm when they see one, but they rely on gossip instead.) And so, normal Merton-style sociologists who see science in action have the choice — to blame Merton's theory as (unwittingly) too idealizing (in its view of researchers as self-critical), or to take as ideal some pathetic informants (who cannot exercise any self-criticism). Why are they eager to take what they report as the ideal? As perfection itself? Because it vindicates them. And why are they not laughed out of court? Because science delivers the (nuclear) goods. And if low-grade science does that, then science proper may be just low-grade cargo cult. How simple. They will say anything to support science as cargo cult. If you must argue with them, attack this, then!

The classical myth of induction, though mostly an incomprehensible cargo cult[35], is at least egalitarian: (allegedly) anyone can perform experiments, and, if one doubts any report, then one can observe the facts for oneself. The current myth of scientists conforming to some paradigm is, by contrast, an unashamed élitism: it is a defense of the (nuclear) scientific establishment, the blatant irrationality of those who promulgate the authority of the science that has already delivered the (nuclear) goods and has thus no need to notice any foolish doubting Thomas. Except that I

---

35 Bacon's writings were very popular in the Age of Reason. He said explicitly that experiment is but the way to show Nature respect in order to make Her yield and reveal Her charms: it is stooping to conquer. (Sexual metaphors comprise an integral part of the popular cabbalistic literature of the time that greatly influenced Bacon.)

am a foolish doubting Thomas — especially when the Establishment tells me to believe — to believe anything at all. Why should I? Believe? Just in order to legitimate their timidity? No way.

Ernest Gellner has explained the low level of fashionable philosophy, sociology and political science, as he characterized them all as mere revalidations of existing doctrines in the intent to revalidate the existing order. This is very useful as an explanation of the breakdown of criticism as well as of the isolation of critics and dissenters: the nineteenth-century physicist Michael Faraday and the philosopher Karl Popper a century or so later, both suffered isolation for decades because they violated existing doctrines and then the timid leadership considered the violation a danger to the existing order. And they were both dangerous, but not so much because of their ideas as because they demanded autonomy of their public above all; this is scarcely forgivable. And so, all the theories the philosopher entertains about his peers' and friends' betrayal, true or not, are clearly irrelevant: Faraday's calling them unphilosophical is somewhat better. As to betrayal, it was bad faith, self-betrayal, the endorsement of mere fashions, pure baseless timidity.

We not only forget. Without notice we distort. In his autobiography Sir Alfred Ayer expresses an early admiration for the philosopher[36]. If so, then he had a funny way of showing it. I remember, in particular, that Ayer showed up at a meeting of the Aristotelian Society — a rare occasion for him, but then it was a rare occasion: it was the philosopher's presidential address to that Society. He opened the debate saying, very querulously, we do not need you to come and tell us that we should be self-critical! He displayed much subtlety, opening himself to a seemingly forceful attack. When the famous native philosopher complained about the respected refugee philosopher coming and preaching autonomy and all that, he may have been unpleasant and even embarrassing, but, admittedly, what he said is possibly (and even actually) true, and hence it is legitimate. The outrage is not that a complaint was lodged but that it played a devious role as it came in the guise of Ayer's response to the philosopher's ideas. Presenting a complaint as if it were a response, as he did, he tacitly declared a) that the speaker had nothing to offer but a sermon; and b) a redundant, offensive one. He put the accent on the seemingly nasty second point, but the real nastiness is in the first: whether the sermon was good or apt is beside the point; the point is, was what the philosopher had to say true? Or did the philosopher have nothing to say and he presented a sermon instead? Did he pretend that his sermon was a lecture, that it was

---

36 Speaking of *Logik der Forschung*, Ayer said, "I had read and admired the book and was pleased to meet its author" [Ayer, 1977, 164]. He reports that Popper was open to criticism and he describes him as "celebrated" (129). Nevertheless, he joins the choir of those who oracularly declare excessive Popper's expression of dissent from the "Circle".

the presentation of an idea? What a way Ayer had to show admiration. The philosopher's presidential address to the Aristotelian Society, incidentally, was the lovely, profound, breathtaking and trailblazing "Back to the Presocratics" [Popper, 1963] (not so surprising now; certainly an important study on the change of attitude to the Presocratics in the last century is awaiting a spirited researcher). Should my claim for fame rest on my having helped the philosopher decide to write for the occasion on this theme, then I will be content. He had touched upon it in his *The Open Society and Its Enemies* and what he had said there whetted my appetite. I had an interest in his choice of theme: I wanted to hear his ideas on it. Briefly, he describes there the origins of Greek thought as the invention of dialectics, and the novelty of dialectics as of knowledge without foundations (to use the title of Feyerabend's lovely book [Feyerabend, 1962]); but do read the original! I do not mean to advocate its ideas; I have even criticized it as a myth of origin; but we need not go into that: if you still cannot allow for appreciation without endorsement, then you may be reading the wrong stuff right this moment and you may save time letting go of it. (Yet earlier writers, from the great pioneer of modern studies of ancient Greek thought, John Burnet, to contemporary authorities such as Kirk and Raven, whom the philosopher was, alas, nastily attacking in that address, they had all constantly struggled with the hopeless need to agree with all the contesting Greek thinkers. Regrettably, the philosopher's nasty attack on Kirk and Raven shows that he too did not absorb the lesson that criticism is appreciative: even though he said so repeatedly, he did not internalize it. In self-righteous mood, Lakatos emphatically declared it mere lip service and dismissed it with contempt as sheer hypocrisy.)

The philosopher had nursed an old grudge against Ayer, and perhaps Ayer knew about it. All that Ayer had ever borrowed from the philosopher is a term: "basic statements". They both used this term to denote observation reports. What the philosopher says of it [*LScD*, 35, n*2] is that though his new term came to denote a new view of observations (as tests), others (*i.e.*, Ayer) have adopted it to serve as a synonym for an old one (as valid); what annoyed him, let me report, is that Ayer willfully let go of an opportunity to mention him. Was that so wrong? I do not think so.[37] On the contrary, I oppose the philosopher's habit of acknowledging with pomp and circumstance terminological inventions[38] to colleagues he did not appreciate as thinkers but befriended all the same: words, as he

---

37 This is no comment on Popper's note on Ayer's distortion [Schilpp, 1974, Replies, n13].
38 [Munz, 2004, 16-17] expresses displeasure at Popper's making empty acknowledgements to him. This is indeed rather objectionable, but to maintain a sense of proportion we may notice its popularity. "Agassi reminds us", says Robert K. Merton, of the meaning of a famous but trite neologism [Merton, 1972, n11]. For important examples of this unpleasant practice, see [Agassi, 1963, n42ff.].

repeatedly said, do not matter. (When I composed the first subject index for *The Open Society and Its Enemies* I noticed that he praised Carnap there a few times, always for verbal innovation. When I mentioned this to him he smiled.) He also made other funny (and in my opinion, for what it is worth, even insulting) acknowledgements, such as, David Miller has corrected him on some ascription of some verbal innovation.[39]

I remember the famous John Kemeny (co-inventor of BASIC) coming to pay homage to the philosopher and express gratitude to him for all that he, Kemeny, had learned from his writings. Though I only knew of one famous paper that he had co-authored (I will not trouble you with its contents: it is now rightly forgotten), I surmised from the manner of the visit that he was famous. He humbly told me that his reputation began when he was assistant to Einstein — which, he added, was due more to luck than to talent. He impressed me greatly and I was eager to hear the response of the philosopher to his expression of gratitude. I thought he would be delighted, yet his anger knew no bound and he nearly asked him to leave: why do you not mention this fact in print? Good question. Kemeny left in a hurry, rather shamefaced, never to return or otherwise resume contact. He utterly forgot the whole incident.

I cannot deny the charge. Kemeny soon had a book out, *A Philosopher Looks at Science* [Kemeny, 1959]. I have reviewed it favorably and still recommend to students and colleagues; it had this funny defect that in time fades out: it misspells the philosopher's name as "Carnap". This soon became a popular custom.[40] He also soon joined Bar-Hillel's defense of Carnap — and likewise set a nasty trend.

Bar-Hillel had claimed Popper misunderstood Carnap — after the latter had written a five-hundred-page treatise of mere clarification. (He had invented a kind of mock-clarification and a name for it: "explication".) There are two kinds of probability (= fields abiding by the axioms of probability), Carnap proposed: statistics and generalized-deducibility-which-is-confirmation; moreover, the two are of equal value, he added: the probability of an event equals that of the statement that describes it. There are two kinds of confirmation,[41] Bar-Hillel now revealed: the theo-

---

39 Some of Popper's acknowledgements to me seem to me regrettable. He granted me one that looks like a paragon of careful, conscientious acknowledgement [Popper, 1956, 255]. It is a distortion of an insignificant detail in an insignificant dispute: a correction of an omission by Carnap that Popper made much fuss of after I showed him that he could use it to prove Carnap inconsistent. That proof provoked much scholasticism, to which Bar-Hillel and Kemeny contributed, though they had no trouble recognizing a contradiction. Carnap wisely rejected their defense of his views in the preface to the second edition of his book [Carnap, 1962] where he withdrew his intent to publish a second volume. Alas, he later tacitly withdrew this rejection [Schilpp, 1963, Replies]. (Carnap's scattered autobiographical assertions are usually too inaccurate.)
40 Worse offenses followed suit. See Chapter Six below.
41 The song and dance about the shift of emphasis from the theoretical problem of induction to the pragmatic one is baffling. Traditionally, rational action rests on rational thought, and rational thought is faith in the right theories. This is an obvious error: refutation destroys credibility, not applicability. And

retical and the pragmatic, and their numerical values differ. (Apologists for induction are now increasingly ready to view the theoretical and the pragmatic problems of induction as distinct; this is admittedly some measure of progress.)[42] Popper's characterization of theoretical confirmation as a result of efforts to refute is by far superior to anything that may compete with it. Bar-Hillel did not challenge this but declared that Popper was confused when he ascribed to Carnap the undertaking of the same task: Carnap was concerned with pragmatic confirmation, not with theoretical confirmation. Moreover, continued Bar-Hillel, the fact remains that even Popper agrees that confirmation depends on measures of probabilities, though he denies that it is a probability measure. Kemeny could not honestly endorse this sort of obfuscation, but he endorsed it anyway. Carnap himself valiantly rejected the defense his two admirers had offered [Carnap, 1962, Preface to the second edition], and he withdrew his original intent to publish a second volume. In his autobiography [Carnap, 1963], however, he returned to his old stance: there was no real dispute here, he said: he and Popper were merely talking at cross-purposes. (First it was a family squabble that the philosopher blew up in self-aggrandizement; now the family does not even quarrel: it is merely at cross-purposes with itself.) Naturally, this sent Carnap straight back to his aborted project. He worked on it till he died. After he died, Bar-Hillel rushed to Los Angeles to look at the final manuscripts of his beloved master. He saw at a glance, he told me later, that the project was stuck badly on a reef. And then something snapped and he went looking for another and more trustworthy master.[43]

---

so applicability is the less problematic of the two, Duhem has noted. The better philosophers, from Hume to Russell, saw the theoretical problem objectively, as regarding learning about universal theories from particular experiences. Hume asked, how is this learning justified? Popper broke new grounds in his researches, Bartley and Wettersten observe, when he developed an alternative that led him to giving up justification altogether. He took the first step in this long and arduous journey as he asked, what inference from particulars to universals is valid? For my part, the story is very different: corroboration for pragmatic ends wants a theory different from that for theoretical ends: the practical corroboration of some refuted theories (such as Newton's theory of gravity) is high and that of some highly corroborated theories (such as Einstein's theory of gravity) is low. See next note.

42 Practical corroboration testifies to the measures taken to prevent some unfortunate applications [Agassi, 1985]. Theoretical corroboration should indicate intellectual value, and different kinds of this are extant. For example, some practical problems arise within theoretical discourses from which they gain their importance, and some relate to education [such as the Atwood machine] and to tests. Similarly, some theoretical problems concern practice and pertain to it and gain their importance from it. And some problems gain importance because they challenge mathematicians. And so it goes.

43 To maintain a sense of balance let me mention that the admiration that Quine displayed towards Carnap never wavered [Quine and Carnap, 1990, last page]. No doubt, Quine never agreed with Carnap about metaphysics (see his letter to Carnap of Feb. 15, 1938): he criticized him publicly repeatedly. Yet he admired him, as he viewed him, so he told me, as the prime follower of Russell's grand plan to apply logic to philosophy. The honor for this, in my view, goes not to Carnap but to Wittgenstein, whom Quine never respected. He said to the last, Carnap was an honorable man. Unlike Lakatos, he said this with no shred of cynicism. Although Quine admired Carnap in dissent, it was dissent that led Bar-Hillel to disappointment. Surely, we may admire people for their character or for their ideas, and the admiration of ideas and assent to them may diverge; it usually does.

I never doubted the sincerity of Bar-Hillel; no one did. Yet he admitted when we met in London that he saw more agreement between our respective masters than he would say publicly, and he admitted to me in our last conversation (in Boston) that he was awaiting the final clarification as long as his master was writing. Come to think of it, were he (or better, Carnap) to say not that a constant misunderstanding had divided Carnap and Popper, but that Carnap agreed with Popper about theoretical science — that scientific theories are refutable — and dissented from him about practical matters, then the whole sorry recent history of the philosophy of science would have been different. Why could they not say so? Why could they not admit in the heat of the debate that they were aware of defects in their philosophical system? Because they denied that they had any system, of course.[44]

To return to the conduct of the philosopher, he usually lost control only when defensiveness was too much for him to respond to otherwise. He always became very friendly with anyone who would not be defensive, like Peter Medawar, who used freely ideas he had found in Popper's *The Poverty of Historicism* in his interesting and prestigious series of radio talks, later published as his *The Future of Man* [Medawar, 1961]. Quine too is indebted to Popper, and he is characteristically very cavalier about it in his punctilious manner. Unlike Kemeny, they were frank and open when he challenged them, and this enhanced his friendships with both.

A rejection similar to that of Kemeny occurred when Isaiah Berlin came to the London School of Economics to deliver the 1953 Auguste Comte Lecture there[45]. He spoke against the theory of historical inevitability (namely, of destiny). After the most impressive torrential barrage that his lecture was — as all of his lectures are — he rushed to Popper, seemingly going to embrace him. I could not believe my eyes: the philosopher met him with an arctic blast: he all but called him a plagiarist. The lecture did not refer to Popper. Rather, in its conclusion it referred to the most popular philosophical authority of the time: Ludwig Wittgenstein. Berlin's lecture would have been proper were it not Popper but the other fellow who had written two impressive and synoptic books on which he was relying in his lecture.

Let me dwell on the difference between the two. The philosopher

---

44 For Bar-Hillel's last published assessment of Popper's output, see Appendix to Chapter Ten.
45 Isaiah Berlin's lecture soon appeared [Berlin, 1953]; [Berlin, 1969]. The published version differs from the one he delivered; it mentions Popper approvingly, makes no mention of Wittgenstein, and has a new conclusion. At the time, the very mention of Popper made quite a stir, and some commentators mislabeled Berlin a Popperian. That he was more of a follower of Wittgenstein, as explained in the text here, mattered much less, and for two reasons. First, it was a novelty. Second, names are easier to notice than the import of ideas. I suppose he blamed me for his reputation as a Popperian, and perhaps with reason; in any case from then on he disliked me.

had gained the title of the only philosopher in England (not in Scotland, incidentally) who was stubbornly rejecting Wittgenstein's authority as the two had a dramatic clash soon after Popper had settled in England (it was after the war). This reputation was unfair to Bertrand Russell, as he was then equally unimpressed with Wittgenstein's alleged philosophy. There never could be a clash between these two, however: Russell was the only living person whom Wittgenstein revered, and so in his presence he was restrained and even silent. Nevertheless, the unphilosophical public that peddled Wittgenstein's wares dismissed Russell as already too old. In his autobiography he confessed beautifully that this made him uncomfortable — which is understandable. The two classic papers against the Wittgenstein lore prior to Ernest Gellner's *Words and Things* that started in the late fifties the reverse tide that terminated the Wittgenstein era[46], were Russell's magnificent "The Cult of Common Usage" [Russell, 1956, 154-9] (see also [Russell, 1959, 217]) and Popper's "The Nature of Philosophical Problems and Their Root in Science" [Popper, 1963, 66-96] — the very paper that had brought me to his seminar. I suppose this paper (his chairman's address to the Philosophy of Science Group of the British Society for the History of Science, 1952) was the result of his 1946 encounter with Wittgenstein, which has been part of the philosophical gossip in England ever since.

The story of the encounter (Wittgenstein scholars call it "the Popper Poker"[47]) is recounted in Popper's autobiography as well as in at least two other writers' reports that agree with it. There is a printed record that two eyewitnesses denied it (publicly but not in print) with no detail and no alternative version in sight. As Stephen Toulmin, who witnessed it, is hardly a Popper fan, I rely on him in my endorsement of the factual aspect of Popper's version.[48] My version of the story is this. Popper was invited by the Cambridge Moral Sciences Club to speak — an initiation rite, I take it — on the non-existence of philosophical problems.[49] The inten-

---

46 The reluctant praise that Gellner received [Quinton, 1961] shows the force of his argument.
47 This title has been changed to "Wittgenstein's Poker" due to the appearance of the best selling *Wittgenstein's Poker: the Story of a Ten-Minute Argument between Two Great Philosophers* [Edmonds and Eidinow, 2001]. This was not the only case of Wittgenstein waving his poker expressively [Annan, 1990, 301]; [Munz, 2004, 68].
48 Much more corroborating evidence is now available to support Popper's version [Edmonds and Eidinow, 2001]; [Smiley, 1998 and 2004]; [Munz, 2004]; [Grattan-Guinness, 1992]; [Gregory, 2004].
49 Popper's version regarding the choice of title for his Cambridge talk is contested [Grattan-Guinness, 1992, 6, 7n5]. This matters little, since the atmosphere was tense in anticipation of antagonism. Smiley refers to Wittgenstein's unusual rudeness: as the host and the chair of the meeting, he had no right to leave abruptly. Other commentators refer to his generally queer conduct, possibly in tacit admission that he was generally impolite, thereby reducing the import of the occasion. Wittgenstein too made light of it: in a letter to Rush Rhees he wrote, "Lousy meeting — at which an ass, Dr. Popper, from London, talked more mushy rubbish than I've heard for a long time" (Austrian National Library, Autogr. 1286/18-5). The occasion was special all the same, and this explains the larger than usual doses of rudeness, as Russell's letter to Popper points out: "I was much shocked by the failure of good man-

tion behind the invitation seems to me clear enough: he should not have come except to capitulate. He came anyway, fought and won — only the debate, not any approval, save that of Russell. He argued that meaningful statements outside the domain of science are at times significant — also for research. He had already said so in his classic *Logik der Forschung* of 1935. Wittgenstein, who was poking the fire gloomily (thus proclaiming, by ordinary English usage, that he was at home in Braithwaite's residence where the meeting took place), finally responded: shaking the poker angrily he demanded an example of a metaphysical proposition.[50] "Do not threaten visiting speakers with pokers" came an example in swift response. Wittgenstein threw the poker into the fire and stormed out. To be fair, one should note that Wittgenstein was handicapped: we are told by different witnesses that he always monopolized debates, especially in the Moral Sciences Club, except in Russell's presence, when he would be silent, and here he was, expecting capitulation, meeting a surprise attack which he could not brush off, and in Russell's presence, which tied his hand anyway.[51]

All this is of no small interest for students of Wittgenstein, since it constitutes most dramatic evidence against the currently received lore according to which he withdrew the notion, peculiar to his first book, of the non-existence of philosophical problems. In his second book that appeared posthumous (and competed with his first for top popularity) he gently ridicules his first, and hints (in its preface) that possibly he disowned it, reporting that it indicates a change in his way of thinking (*Denkweise*). A little attention reveals that the second book endorsed the main message of the first, and ridiculed only the style it employed and a certain narrowness of outlook it exhibited: philosophical problems do not exist, both books declare, since philosophy is outside language. (There was a great difference between the two books, though: the rationale offered in the first book is the theory that the language of science is identical with language as such, namely, with the ideal language, thus forcing everything outside science, philosophy included, to stay outside language as well, with logic and mathematics allowed in to the margin by sheer

---

ners", he wrote, adding, "in Wittgenstein this was to be expected but I was sorry that some of the others followed suit" [Grattan-Guinness, 1992, 15].

50 Popper's first response to Wittgenstein's challenge was to mention the problem of induction as an instance of a philosophical problem. Wittgenstein had endorsed a solution to it [Wittgenstein, 1922, 6.363] that is highly problematic. He nevertheless dismissed the example as more scientific than philosophical and demanded another. Popper complied. His other example was from ethics. It was harder to dismiss, as Wittgenstein had declared all imperatives not properly worded, metaphysical sentences [Wittgenstein, 1922, 6.421-1]. He should have withdrawn it. He never did. It was not very wise; was it?

51 Some of Wittgenstein's apologists report an added handicap: at the time he was in personal trouble. This is obviously (but tacitly) an added mitigating circumstance. Hence, they tacitly admit defeat.

courtesy. The second book has no rationale;[52] its author displays hostility to philosophy not through a rational exposition of some reasoning and arguments; rather, in the style of a sage he illustrates his views in aphorisms and maxims and sketches and anecdotes; except that traditionally sages are never hostile, come what may.) It seems that the late phase of the positivism (= hostility to metaphysics) of Wittgenstein was different from his early one. Berlin judged the late phase immune to the criticism that Popper had leveled against the early phase of Wittgenstein's views. Berlin also judged the later Wittgenstein and his English followers better, since they were more humanist in mood, than the earlier Wittgenstein and his Viennese followers, on account of Wittgenstein's early identification of language with formal language and later with ordinary language. (This is why Russell's critique is so biting: what these people call ordinary discourse is obviously far from the ordinary, he observed, being ordinary Oxford-college common-room discourse.)

Berlin was less impressed with Popper's arguments than with Wittgenstein's reputation (among the élite and the general public alike) and with his personal magnetism and mannerism. Appealing to his authority, Berlin suggested in his 1953 lecture in the London School of Economics that technical terms should be permitted only in the natural sciences, not in the social sciences, where they should be decreed devoid of meaning. And then the very term of reference for historical inevitability (or destiny) will be barred as illegitimate. This is typical of the Wittgenstein legacy: rather than attend to the idea in question and discuss its merits and defects in the search for the truth, decree a word or rather a concept ("destiny" or its cognate) illegitimate, refuse to comprehend it (even while being able to identify words cognate with it), and one more problem will have vanished. (How this exclusion of the word "destiny" helps prevent one from combining the two legitimate words "historical" and "necessity" and thus obtaining the cognate of "destiny" I cannot tell.) This attitude will not be entertained nowadays, as Jagdish Hattiangadi observes apropos of the most fashionable current book on the writings of Wittgenstein, a book by the celebrated Saul Kripke, which presents a hypothesis answering the question, which problem did Wittgenstein try to solve? (The book is a best-seller.) The true answer is obvious yet systematically ignored by devotees: he never withdrew his presumption that intrinsically philosophy admits no articulated problem, that all problems

---

52 The question is very popular as to how much did Wittgenstein change his opinions. The major change was this: Wittgenstein lost the rationale for his anti-metaphysics. This fact is utterly absent from the literature. Munz says [Munz, 2004, 55], "Wittgenstein had changed his mind, but would never explain". He suggests that it is worthwhile to examine this matter. Well then. Did Wittgenstein revoke his ban on metaphysical discourse? The evidence that the poker incident provides is negative. Munz, who was present, does not deny this. He deemed the incident a colossal error on both sides and offered a friendly dream-version of how it should have happened [Munz, 2004, 101-108].

that are open to proper verbalization belong to logic or mathematics or science or anything else, but not to philosophy. In his early days Wittgenstein had a theory of proper verbalization, and Russell criticized it severely [Russell, 1922].

Now Berlin's suggestion to taboo certain terminology does sound somewhat in line with Wittgenstein, and it is even harder to implement: considering the term that refers to social class or the one that refers to trade unions (not to mention terms denoting fixed real prices, the gross national product, balanced budgets and the elasticity of demand), is such a term common or technical? Berlin himself was not decided on such matters. A few days after his lecture he was kind enough to grant me an appointment and see if he could help me in my financial constraints; he asked me then for my response to his lecture and I asked him if he considers terms like "social class" technical. The published version of Berlin's lecture has no Wittgenstein, no proposal to outlaw any sort of terminology, and a footnote praising Popper. As a result commentators still couple the two.[53] Serves them right — both of them. My students at a seminar that discussed this matter were quick to observe how diverse these two are: Popper was criticizing theories and thereby defending liberty; Berlin was criticizing concepts, leading nowhere in particular. The philosopher does not need my observation on Berlin: he was always quick to notice the difference between a conceptual difference and a ("substantial", "real") disagreement: he always insisted, contrary to essentialism, that conceptual differences are always legitimate, and so he dismisses all conceptual analysis as redundant — except when it involves a complaint. He went on complaining, they say.

Perhaps I should be a bit pedantic and devote a paragraph to a report on Berlin's point in some detail and with more precision. The final 4-page-long paragraph of his Historical Inevitability seems almost Popperian. In it he dismisses the doctrine of historical inevitability as pseudoscientific (not as meaningless): there are no empirical tests for it. The dismissal of all metaphysical systems as equally hollow, he bravely admits, may be invalid. He then goes on to explicitly permit technical ter-

---

53 Berlin was a staunch liberal but not his famous idea that there are two kinds of freedom, positive and negative [Berlin, 1997]: it is even the standard illiberal response to liberalism (stemming from the writings of Hegel): it is the complaint that the liberal concept of freedom is defective as it is merely negative or merely formal. It is the demand for a positive concept of freedom, meaning, presumably, the freedom that free people regularly exercise, namely their voluntary adherence to tradition. (Russell says, Hegel recognized only one freedom: the freedom to sacrifice oneself for the glory of Prussia.) Berlin sent Popper a copy of the lecture and he responded in an overly friendly manner, suggesting a liberal reading of positive freedom. He even ignored Berlin's Wittgenstein-style translation of Hegel's banality to a verbal form, making it a distinction between "freedom from" and "freedom to" [Berlin, 1997, pages x, xxxii, 203-6, 233-4, 237]. Lovers of verbal distinctions for their own sake can put after the word "freedom" any preposition that the English language tolerates (for, in, at, of, by, before, after, so-that, in-order-to), add distinctions between their different usages, and conclude with its preposition-free use ("the land of the free").

minology to the various social sciences. Though he does not say so, he presumably will allow the same luxury to historians of these sciences. But historians, he adds (meaning political historians), "can scarcely do that": they cannot avoid all use of ordinary language and they thereby endorse the moral categories built in it, he observes. This is an echo of Wittgenstein, but it is not Wittgenstein. (In particular, Wittgenstein never demanded the endorsement of all ideas implicit in ordinary language. Heavens Forbid!)[54] Yet many commentators see this as the unmistakable mark of Berlin and the more perceptive see this as his attempt to synthesize Wittgenstein and Popper — which surely has glamour. Others were also lured by this project (namely of compromising between Popper and Wittgenstein or Carnap or), notably Jaakko Hintikka and Paul Feyerabend, and, of course, Imre Lakatos, who organized a whole international conference for this noble purpose and lured both Popper and Carnap to it, each hoping to see the other accept his own view of philosophy and of the difference between them: Carnap wanting to see Popper admit that a mere misunderstanding rested on Popper's exaggeration of the difference between them and Popper to see Carnap admit that there was a genuine and important disagreement between them and that Popper, like a sheriff in a Western movie, was still alone, yet a holy terror. It was no good: no one yielded.

And so I come to what seems to me to be my major crime. A reject, I became a leading philosopher (I will return to the theme of leadership, don't you worry) — thanks to the tremendous investment of the philosopher in myself. There was — so one may suggest — a tacit give-and-take: I had to contribute, some time later, what I could, so as to rectify the wrongs done to him, at least continue for one more generation to protest against the compromisers. To that effect I was tacitly appointed the philosopher's biographer, I understand. He told and retold me the story of his career — a story which I still find fascinating. He made me repeat some details of the story to him, possibly because this way he could insure that I got the details correctly without dictating it to me. I got the drift of this design, of course. I knew it was not for me. (His autobiography confirmed to me again the unreliability of my memory.) I said, Karl, I am going to disappoint you. He said, leave it to me; I have no time to discuss this with you. He was convinced that he could mold me to become a pedant even though he had failed to persuade me by arguing that pedantry is necessary for true scholarship: I had broken with the Talmudic tradition

---

54 The argument regarding the bias of ordinary language (Wittgenstein-style) in favor of received moral ideas is all that Berlin's *Historical Inevitability* offers. This bias is ubiquitous but harmless: we need not admit what language suggests. (Galileo's opponent, Cardinal St. Roberto Bellarmino, used language against Copernicus.) Appeal to ordinary meanings of locutions, David Pole noted [Pole, 1958], is inherently backward looking.

pledging to avoid all forms of apologetics, pedantry included. He took recourse to measures that I found objectionable. When he began to succeed in manipulating me, I decided to change our relation drastically. Later he expressed to mutual friends his resentment of my having stayed with him as long as I stood to gain and then thanklessly left. This charge is reasonably near the truth. It deviates in two small details, however: while I served, I served best, and thanklessness is not in my style. And if this melancholy account seems thankless to you, then view it, please, as my poor control over my pen, nothing more than that.

The need that the philosopher had to receive the recognition of his peers diminished in time as they were all dying out and usually sank fast out of sight: memory in matters fashionable is short and scholarly memory is delegated to serious historians who ignore fashion or present it as fashion.[55] His central target, the positivist theory of meaning, is now as dead as a doornail, either due to his deadly arguments, as his autobiography claims, or on the authority of Quine who had no good word for it. His constant effort went to the fight against the fate of oblivion. (I once observed him glance at a long shelf filled with the collected works of a person from well beyond his horizon. He stood there looking at the shelf for quite some time, thunderstruck.) In accord with the myth that in the arts but not in the sciences one may be rediscovered significantly, and in accord with the idea that in this respect the fate of ideas in science is shared by rationalist philosophy or scientific philosophy, or the philosophy of science, or the philosophy that approves of science, the feeling mounted that it was now or never. And if never, then all this terribly hard work was in vain. I once told him my opinion on his place in the history of philosophy. I told him I thought that it was secure, that he could do little in the evening of his life to change his position overmuch. He did not ask me why, and I am grateful to him for that; I think he could not bring himself to respond to the criticism of his disciples, and that this, or else ventures into new interests, is what could possibly further raise then the (already high) value of his total output.

He was working, I understand, harder than ever. He insisted that I wanted him to stop writing; he insisted that I wanted him to make room for the younger generation. I do not understand this: I do not see how it can possibly be denied that there is room for everybody, that everyone can have their say; for example, had he openly criticized us, me or any other of his former students, we would all be dancing at the top of the

---

55 As the aim of the output of Friedrich Stadler is to legitimate continued research into the output of members of the "Vienna Circle", and as now their verification principle is *passé*, he insinuates that they never laid much stress on it. This conflicts with their *Manifesto* and the early volumes of *Erkenntnis* and other publications — of Schlick, Carnap, and Reichenbach, not to mention Ayer. It also conflicts with what the public understood them to say [Joad, 1950] that was then the source of their appeal. Stadler does not mean to fool anyone. He simply decrees orthodoxy like a Czar. See also next note.

heap. He wrote to me to Hong Kong that his additions to his theory of verisimilitude (of scientific theories as series of approximations to the truth; I will not discuss it here as it is too involved in technicalities) includes a response to my criticism of a part of his theory of corroboration (the part that says, science without positive evidence is impossible, the part he calls "a whiff of inductivism"; I say, if so, then the theory of science as conjectures and refutations is false and should give way to the theory of science as conjectures, corroborations and refutations). Had he said so in public too, the situation would have been much clearer: he would have been able then to explain all the purposes of his development of his ideas; the whole picture then would have been better understood, and so on. Instead, he resented my alleged suggestion that he should not work hard; he even reported that I had told him he went on working hard because he was greedy. Though I do not remember ever having said anything even remotely resembling this, perhaps I should accept his report: we say what our audiences hear us say, he always insisted, not what we intend them to hear. So I am ready to accept the charge and here unhesitatingly withdraw what he reported that I had said: as far as my preference is concerned, I had no preference ever for him to stop working hard, and I always wished him success, wanting him to live long and prosper. If one can apologize for what one is reported to have said even while being incredulous about it, then here is my apology to him. I do not know if the antecedent of the last sentence is true, though, and he was right to reject conditional apologies as scarcely apologies. I am at a loss.[56]

A point about recognition, in private and/or in public, contemporary and/or in history. Historians not prone to conformism repeatedly improve standard textbooks[57] and at times even revolutionize them and this includes the standard history textbooks, which, revolutionized, may record and legitimate some important heresies — such as Faraday's theory

---

56 It is hard to reconcile Popper's view that we say what people hear us say with his constant complaint against distortion. Fortunately, he also recommended viewing distortions, even willful ones, as indications of some ambiguity or some lacuna or even some error. This is clear in the case of science. Take, for example, the standard distortions of Einstein's ideas. Some, like the twin-paradox, are hard to eradicate because they rest on operationalism and in his early days Einstein was an operationalist (only to some extent and very interestingly so). Other distortions, such as that he dismissed quantum theory as false, rest on the inability to distinguish between truth and completeness. (The dispute is clear now: Bohr claimed that quantum theory is complete in the very sense in which Einstein said it is not. Einstein showed that Bohr's view leads to quantum entanglement, which looked incredible, and Bohr said, no experiment can ever display it. To everybody's surprise, it did turn up.) Now the distortions of the ideas of Popper, being political, greatly differ from the distortions of the ideas of Einstein. Some distortions vanish with no further ado and so we may safely ignore them. This indicates that the application of willful distortion that runs in the face of easily available evidence is a counsel of despair. This is good news. Popper could have waited, then, were he more patient. This is what Faraday recommended: have faith and wait patiently. Except that the Establishment waits for you to die first.
57 Momigliano said [Momigliano, 1990, 56-7], practically all improvements of standards of historical writing by boosting the critical attitude are due to contributions of some skeptics. For, the standard of others is unattainable, and so it is also inapplicable.

of electromagnetic fields. (The philosopher was very excited when I told him about Faraday, and he greatly encouraged me in my research.) Otherwise Faraday, or Spinoza, for a more conspicuous example, could never be brought out of relative obscurity. True, he never was utterly obscure; but then neither is any modern thinker who was later allegedly rediscovered: the missing official recognition only masks the little recognition that is there: an idea with no recognition is lost for good.

This is the hypothesis of a two-tier intellectual system, one level carried by the established leadership and the other level by the *avant-garde*, by the underground — in the American sense of "underground" — and both merge in the standard vision of the past. The hypothesis may be viewed as a part of Gellner's theory of validation: the leadership validates and revalidates the real goods, often produced by the underground or by the *avant-garde*.[58] Any real contribution by the underground or *avant-garde*, to elaborate on the present hypothesis, sooner or later merges into the mainstream or else it gets lost. Hence, at best, it takes time for an innovation to be recognized unless it is originated by members of the established leadership or in a leading center of learning. This is a repeated sociological observation. Hence it is a repeatable, a *bona fide* empirical (namely, refutable) observation-report ([Russell, 1945, Book 3, Part 1, Chapter 14B], [Caplow and McGee, 1958], who speak of the aggrandizement effect and of the grapevine, (Ch. 5), and my essays on cultural lag in [Agassi, 1981]). The time lag is explicable, then, by the hypothesis of a two-tier intellectual system. Assuming that the present hypothesis is true, we may ask, then, why is this highly inefficient arrangement maintained?[59] Why does the learned public wait for its leadership to tell them what innovation is significant? The only reason is that the leaders deem the led — the learned, the professional, the rank-and-file researchers —

---

58 The novelty of Gellner's treatment of the intellectual leadership is not in the view of it as on two tiers but in his detailed observations. The chief difference between them concerns progress: the one revalidates it and the other seeks it. In Gellner's view they concern intellectual progress in entangled and overlapping ways, serving different functions and employing different social mechanisms. (Popper's theory, then, invites completion. It should then be a proper analysis of both [Jarvie, 1994, 226].) Gellner explains the time-lag observation: the task of official leaders is to sanctify the better ideas of the counter-leaders; and this demands procrastination: they should be cautious or cowardly (as you wish). This is common in halfway progressive societies like ours. Unluckily for Popper, the counter-culture lent momentum to procrastination. Gellner suggested that the general recognition of the validity of Popper's criticism of Marxism plus the refusal to take it as conclusive made them hold on to Marxism and this made them irrationalists. This, let me add, was because the leaders of the counter-culture used Marxism as a mask for their reactionary views that justify their political impotence. This is strange, as the counter-culture excelled at activism, helped accelerate the end of the Vietnam War and helped the civil rights and the women's movements. The sense of helplessness still prevails in left-wing circles, perhaps because helplessness and utopianism (especially Marxist) reinforce each other.

59 The poor quality of some criticism reflects poor training. The support that the establishment lends it forces the counter-establishment to improve education. This is highly inefficient as the establishment is in charge of education. The counter leadership should come up with a semi-official document offering helpful minimal guidelines for scholarly autonomy. See also next note.

hopelessly heteronomous, hopelessly dependent on others to tell them what to think, and the leaders then revalidate the poor view that the led have of themselves. If so, then, the construction of a good underground communication system will force the leadership to improve matters. At least in the more competent professions the underground takes a shorter time to have its messages reach the public; and so, it is not just the fault of the leadership that change is sluggish, since the underground can steer a campaign against incompetence but does not. Now the availability of the Internet offers a challenge and a golden opportunity for the *avantgarde* to force the Establishment to improve.[60]

The two-tier intellectual system is inefficient; even the Establishment admits this, yet while placing the matter in the distant past. We may consider the rather emotional letter that Galileo wrote to Kepler many years ago (before 1600) in which he complained about the crassness of the multitude of scholars and their leadership, while considering himself and Kepler to be the underground. This view of Galileo is standard: historians declare the established leadership of that period incompetent, and they describe the underground of that period as if it was the true leadership. Is the intellectual crowd today different from what it was in the days of Kepler and Galileo? The generally received answer — the Establishment's answer — is reassuring: Galileo and Kepler lived in the days of the rule of religion over science: now, we are reassured, now we are all autonomous. (Alternative version: now science is autonomous, as we are all the devout followers of the established, autonomous leaders (the paradigm-makers).) And when one conjures a later example — the tragic story of Semmelweis (in the mid-nineteenth century) is my favorite — the field of this example (medicine) is declared one that has emerged from the middle ages (from the pre-paradigm era) only yesterday. Ask for evidence for that and they will use your own evidence to support this account. Ask what example will be a refutation rather than support of this account, and you will be invited to offer any contemporary example. Offer one, and it will be dismissed authoritatively. Alternatively, your contemporary example will be declared an exception: it is unreasonable to expect perfection. (The paradigm case here is that of Nobel laureate Barbara McClintock: now that she has won recognition, they suggest, the problem is solved. Not so; now that the damage due to a thirty-year delay is admitted, the problem is, what is the best safeguard to institute against its repetition?) In brief, the established leaders know they must win this debate else chaos will reign. They see only two options: they rule or

---

60 The communication highways are already forcing establishments to improve, but too slowly. The new situation raises many new problems, mainly of new complementary modes of filtering. The richness of available information renders these problems challenging, and we should help making them enhance individual autonomy.

chaos does; all Establishments think this way.

I tried to air all this in the presence of the philosopher. I could not. He was greatly annoyed at discussions like the one presented here. I offer socio-analysis instead of criticism, turning peers into objects of study rather than viewing them as partners. He found this objectionable. He demanded that we always take our opponents seriously. I heartily disagree.

His taking seriously peers whose job is to be unserious (to pretend to be serious) is an error, one that prevented him from taking his former students and disciples seriously. He should have discussed Bartley's criticism; he should have taken Feyerabend's criticisms and mine as if they were offered by perfect strangers. Our comments are not less serious than those of the least positivist around. There is here a "myth of a framework" which the philosopher is a victim of, an opinion that he endorsed, not noticing that it falls under the same heading of a "myth of the framework"; it is a myth against the framework. That he could communicate with everybody was his conceit.[61] He allowed himself this conceit because it placed a tremendous burden on him: it forced him to work day-and-night in the endless task of increasing the clarity of his discourse. Any communication-barrier — and even the wisest cannot avoid them every time — then he had the choice between giving up his conceit and blaming. And blame he did; he blamed in abundance. For years I tried to find the logic of his incessant blaming. Perhaps I have found it.

I am applying here one of the things that impressed me in the introductory courses of the philosopher. It was his advice: as we are all prejudiced to this or that extent, beware of those who declare themselves free of all prejudices: they evidently have lots and lots of them, he added. He did not claim, then, that he was free of prejudice; rather, he pronounced that, taking his opponents very seriously and working at it very hard and very sincerely he was best able to fight his own prejudices. But this very pronouncement is a (meta-)prejudice [Magee, 1997, 232].

On what ground did he rest his pronouncement that he was sincere

---

61 Popper refused to communicate with Feyerabend or with me about our view of the shortcoming of his theory of explanation. In his *magnum opus* he said, theories explain the truth of observation statements. But, he said elsewhere, explanations might indicate that the assertions that we had sought to explain are inaccurate and replace them by ones that are more accurate. These cases lead to crucial experiments. In cases in which corroboration goes to the more accurate assertions, the ones that we began with end up explained *not as true but as mere approximations*. This is an innovation of Einstein and Popper that conflicts with the traditional theory of explanation as deductive that Popper initially endorsed (with modifications). Popper has quoted Galileo's expression of admiration for Copernicus for his refusal to accept some astronomical data that conflicted with his hypothesis. Hence, says Feyerabend, Popper should admit refutation of his view that observation always wins when conflicting with theory. Popper could admit all this with ease, as he said of any move in the game of science that as long as it does not block the continuation of efforts to produce conjectures and refutation, there is no need to complain. His theory is more powerful than he had noticed. It renders redundant his demand to give priority to observation over theory. All this may be false, of course. If I am in error, he should have corrected me. He did not. Instead, he put pressure on me to desist.

in his love of criticism? Two responses were frequent in his repertoire. One: I work very hard and in response to critics and I try my best to put their criticism in the best light as I respond to it. Two: never mind the grounds, just try to criticize my view, and if you succeed I will be grateful, as I love criticism.

There is a catch here. He knew that in my view he disliked criticism. He said to me, quite rightly, this put him in a catch: he was in the wrong if he defended himself, since he was thereby rejecting my criticism, and he was in the wrong if he did not, since he was thereby indifferent to my criticism. Now a catch proves nothing, since its logic is the same if the opinion in question — that someone dislikes criticism — is true or not. So much I agree. But the same holds for the two responses from his repertoire cited (they are near-quotations) in the previous paragraphs: all of his responses are question begging in a slightly sophisticated way. They should all be dismissed as suspect: both my criticism and his response are invalid just because, regardless of the facts of the matter, they are acceptable if they meet with acceptance and objectionable if they meet with objection.

What is not acceptable, in general and regarding the specific requirement to take everyone as seriously as possible is his conviction that any criticism launched against him in public by his former students and disciples is a mark of a betrayal and so he should not have taken it as criticism at all and so he was exempt from examining and answering it as if it came from utter strangers. Why, I ask, should he have treated criticisms by traitors less seriously than criticism by strangers? To this his answer is that his former students' public criticisms happen to be very poor. And *a priori* at times this surely is true, but at times he may have misjudged it.

This last sentence of mine would constitute serious criticism were it not question begging. Yet quite obviously it is: why should my assessment of my criticism be preferred to his assessment of it?

Something has gone wrong here. It is so just because we took matters on the personal level: criticism is a matter of a critical tradition, as the philosopher has repeatedly and so rightly emphasized. This is not to say that we must endorse public opinion and declare a dogmatist anyone reputed to be a dogmatist. To take an extreme and simple example, Dr Joseph Priestley, the famous discoverer of oxygen and the stubborn phlogistonist, was reputed as a dogmatist despite the testimony of, say, Sir Humphry Davy [Agassi, 1963, 46 and n. 128]; the same holds for H. A. Lorentz of the Lorentz transformations who stuck to classical physics, who was reputed as a dogmatist despite, say, Einstein's testimony to the contrary [Einstein, 1954, 73]. The reputations of these researchers as dogmatists are open to explanation without endorsement. But even there, the details that signify most are their contributions to our critical tradition as we understand it. There is no need to study their psychological makeup. The better question, even on the individual level, concerns the assessment of individual performances in the public arena.

The arrival of Bartley on the scene marked a new era: he came from Harvard on a prestigious Fulbright fellowship as a pilgrim to a reputed guru. More than that: Bartley came in the teeth of opposition from his teachers. He was a true convert. He was, in addition, quite delightful: quiet, witty, learned, and above all an extremely mature and serious scholar. His authority was accepted from the very start. But for the appearance on the scene of Imre Lakatos, my nemesis and his, he would have been the unquestioned leader of our crowd to his dying day.

Bartley excelled from the start as a critic of the philosopher on a most central issue. The philosopher presented to his students a terribly interesting theory of levels of comprehension, from the lowest, in which one can barely follow a discourse, to the highest, in which one can criticize [and critically assess] a theory. Bartley had intelligent criticism of Popper's theory of rationality as expressed in his *The Open Society and Its Enemies*, Chapter 24. This should invite respect. At the very least he knew where the action was: the theory of rationality, he said from the start, is going to replace the theory of knowledge (= the philosophy of science = epistemology) in order of priority. How insightful! At the very least he comprehended the theory very well.

The literature on rationality has grown by leaps and bounds; it is now too vast to survey. I shall outline here the drift of it. On the authority of Plato and Aristotle, rationality was equated with proof despite the skeptic argument that all proof is question begging. Popper proposed that rationality is not proof but readiness to accept disproof. Simple. He added, this readiness itself is not rational, but it is the minimal amount of irrational commitment required to allow for rationality. Bartley said, in this move Popper erroneously accepted criticism applicable only to the old criterion of rationality, not to his new one. On this Bartley is obviously right. It is a criticism of a move in a debate, and it is unanswerable and the philosopher attempted to answer it only once (see below). This is the end of one chapter in the rationality dispute — until some one refuted Bartley's refutation, at the very least.

Bartley went further and offered an alternative, which he presented as his main claim for fame: apply the old criterion to itself and it breaks down (this is the skeptical move); apply the new one to itself and all is well: I, William Warren Bartley, III, am willing to subject the new criterion to itself in that I solemnly undertake try to criticize it and to drop it as soon as it is validly criticized.

I have no stake in the dispute over the status of Bartley's new criterion, since I have validly criticized it. (Later!) I admire Bartley, however.

In the year 1963 the philosopher gave a talk on the radio station of the University of Illinois in which he expressed the highest appreciation of Bartley for having presented the philosopher as a revolutionary thinker, as one who has rid us of the need to justify our views: having ascribed to the philosopher non-justificationism, Bartley explained to the philosopher himself why his peers had found his proposals hard to entertain. I arrived in Illinois a few months later from Hong Kong, and I heard this with great pleasure as that lecture was replayed on the local radio station. I could not procure a copy of it for reasons of copyright, but I paid for one to be sent to the philosopher just before the original was erased. He promptly lost it. Pity: I could have saved so much heartache.

This was no premonition: everyone could see trouble coming. During the whole period of my apprenticeship, the assigned logic and scientific method two-year lecture course was the only one the philosopher gave — with the exception of one special course to a senior and graduate audience. (The same holds for seminars: except for his celebrated Tuesday Afternoon Seminar, he also once held a seminar on the foundations of logic, tailored to Lejewski's needs; it had a very small membership and met for a few sessions.) In that unusual course he spoke most unusually: he chose to begin the course with an attack on both Bartley and Jerzy Giedymin.[62] (Both were pilgrims and Giedymin was the senior; he had come from Poland, where the isolation was a great obstacle to his progress. He has a favorable mention in Popper's *Conjectures and Refutations*, where Popper cleverly pitches one of his subtle ideas against my view on corroboration. I will not drag him into this story though: at issue is not my view of corroboration but my criticism of Popper's. The criticism of one alternative, however valid and important, does not deflect any criticism of another: two wrongs do not make a right.)

The philosopher began his seminar with an attack. He said, these two people's critical comments on his own ideas are really poor, and even attempts to improve them before responding to them cannot be successful enough. Nevertheless he would try to respond to them — because of his great love for criticism In 1960, when I was away already, I had heard

---

62 Popper's opening gambit was surprising, disconcerting, and disappointing. We all looked at Edward (Ted) Goodman, the oldest and most highly regarded participant. He was pleased to play the humble part of one of the group of graduate students although he excelled as the secretary of the Acton Society and as the Director of the Joseph Rowntree Reform Trust. He authored an important though little known work that showed his originality as a liberal economic reformist [Goodman, 1951] and a book on the use of overheads [Goodman, 1969] that anticipated a discussion by William Baumol. I do not know what had brought him to the Popper entourage to begin with: he was always there and we were close friends. (He also tried to find some financial support for me. His trust offered me a grant, but as it arrived after my appointment in Hong Kong, I gratefully declined.) As Popper continued with his complaint, we all felt increasingly embarrassed and lost for response. As we stared at Goodman, he was distressed and fixed his gaze on the wall. Uncharacteristically, Popper continued doggedly with his onslaught in utter disregard for the atmosphere and the embarrassment.

that he had convinced Bartley that his own classic text should be read in accord with Bartley's view, so that he (Bartley) was not criticizing but only clarifying that text. In 1962 I discussed this with the philosopher: I said, as I think Bartley disagrees with you, you cannot agree with both of us. He brushed this aside as unimportant. In 1966, with much tricky maneuvering by Lakatos, the two fell out publicly, embarrassingly, with the philosopher censuring Bartley as an ungrateful go-getter. The big row was redundant and regrettable, as it caused much damage, particularly to Bartley's progress. They finally made up; they then abstained from reference to each other's views on rationality, even where this was required. So much for the sincere love of criticism.

I cannot finish here. Let me add a happy coda. Bartley has won the gratitude of all of us, the admirers of the philosopher, for having edited and seen through to the press the master's long awaited *Postscript*. This act clears the air somewhat as it is now possible to refer to an open text rather than to unpublished galley sheets accessible only to the elect. He was working on a life of the philosopher before he died. His life of Wittgenstein is superb and if there is anything on the philosopher, sufficiently clear to be publishable, I am certain it should be salvaged.

Bartley's disagreement with the master, though not one I share, seems to me very significant on at least two counts. It opens the way to the study of Popper's predecessors in the study of the matter of commitment to rationality, from the end of Plato's *Phaidon* through Heine's marvelous Religion and Philosophy in Germany to Weber's desperate fideist rationalism. It has also created a new kind of dispute between the two (Bartley and Popper): what problem is more important/basic/central, that of the demarcation of science or that of rationality? The study of this kind of question should enrich the logic of questions immensely. We need studies of the problem, how do we set our agenda?

This question is very general: every research is a social phenomenon, we have learned from Popper. And every social grouping has its institutional aspect that merits study, we have learned from Jarvie [Jarvie, 2001]. And so every research group has its agenda — is even determined by its agenda. How? Who by? This is a new and exciting research problem for the sociology of science. It interests particularly those who argue that the current agenda of their peers is erroneous to the extent that it can alter only by a revolution. (This is one of the few ideas that Kuhn has advocated and that seem to me both correct and important.)

CHAPTER SIX: THE PHILOSOPHER AND HIS WORKSHOP

> Our successors will one day be amazed at the things we quarrel about and even more at how excited we grew in doing so.
> (Ernst Mach[1])

The picture I have thus far sketched is, I am afraid, a gross distortion. The distortion is the result of the omission of the most significant characteristic of the philosopher — that intense seriousness that forced him into strong, inspiring interaction with almost anybody around him. This was too much for some who would gladly interact with him less intensely and were obliged to do without his company, and this is how he regularly attracted excellent people who were willing to tolerate his manner only because they judged interaction with him most profitable for themselves. They were right in their judgment: he did have something to offer: his intellect: the ideas he had developed over the years of intense work and the ability he had developed for extemporaneous thinking — always on what they cared about, provided they cared about it passionately. He could discuss almost anything — but only very seriously.[2] He elevated the most frivolous remark thrown at him to the highest level he could.

I will describe later an amusing instance of the philosopher elevating a trite remark to a high intellectual plateau. Before that, however, I allow me to linger on this point — his combined forte-and-foible: his utter and relentless and intense seriousness day and night until exhaustion imposed a few brief hours of sleep disturbed by chronic illness.[3] The seriousness is not at the expense of a sense of humor; many individuals are devoid of a sense of humor and this does not make them in the least serious — especially when they replace the fun of humor with derision and contempt and that expression of superiority that Henri Bergson declared the heart of humor[4] (to which Russell responded by saying that were

---

1 [Mach, 1976, Ch. 16, n. 6]
2 Popper was as serious as Kierkegaard, who found going to church casually an affront to the Lord, since one should approach Him only in fear and trembling [Agassi, 2003, 154]. This goes along with the traditional view of research as worship (that Robert Boyle ascribes to Philo Judaeus). Both Kierkegaard and Popper ignored the traditional promise of equanimity as a reward for proper conduct, perhaps because they disliked equanimity and perhaps because tradition equates equanimity with inaction. Spinoza delightfully differed. They (Boyle and Popper) disliked him for his determinism. Popper should not have, as he played down determinism that does not oust commonsense indeterminism [OS, ii, 85] — and Spinoza complied with this famously, as he put great stress on the freedom of the will. Popper expresses regret that he was late to read Spinoza's correspondence [Autobiography §6]. He does not elaborate. One may read this as perhaps an admission of having been somewhat unfair.
3 Popper suffered from a punctured diaphragm (hiatus hernia) and from chronic headaches and stomach ailments; he was very susceptible to head colds and to stomach upsets.
4 "In laughter we always find an unavowed intention to humiliate and consequently to correct our neighbor." "Laughter is the perception of the substitution of mere mechanism for adaptive pliancy."

Bergson's view of humor true, then, the poor fellow who made us laugh as we notice him slip on a banana peel would make us split our sides when we discern that he has broken his neck: witnessing a fatal accident should boost immeasurably our feeling superior)[5]. The philosopher had a terrific sense of humor, clean of contempt, derision and all sense of superiority: his writings are full of subtle and lovely jokes; he told terrific jokes in his lectures and he invented them with ease in his (admittedly rare) friendly relaxed conversations. And they were all very good and very apt and very serious. I have never heard him tell a bad joke, a sex joke, good or bad, or a joke with no direct relevance to the ongoing conversation; which means that most of his jokes were ad lib (and so can hardly be reproduced here). The systematic absence of sex jokes in his conversation was remarkable, since in most societies the rule is that sex is a matter to handle with some levity.[6] His jokes in general, come to think of it, never express levity, being all witty and humane and good-natured; very proper and very serious in intent.

I never heard him make any vulgar remark with the exception of his already mentioned anti-Jewish and anti-Zionist remarks that were rarely expressed and then very briefly.[7] To the latter I responded saying that for me the Zionist venture is a fait accompli: taking sides in the dispute over Zionism equals taking sides in the War of the Roses; he responded by suggesting that the U. S. should grant free entry to all Israelis in order to reverse the process. This way he expressed both his general, utter indifference to the national sentiment,[8] as he always did when dis-

---

5 [Russell, 1985, 37, The Professor's Guide to Laughter]. Compassion with the injured victim of a banana peel obviously stifles laughter. But then we feel compassion even if the victim of the banana peel is not injured. Indeed, the best humor combines poking fun at people (including humanized animals or even humanized sticks and stones) with the compassion with them that should free humor of contempt. Not so sarcasm, satire, caricature and parody, or any other form of *exposé*: they are destructive weapons, meant to be savage. Bergson's idea may be more adequate for them, but even then with a proviso: the best and most devastating of these weapons allow compassion to infiltrate stealthily, showing that all weaknesses notwithstanding, their target still is "human, all too human" (Nietzsche).
6 I never heard anyone tell Popper an off-color joke. In the very rare event of even the slightest use of improper words in his company, he responded sharply and very seriously.
7 Regrettably, Popper's hostility to drugs, pop-culture and the media made him a paternalist, as he explained in his letter to Thomas S. Szasz of July 20th 1961 and elsewhere. (The consistency with which Szasz adhered to his liberalism is admirable: opposing all laws forbidding the use of drugs, he advocates reasonable government control over it.)
8 In his early days, Popper showed compassion as he advocated the diversion of the national sentiment to positive channels [Popper, 1927]. This is insufficient as it seemingly condoned the manipulation that he otherwise rightly condemned [*OS*, i, 199]. Objectionable popular ideas invite a sincere search for improvements that show concern for the needs that make these objectionable ideas popular. And then one may read into the old objectionable idea its improved version. Popper did this in a letter to Isaiah Berlin, where he offered a friendly interpretation for Berlin's objectionable idea of freedom. Later on Popper dismissed ideas that cater for the need to belong to the closed society with the demand that we bear the cross of civilization. Here he is in conflict with the traditional liberal recommendation to make all law-abiding as agreeable to the public as possible. (He endorsed these ideas, of course, but not all the way.) The needs wrongly expressed as the need to return to the closed society are easier to meet in the open society. But this takes some planning.

cussing any matter pertaining to nationalism, and his specific, strong distaste for the Jewish national sentiment, thus making lavish acknowledgement of his deep Jewish bond, his protestations to the contrary notwithstanding[9]. (His systematic affirmation of his being Jewish to anti-Semitic audiences is merely an expression of his distaste for discrimination.)

I should also mention here that he did write at least one paper [Lakatos and Musgrave, 1970, 51-58] that I judged poor [Agassi, 1988] and it was for me an unbelievable surprise. Though this paper, as well as the book in which it appeared, had met with great success, I found it very poor as I explained in my review of it. I suppose that the poor level of the volume has something to do with the unusual appearance of a poor paper from Popper's pen.[10]

I have never heard him make an unintelligent and unfair remark except what is mentioned in the previous two paragraphs and some of his discussions of his peers' attitudes towards himself — which is regrettably a running theme of this record. And he almost never discussed sex or even alluded to anything directly connected with it [Munz, 2004, 19]. His adherence to this rule was so regular that the exceptions stand out and they all happened in contexts that clearly demanded seriousness. I never heard him so much as mention sex except on very rare occasions, and then only when it was unavoidable, and then he handled it at a great distance — with the caution of a physician looking down the throat of a patient with a very severe head cold. He only once, and under great pressure, expressed to me his view on sex, and then sex was not even mentioned. Let me tell the story in some detail; I think it will provide the

---

9 It is customary to measure the degree of self-identification with a group as that of the pride in belonging to it. This is shortsighted: shame is no less significant than pride [Agassi, 1999c, Chapter 3.1]. For example, Popper said, as a Jew he was ashamed of the inhuman parts of Scriptures. Also, in an interview [Urban, 1993, 213], to his condemnation of the "mentality of mindless killing and sadism" he added the regrettable assertion that this mentality "stems from certain elements in the Old Testament and from distorted interpretations of the New Testament". This is a juxtaposition of an uncharitable reading of one text with a charitable reading of another. This is Popper's bias against Judaism coupled with his off-hand dismissal of all criticism of Christianity. As to historical Judaism, Popper was ignorant of it (see, for example, [Popper, 1998, 44]). But he must have at least browsed through Russell's *History of Western Philosophy*. He could not fail then to see then that Russell held a friendly view of Judaism. And he was well aware of Russell's rejection of Christianity. I do not remember disregarded views that Russell had expressed except for his views on religion. But in saying all this I am missing the point, as it is not factual but moral. He said that much in his tense response in his seminar, to my criticism of his distortion of the Sermon on the Mount [*OS*, ii, Chapter 11, iii]. It was not to my liking. And at the very least Popper could have learned from Heinrich Heine's enlightened refusal to condemn religion or to preach to its practitioners or to romanticize it in any way — not even while viewing it nicely and with the greatest of sympathy as the mere folklore expressing naïve folk sentiment.

10 The nadir of Popper's output is his cooperation with Lorenz that ends with a distressing sermon [Popper and Lorenz, 1985]. But, as Maugham repeatedly said, writers have the right to be judged by their very best. And Popper's best conflicts with his worst, as it shine with his profound humanism. Nevertheless, let me notice that Popper said Lorenz had apologized for his Nazi past and he accepted that apology. This is far from satisfactory [Bischof, 1993]. In his *Civilized Man's Eight Deadly Sins* of 1973, in the chapter on genetic sins, Lorenz repeats his admonition against extremism, giving as examples of extremism Auschwitz and American democracy. This is really too hard to swallow.

background to the atmosphere in his company, in his study, in his office, and in his seminar. (As far as I know, he avoided participating in committee meetings, except when he had to defend a graduate student's cause.)

I was writing a piece against Wittgenstein at the later period of my apprenticeship — before I left London — when the philosopher was urging me to write and repeatedly and in various ways expressing extremely kindly and very solemnly great readiness to help me write and to help me improve my writing technique. (Indeed, despite the great variance between our styles of writing, I owe my writing accomplishment, such as it is, almost exclusively to him.) The proposal that I write about Wittgenstein particularly tickled him: Wittgenstein was the most popular — the only really popular — leader of philosophy in England at the time, professional and amateur alike. He was, I suspect he remained, the philosopher's only bête noire:[11] there could be no greater public expression of loyalty to him than to lunge at Wittgenstein. I knew this all along, of course, yet preferred not to discuss Wittgenstein in public, thinking (as I still do) that despite the important minor technical contributions of his first book to the standard stock of philosophical knowledge, philosophically his whole output is rather trite: it is neither interesting nor inspiring. And then the occasion arose (the publication of [Wittgenstein, 1956] that is blatantly anti-rationalist)[12] and I tried my hand at discussing his philosophy (and its application to mathematics, to show that the book's repulsive qualities were no mere accretions); the philosopher was flatteringly supportive and curious. In my next visit I brought the draft. No sooner did I read my opening remarks to him than he condemned it all. (I dropped the project, though I still cling to my old view of that book; this is no loss to learning: the irrationalism of Wittgenstein's view of mathematics is by now public knowledge. I have since had sufficient occasions to publish

---

11 Popper was friendly with many Wittgenstein fans [*Autobiography*, §39]. In comments on an interview [Stadler, 2004, 474-97], the interviewer, Friedrich Stadler, the official keeper of the fading memory of the "Vienna Circle", uses Popper's friendship as evidence that he had not distanced himself from the "Circle" as much as he wanted his audience to suppose. This echoes Schlick's nasty remark that Popper was not as different from them as he was claiming. This is to say, Stadler viewed personal and intellectual distance as going together. Now, in Stadler's opinion, could Popper honestly befriend an advocate of the return to the closed society? If he says yes, then he dismisses Popper's passionate advocacy of the open society as exaggeration. If he says no, then he ignores Popper's friendship with Lorenz (see previous note) and he limits seriousness to the *doctrinaire*. Similarly, Feyerabend claimed that as Popper translated his intellectual distance from him into a personal one he was a prig — as were Popper's followers, myself in particular.

12 The dismissal of Wittgenstein's book on mathematics is common. See [Wrigley, 1977] on the reviews of it by Michael Dummett and by Paul Bernays, as well as the one [Agassi and Jarvie, 1987, 51-68] by J. O. Wisdom. These days no matter the prestige that Wittgenstein may still have, any prestige that his book on mathematics may once have had has completely drained away, partly on the authority of Gödel [Feferman ,2005]:

"One of the gems is Gödel's put-down of Wittgenstein's book on the foundations of mathematics (30 October 1958): 'I also read parts of it. It seemed to me at the time that the benefit created by it may be mainly that it shows the falsity of the assertions set forth in it.' As a footnote he added: 'and in the *Tractatus* (the book itself really contains very few assertions)'."

my opinions on the philosophy of Wittgenstein (and of Carnap). I would have loved to show the results to the philosopher as marks of my still strong loyalty, but I resigned, of course, to his refusal to glance at my output.[13]) He was understandably somewhat embarrassed at the failure, and then he did what he usually did when embarrassed: rather than explain the embarrassment or try to clear it, he usually made his company a great gift.[14] And the most precious gifts he could give were always some wonderful ideas, and with a whole range of preparedness, since some of them impromptu and others were expressions of some long gestated thoughts that burst to come out but were somewhat inhibited.

He had a few such long gestated ideas; the most conspicuous of them is his speculation regarding the rise of Western polyphonic music and its effect on Western music in general and even on Western culture in general. He published some of them in his autobiography (§11), and (as he said there) he hardly ever gave them airing before. Some of the greatest authorities in the field considered his ideas in the history of music remarkable, though at first they naturally said so only orally. Tomas Kulka has won public gratitude by having shown how interesting they are for students of aesthetics [Kulka, 1989] (see also [Blaukopf, 1994]).

To return to my story, the idea he discussed with me at the time has to do with his profound hostility to both Freud and Wittgenstein. Their similarity lies in Wittgenstein's repeated claim that "to ask philosophical questions is sick": the task of philosophers is to cure sick minds by clearing the air of metaphysical fog, he said (on the basis of his (adolescent) metaphysical view that metaphysics is all fog, so that nothing remains when the clearing is over). In Popper's opinion, both Freud and Wittgens-

---

13 My hope that Popper would like anything by me (on Wittgenstein or on anything else) was an illusion. When I wrote this, I forgot that I once had shown him something of this kind [Agassi, 1959]: two very brief letters to the editor of the *Times Literary Supplement*. Although they concerned not Wittgenstein but etiquette, they were heretic, as they defended David Pole's book against Wittgenstein. The anonymous reviewer argued against Pole by appealing to Elisabeth Anscombe's personal reports about Wittgenstein. My claim was that these are irrelevant. The reviewer dismissed me as a disciple of Popper. I responded, saying that this again was personal. Worse, as he hid behind the veil of anonymity, he was also unfair. In my youthful zest, I was proud of my success, all the more so since Anscombe openly agreed with me. (Not that the reviewer admitted defeat or that he suffered from it. As it turned out decades later, he is an Oxford high muck-a-muck.) I thought Popper would offer a kind response. It was surprisingly hostile, though he expressed clearly only dislike of my use of an expression that he deemed too colloquial. (The expression was "I guess".) This upset me. I said, perhaps I should not publish anything then. He told me years later that he took this as censure of his having taken his time sending his *Logic of Scientific Discovery* to the press. This made no sense to me except as displaying his character as a glutton for punishment. But perhaps I am too shortsighted. My response was to his pedantry, and it was indeed this pedantry that kept him from sending the book to the press. And so, one may perhaps bring together my assertion and his far-off reading of it. Nevertheless, this is very upsetting because my communications are as direct and open as I can make them and he always sought hidden messages. Of course, this mode of conduct enabled him in conversations to feel responses to his assertions and to respond to them without hearing them. This is uncanny but not advisable to emulate.
14 On some friendly occasions, Popper insisted on giving me some old books of his – as expressions of affection. From his autobiography (§2) I learn that this was a great sacrifice.

tein are right on one central issue, and this, paradoxically, condemns their ideas most. Freud was right in his conjecture that all sublimation is neurotic and all acceptable curiosity is sublimation of early infantile curiosity [about sex]; and this includes all curiosity — religious, cultural, artistic, scientific, or other. Hence, the cure of neurosis, in the style of Freud or of Wittgenstein, is the end of civilization as we know it and is thus objectionable as too costly.

(I once met an Irish diplomat, Brian Merriman by name, who presented himself to me as a Popper follower. He had a book published that was a light-hearted fantasy [Merriman, 1976]; it is not striking in any way except in that it describes sexual conduct as evidently less problematic and less interesting than food intake. This exercise is much more interesting than a casual glance will reveal. He had visited the philosopher, he told me, only to receive admonition for his permissive attitude to sex as unproblematic.)[15]

At first I was disappointed with the gift, with the idea that we need neuroses in order to progress (or even merely to stay in place): the idea is neither good nor new. Obviously, it is a misanthropic error rooted in the Judeo-Christian theory of original sin and the Greek myth of the Beast in Man (that Plato enthusiastically endorsed, *Rep.*, 588c). It contains the assumption that by nature we are lazy, so that natural and healthy human curiosity is idle — contrary to Popper's own insistence [*OS*, i, 70, 78-9, 241, 262] that we have no knowledge of human nature and contrary to his appeal to the animal kingdom for evidence of our ignorance (as even puppies are observed to show regular curiosity about their environment).

The author best known for the dissemination of this misanthropic idea is the Platonic-Judeo-Christian St. Augustine, whose leading modern heir and admirer is none other than Freud. The *locus classicus* for it is his moving *Civilization and Its Discontent* (1930), which is his expression of his own ambivalence about ambivalence; he even opted for the idea that we should remain neurotic in some measure.[16] He wanted to cure only neuroses that interfere with work: only they need and deserve treatment:

---

15 Adam Gopnik reports [Gopnik, 2002] in detail Popper's appreciation of *The Catcher in the Rye* by J. D. Salinger as realist — perhaps because it contains a description of problematic sex. (Its adolescent hero goes to a prostitute but only talks with her.) To my regret, Popper never talked with me about Mrs. Gaskell, whom he admired despite her relatively open attitude to sex [Pryke, 1999]. Nor did I ask him ever about Oscar Wilde, whom he never mentioned except when he borrowed from him these powerful lines: "And all men kill the thing they love, / But let all this be heard, / Some do it with a bitter look, / Some with a flattering word: / The coward does it with a kiss, / The brave man with a sword." He found it thrilling that the best expression of loyalty may be an act of treason and that a beloved master is not exempt. This is only a supposition, however, as he refused to discuss such matters. I do not even know if he had read the whole of Wilde's poem, let alone if he was familiar with Wilde's "The Critic as Artist". It is hard to assume that he was; which is a pity: he would have loved it.

16 [*OS*, i, 295]:
"I suppose that what I call the 'strain of civilization' is similar to the phenomenon which Freud had in mind when writing his *Civilization and Its Discontents*."

the puritan philosophy is the mistrust of the individual and the use of overwork as the sacred cure for all of the individual's ills. The ability to work, to act as productive members of society, in Freud's words, is the only thing that by his puritan standards is of real value. There is an exception to this, though: Freud also wanted healers perfectly cured and hardworking and creative; but perfect cure is a joke anyway, especially by Freud's own canons, as Popper observed [*OS*, ii, 352]; that the Freudian community nevertheless passed over this critique unnoticed cost them dearly: they still have to sort out for themselves the purpose of training-analysis [Wisdom, 1984]. In any case, there are legends about Freud's neuroses about his originality, about his critics, and about his disciples — not to mention the one about music. (As an infant prodigy he prevented his sister from learning to play the piano as this disturbed his thinking. She did not forgive him to her dying day, years after he died. This explains the bizarre fact that he was both passionately interested in music and self-proclaimed tone-deaf.) Freud's compromise solution, then, was, neurosis is good or bad depending on whether it helps or impedes production; it is in line with his other moving masterpiece, The *Ego and the Id* (1923), where he says this, and where he complains that we are doomed if we listen to Father and we are doomed if we do not, and that Father's effort to tame is at times so successful that it utterly incapacitates. For, in that masterpiece, too, Freud remained faithful to Father and admitted that his infantile desire for Mother was plainly wrong, except that he appealed to Father to be a bit more compassionate and a little less strict. The appeal is in no way contrary to the strict and traditional puritanical attitude that Freud wished to soften; it may be claimed that the kind of softening of the tradition that he proposed is already part and parcel of that tradition: whenever the traditional code of education leads to suicide, rather than blame it and attempt to alter it and prevent further suicides, as we should, we traditionally and systematically and hypocritically and cowardly blame some stringent application of the traditional code by too strict a Father or too strict a father substitute, and blame that on some excessively harsh attitude — at least for the case of a tragic suicide that proves in retrospect to have been a case of an unusually sensitive individual who had merited exceptionally gentle treatment but due to some regrettable oversight received too large a dose of discipline. (This is crocodile tears: regretting the damage and blaming an accident or a scapegoat or anything else just in order to prevent blaming the system and thus blocking all attempts at its improvement. All this is nothing new, as it is the topic of Kafka's *Letter to His Father*; Kafka's conservatism is a constant cause of embarrassment to those of his commentators who present themselves as radicals; they hide his conservatism and their embarrassment about it behind thick metaphysical fog, insinuating that he liked it.)

I do not know what was Popper's opinion on the need for curing individuals who are so incapacitated by strict education (or by other causes) that their neuroses are so severe that they are unable to work or that they live on the verge of suicide. The philosopher told me about a tragedy of this sort in my early days: before my arrival in the London School of Economics a brilliant young economist there who was of that ilk ended up in a mental home where he committed suicide. Before he went to hospital, he was undergoing some sort of analytic treatment. The philosopher had warned that young economist and told him that the analyst was contributing to the disaster and that psychoanalysis is pseudo-scientific, he told me, but to no avail. This may amount to a dismissal of my question, should we institute treatment to cure utterly paralyzing neurosis? For, the dismissal of my question would rest on the tacit claim that since psychoanalysis is no cure, my question has no point. If so, then this would not do: even if we have no cure at all, the question remains, is cure desirable? This question signifies both theoretically and practically — as a motive force for research into the possibilities of such a cure: we know of no principle that can help us discern the curable mental illnesses from the incurable ones. And so we do not know if such cures as we may one day find are commendable or not [Fried and Agassi, 1976 and 1983].

What then was the philosopher intimating to me when conceding that the analyses that Freud and Wittgenstein offer of curiosity as neurotic are correct at bottom yet the aim of curing it is reprehensible? I do think that here a strange phenomenon may confuse and thus block efforts at a simple answer. The great strength and weakness of the philosopher is his dedication to hard work — and to work of the highest quality as judged by the highest of standards. It took me time to see this as a weakness: I was repeatedly disturbed, but also intrigued, even thrilled, when he said in his seminar, every time someone appealed to philosophic public opinion or received standard, even to the best of it: we should not swim with the current as we are the makers of the current. And he always did attempt to address the best problems in his field of vision, and by the highest possible approach: to repeat, he almost never argued cheaply. You can imagine how I had felt towards his output as long as I could not find one page of his that was not highly significant (whether I endorsed it or not), and how I felt towards his person when I noticed that he almost never said anything trite or vulgar, and that almost everything he had noticed is of the highest quality, almost every reference of his to any work of literature, a painting or a piece of music. This was not a pose: it was consistently his character: his spacious home and garden, his furniture and the reproductions on his wall,[17] the pieces he played on his grand piano (for Feyera-

---

[17] A well-framed reproduction of *The Man with the Golden Helmet* dominated the huge, bay-windowed living room on its grand piano and armchairs of Popper's own design (in a toned-down Bauhaus-style).

bend and me) and the books in his bookcase — everything to do with him was of the highest quality.[18] And yet this highest of standards and this super-high-brow attitude also always disturbed me. His intimation of his view of the indispensability of neurosis for civilization, though in itself it was a disappointment, has made me see some integrated picture; I had a prejudice that needed correction, and this was it.

The prejudice in question concerned the peculiarity of philosophy as an intellectual activity, involved with a complicated history, to do with the idealization of intellectuals, with the assumption that all serious scholars must possess broad interests that reflect progressive, broad, open minds. This idealization can be entertained only by the extremely credulous, as many scholars who were extremely limited in scope and political outlook nevertheless rightly won the gratitude of their peers and of scholars in general and of the general public too. This assumption is still often entertained, as it has deep roots in our tradition, as it was a medieval idea that was turned into the ideal of the Enlightenment Movement of the Age of Reason, though even at the height of its popularity it was too lofty to be seriously entertained so that its endorsement as a matter of course was often mere lip service. Nevertheless, until the eve of World War II most Western scientists tacitly accepted the idealization of the serious intellectual as a part of their heritage. It was then a shock for many of them to learn that practically the whole of the German scientific community were ardent chauvinists and that most of them did not oppose the most brutal and irrationalist ideology ever. It was a shock to them that the great thinker Werner Heisenberg (of the Heisenberg inequalities) was not very good as a philosopher and historian of science, even though he did repeatedly publish something or other in that field, and that he had strong sympathies toward this ideology at the very least (I have argued that he sympathized with this ideology sufficiently culpably even after the war [Agassi, 1975, 416-17]). This shock is now almost entirely suppressed, but the idealized image of the scientist as open-minded, as a seeker of progress through enlightenment, cannot be revived, I surmise, as the regrettable and unintended consequence of the process of scientific specialization. Thus, in the writings of Aldous Huxley (*e.g.*, *Brave New World* and *Suddenly Last Summer*), specialization is described as the loss of this ideal, presented with nostalgia for the Enlightenment past in C. P. Snow's

---

18 Popper found distasteful Einstein's practice of playing the violin after his public lectures. This surprised me at the time, but now it seems to me a part of his allergy to anything not of the highest quality, perfectionism run wild. (Similarly, when Charlie Chaplin realized that he would never play the cello to perfection, he stopped playing. Incidentally, Popper admired his *Gold Rush* as well as the scene from his *The Great Dictator* that he mentions in his *The Open Society and Its Enemies*, Chapter 12, n35.) The odd thing is that the starting point for Popper's deliberations on music, as he told me, was his contrast between the romantic view of it as destined for greatness with the older, friendly view of it as sheer fun. Alas, his published text on music [*Autobiography*, §11] has no reference to fun.

celebrated *The Two Cultures* that I have criticized elsewhere [Agassi, 1981, Chapter One]. Knowing all this I still clung defensively to remnants of the idealized image of the intellectual even after I had learned that the great polymath Bernard Shaw had a good word to say for the political leaders who later proved to have been the worst ever (even though only in the early days, of course, and even then he was critical of them: he said of the German and of the Soviet dictators who had killed their critics, that they had thereby dug their own graves).[19] And even after I learned my lesson and knew that great thinkers of all sorts, even great humanists, are not always progressive and enlightened, even then I still took it for granted that the Enlightenment ideal image is sufficiently nearly true of significant philosophers: philosophers cannot be the universal scholars as the Enlightenment Movement wanted them to be, but in order to be of any worth they must aspire to that position, I tried to insist, and they must at least be progressive in tendencies, and broadminded too. Oh, I did not mean that philosophers have to believe in progress, merely that they should wish it and endorse it as a positive value, and that they should be not so loyal to tradition as to fear great progress.[20]

I always knew that each ideology had its philosophers — the romantics, the reactionaries, the chauvinists, the varieties of irrationalists,

---

19 I was in error on this. Shaw viewed as realists power-hungry politicians of all sorts. This may be excused as disdain of utopianism. Not so his admiration of them. He recommended Stalin for a Nobel Prize. His Endorsement of the Nazi cancellation of the Versailles Treaty in November 12, 1933 is reasonable. Not so his hilariously exaggerated admiration for the beast, expressed in *Observer*, November 5, 1933. He missed many occasions to regain some sense of proportion [Klett, 1988]. His *Geneva* (1938, rev. 1945, 1945 Preface and notes) displays futile efforts in this direction. Anti-utopianism does not require the courtship of tyrants; Russell rightly viewed this failing of Shaw as a character defect. He belittled Shaw's ideas, admitting as important only his ridicule of humbug [Russell, 1956, 77]. This is a trifle unkind, as an oversight of Shaw's daring to express boldly some unpopular sane ideas, such as his opposition to World War I and to the Versailles Treaty. Although he foolishly sang the praise of the social conditions in Russia (in his famous letter to the editor of the *Manchester Guardian* of March 2, 1933) that he never took back, he strongly condemned the Soviet show-trials. His defense of eugenics notwithstanding, he ridiculed Nazi racism and anti-Semitism. (He reconciled this with his admiration for the beast by asserting that it was not serious. After the War it was incumbent on him to withdraw this outrageous contention; he did not.) He criticized democracy thoughtfully (Preface to *The Applecart*) and praised the rule of law [Shaw, 1965, 377]. He sought to safeguard science against "rulers who monopolize physical force" [Shaw, 1965, 487]. He was a great supporter of political activities in committees (illustrated in his famous letter on it to H. G. Wells of March 22, 1908). Moreover, he spoke about "the sacredness of criticism", declaring it the right and duty of government; he said, "Civilization cannot progress without criticism" [Shaw, 1965, 396-7]. He was thus a critical rationalist of sorts [Agassi, 1981, 102]. In disregard for Spinoza's criticism (*Political Treatise*, concluding passages) he advocated some sort of an élitist democracy ("a Democracy of Supermen" [Shaw, 1965, 179]), that will supposedly be the "impartial government for the good of the governed by qualified rulers" [Shaw, 1965, 883]. In sum, he was much more of a democrat than one might conclude from the mere familiarity with his support for every sadistic, reckless, self-centered tyrant in sight. (He even ignored the notorious facts that tyrants often act too cruelly for their own good: he condoned their humbug at its worst.)

20 All this is very mild: I ignored here much worse assertions, such as that of Feyerabend, who declared the beast "an intelligent person, more intelligent than most of the critical rationalists, with a clear view of the complex historical processes" [Feyerabend, 1979, 92]. He could not mean this, and in saying it anyway he could not be nastier.

the Wittgenstein fans[21] the mock-radical and/or mock-liberal leftists, and even the Nazis; (If you refuse to appoint Martin Heidegger to this post, then by all means consider the official Nazi philosophers; in the present context this makes no difference.) I never expected to learn anything from them, though, or from such philosophy professors as Geoffrey Warnock, who at the place and time of my apprenticeship made names for themselves as leading philosophers because they complained loudly and bitterly against the wicked who demanded of philosophers to be broadly philosophical instead of allowing them to sink comfortably into the narrowness of their chosen professional specialization and develop expertise like other academics [Warnock, 1956, Conclusion]. And the same goes for other philosophy professors of that time, like T. D. Weldon, who illustrated the value of the same expertise — the study of words — by showing how deliciously backward looking this may make them. (In his autobiography Sir Peter Medawar says his teacher Weldon had told him to read Popper; in public he never gave the slightest indication that he knew of all this.[22]) The expertise these philosophy professors were speaking of is the expertise of philosophical therapists: they claimed that dealing with mere verbal quirks is a sort of therapy. I found all that hilarious; when Ernest Gellner called such fellows charlatans I did not know if he was right but I cheered him all the same: I found their conduct unbecoming one way or another. It is now shamefacedly overlooked.

In brief, all I knew and all I had learned, whether of philosophers I respected or of others who exhibited some philosophical breadth, tallied well enough with my prejudice that unlike the engagement in science, at least proper engagement in philosophy — to the exclusion of mock-philosophy of all sorts — must at least accompany efforts to conform to the idealized image of the intellectual as a well informed, well rounded, up-to-date individual whose heart is on the progressive side. And Popper fit extremely well this idealized image. So I read him this way all the way too. So most naturally I took his anti-Judaism and his anti-modernism in the arts as merely blind spots, and his anti-modernism at least in part jus-

---

21 To gain recognition as professionals, Wittgenstein's fans had to acquire expertise, even though they preferred to hang about as idle gentlemen old-style, especially in Oxford. In the USA the demand that they should be proficient was stronger. MIT unified the department of philosophy and of linguistics. Yet Harvard hired Stanley Cavell because he had insisted that all the qualification he needed in order to philosophize (Wittgenstein-style, naturally) he had as a native English speaker. This changed when Wittgenstein-style philosophy professors found themselves at the bottom of the academic totem pole. Cavell moved on then to aesthetics. Others moved to the history of recent philosophy or towards the philosophy of science, or, still better, to feminism and to other innovations that as innovations seem safe from older disputes. The latest fad still is the construction of bridges between the analytic (Anglo-American) and the phenomenological (Continental) schools of philosophy.

22 [Medawar, 1986, 53]:
"(... Harry Weldon, also a great admirer of Popper, describes his *Open Society* as the most important philosophic work of the twentieth century even though written under difficult circumstances.)" This Weldon said in private. Publicly, he dismissed Popper. See Appendix to Chapter Ten.

178   Chapter Six

tifiable by the prevalent influence of historicism and of German Expressionism.

(I am uneasy: the expression "German Expressionism" is a name for a historical set rather than an ideology or an artistic characterization proper. The set includes works of many non-German artists. And like almost any artistic movement or trend, the set contains works of different degrees of quality. I do agree, though, that the doctrines and techniques and traditions of sets are not all equally helpful and that those of German Expressionism comprise license to produce intellectually pretentious trash. Our reasons, however, are very different. Popper was simply a traditionalist concerning the arts, and he viewed the experimentalism of the expressionists as historicist and romantic. In my "Art and Science" [Agassi, 2003] I have presented art as an exploration of inner space and this seems to me to be a sufficient response to these charges. My own reason for dislike of most of German Expressionism comes from my agreement with Friedrich Nietzsche,[23] of all people, on matters artistic: good art, he said, no matter how despondent, expresses *joie de vivre*. Popper took even Anton Webern to have held the Romantic view of art as expression, and no doubt the tremendous force of Webern's music[24] is his exploration of the depth of despair; yet it conforms all the same to Nietzsche's requirement in a most paradoxical manner[25] (as do the better

---

23 This refers to Nietzsche's praise for joy, not his demand for it that is joyless and even killjoy. The praise of joy seems to me to deserve better press. It can be a joyful expression of the love of life. But, admittedly, it need not be. Russell justly dissented from the common view that Nietzsche loved life [Russell, 1945, end of Chapter 26].

24 Webern's hyper-romantic Opus 1 as well as some of his arrangements are both superficial and far from desperate. His famous arrangement of Bach's *Ricercar a 6* is a gigantic *tour de force* that is profoundly desperate.

25 Nietzsche appended the essay "Attempt at a Self-criticism" as a preface to the second edition of his *Birth of Tragedy* [Nietzsche, 1996]. It is a comment on Schopenhauer's view of tragedy as resignation. Nietzsche cites there his own idea that paradoxically tragedy too must express joy. This hardly counts either as self-criticism or as a response to criticism: it is no more than an expression of a reasonable dissent. But the general idea behind it is right: the best place to look for a counterexample for the view of art as joyful is to look at the least joyous of the arts, namely, at tragedy — as already Aristotle reluctantly noticed. Nietzsche should have argued, at least by indicating his view as to what renders tragedy joyous, rather than by dismissing Schopenhauer's view to by reiterating his own. He could not share Aristotle's view as he could not see the catharsis as joyful: it goes against his idea that the best life is the resignation to one's fate. Thus, as usual, he was in profound conflict here, wavering between despair (Schopenhauer-style) and *joie de vivre*. This seems true of Freud too, as well as of Wittgenstein and of some other thinkers whom Nietzsche influenced (not Russell, though, much less Shaw, whose *Don Juan in Hell* is a rebellion against all that Nietzsche stood for.).

This is not to reduce the task of art to the task of praising life, nor to see this praise as the common denominator of all art. Consider the difference between *The Song of Songs* and *Lamentations* (ignoring for a while whatever makes them religiously significant): the expression of love for life is most natural for a love song and a staggering challenge for a lamentation.

The praise for *joie de vivre* is often a rider to expressions of contempt for reason and for ordinariness — expressions that are snobbish and often pompous. Nietzsche, as usual, both supported and condemned them. This brought him credit and discredit, influence on Russell and on Heidegger, inspiration to Jack London's forceful *Sea Wolf* and to Hermann Hesse's adolescent *Steppenwolf*.

works of Edgar Allan Poe[26] and of Frida Kahlo[27]). I take Stravinsky's rejection of Schoenberg's music as expressionist and his endorsement and even emulation of Webern's, as intriguing evidence here, even though regrettably the master philosopher[28] rejected all three master composers equally strongly — despite the appreciation of Webern's sincerity that he records in his autobiography.)[29]

I was in error. I should have known better. Soon after I met Popper and before I became his apprentice I met Michael Polanyi, who shared Popper's problems but not his rationalism. Rather, in the existentialist mood of the time, he declared scientific research unique, the way art is: science, like art, is learned in workshops, he said; science masters can no more articulate their views on science than art masters articulating their views on art: to be a good artist one must apprentice oneself to a good workshop; to be a good scientist one had to do the same, he said. In brief, Polanyi was a thought-provoking reactionary philosopher, and he should have taught me that I was priggish in my claim that all good philosophers must have their hearts on the progressive side. Also, Polanyi was a model gentleman, a truly open person. He offered me an assistantship before Popper did. He wanted me to help him as he put his celebrated *Personal Knowledge* in its final shape. I read the manuscript and declined. He remained gentle and most courteous and helpful to the last.[30] Though I do not think I could have worked with him, I do sincerely regret that I was so priggish about it. Were I better familiar with Spinoza's work then, I would know that my indignation was a defensive refusal to be instructed. I now know what I refused to learn: I refused to learn that even a rounded philosopher can be a conservative, even a reactionary, and yet be an interesting and profound philosopher. I hope the dedication of my *Science and Society* to his memory is expiation to some extent.

But I am speaking of Karl Popper, who was progressive in his tendencies, at least by-and-large, yet with a few highly conservative traits, justified by some very high intellectual and artistic and moral standards. I

---

25 This conflicts with the celebrated reading by Charles Baudelaire; it accords with the following, probably due to Jeffrey Savoye, secretary of the Poe Society:
"There's this incredible dogged optimism in Poe ... how could you possibly have gotten up and faced another day under his circumstances without some incredible strength of 'today will be better'? Opportunity was always just around the corner for Poe and he just never quite got there."
27 Kahlo put well her joy in her famous saying, "I am not sick. I am broken. But I am happy as long as I can paint." This agrees with Popper's saying, "We don't mind physical suffering if we are happy" [Munz, 2004, 21].
28 Popper admired Beethoven's ability to express joy despite his personal misery [*Autobiography*, §13].
29 I wonder whether Popper knew of György Ligeti, *Clocks and Clouds* (1973), written under the inspiration of his lecture by that name (1965). Perhaps he was too deaf to listen to it and perhaps he would not or could not like it, gentle and lovely though it is.
30 In the eve of his life, Polanyi agreed to talk to the Boston Colloquium for the Philosophy of Science on the condition that he would have me as his commentator. For fun, he forced me to conjecture its content or to improvise. This became common knowledge at once, and it was great fun, especially since uncharacteristically I visibly read my comments from a prepared page.

now consider his high standards and his unlimited dedication to the cause of philosophy as a trait that is definitely not progressive. But it took me years to arrive at this view, even though it was all too obvious that he fostered puritanism and I disowned it. Yet it was this puritanism, this highest of standards, this relentless effort always to master the best ideas in the field, this relentless dedication to philosophic research, that made the philosopher so exciting an individual to listen to and to converse with that he was regularly the center of an ambulatory workshop.

It is amazing how much Vienna of the time is echoed this way in the personal tastes of Freud,[31] Wittgenstein and Popper. Ray Monk's life of Wittgenstein presents as a spell cast on a portion of the Viennese intelligentsia, an ethos that filled the Vienna air of the time, the compound demand for high standards, utter clarity and a measure of sexual abstinence, linked to a maudlin and anti-Jewish version of mock-Christianity and the conviction that the only way to escape the profound and most painful (Kafkaesque) sense of one's utter insignificance is the tremendous effort to be creative. There was, of course, the liberal part of Vienna too; it is the triumph of the philosopher that he was above the morass that Monk describes, and he kept above it aided by a healthy view of criticism and of democracy, I think, but without totally freeing himself of all of its veneer, as I would have expected in my naïve idealization. The very fact that the Vienna of the time produced so much that is intellectually of great value, even if some of it was pretentious and cheap, is proof enough that I was very naïve; Popper found extremely cheap even the articulation of some of the convictions that some of his Viennese older contemporaries gave vent to. But he made an exception and indicated to me the part of them that he shared, especially the tremendous demands on oneself that the better part of these better older contemporaries of his had required. There is no denying that his excessive work ethic contributed to his triumph over his tremendous handicaps and made life around him so very exciting.

It would be unbalanced to report the great excitement the philosopher regularly generated while overlooking the failure of his methods — the methods discussed a few times already in this record — of taking any opponent as seriously as possible, even though instances of the failure, by its very nature, must strike you as humorous.

I remember once, in his seminar, an American visitor asked the philosopher a simple-minded question. This was the now famous Harvard professor, scholar and philosopher of education Israel Scheffler, who was then a beginner. He was a disciple of Nelson Goodman, then a leading philosopher — perhaps because he was from Harvard and a close collabo-

---

31 "My emotional life has always insisted that I should have an intimate friend and a hated enemy", said Freud with rare frankness (*Interpretation of Dreams* [S.E., 4.5, 483]).

rator of the celebrated Quine, but I cannot say. At the time I already admired Quine, but still showed no interest at all in the work of Goodman. Goodman's *Fact, Fiction and Forecast* [Goodman, 1955] was the rage then. Later on I said a nice word about that book in an essay on laws of nature [Agassi, 1975, 236-7]; at the time I saw no merit in that book and I still think it was popular at the time for no other reason than that its author was famous. Popper dealt incidentally with the same issue as that book in an index of his major book ([*LScD*, New Appendix *X]. It pains me to remember how many times I requested that the philosopher not tell me about that appendix, much less discuss it with me: he disregarded almost all of my requests. In my opinion, any discussion of the laws of nature with no reference to metaphysics is a plain waste of time. This the philosopher would not discuss with me, yet he went on discussing ideas that clearly conflict with it.) Scheffler had written a review [Scheffler, 1958] of Goodman's book in *Science*, a very prestigious periodical that happened to pass my way. It said, the problem of induction is the central problem of the philosophy of science; without mentioning Popper's proposed solution to the problem, Scheffler declared that Goodman's tremendous significance lies in his having proposed a viable solution to it. Goodman had declared induction to be projection — where a projection is but another name for what Popper had called the law-likeness of a generalization — and a valid induction should then be a legitimate projection. The problem of induction thus solved, however, now leaves as still unsolved the problem of projection, Scheffler concluded his review. That this is but the same problem as the one allegedly solved only in a slightly different garb he did not say. Nor did he say that in Popper's view the claim that a given universal statement is a law of nature is a conjecture to be tested. I took it for granted that the review was sarcastic.[32]

I was profoundly surprised to meet Scheffler in person and find him a person too friendly, agreeable and gentle to publish any sarcasm. Upon inquiry I learned that he was a disciple of Goodman (he still is, I understand). I assumed that he was rather a simple-minded philosopher — by choice, not by a given limitation; indeed, he is a qualified rabbi and as such he surely is as capable of sophistication as anyone is, but in the fashionable style of the day he preferred to be a simpleton. It was only the perceptive Bertrand Russell who noted (*Portraits From Memory*) — in passing, perhaps only by allusion — that Ludwig Wittgenstein, then the

---

32 [Schwartz, Scheffler and Goodman, 1970] is one more patch on Goodman's theory of projectability. It clashes with testable hypotheses like "all young emerald are greenish and all old ones are whitish". Mercifully, the theory of projectability dropped out of sight.
  Obviously, my having misread Scheffler's essay as sarcastic is due to my shortcoming. O. K. Bouwsma's review of Wittgenstein's *The Blue Book* [Bouwsma, 1961] looked to me sarcastic too, in my ignorance of his membership in Wittgenstein's inner circle. Hearing this visibly upset him; more so it puzzled him: he failed to see what in his review could give me such a strange idea.

leading authority on (against) philosophy in England, was expressing no more than the yearning, characteristic of a whole irrationalist tradition and shared by Tolstoy and Pascal, for a lost childhood simplicity. Childlike simplicity, incidentally, does not lead to any specific philosophy: it is common to any philosophy held by anyone who had never met with universal doubt. This yearning was therefore expressed in parables — by Wittgenstein in parables about forms of life. This yearning, finally, is explicable by the claim that the likes of Pascal, Tolstoy, and Wittgenstein deemed universal doubt a difficulty utterly insurmountable, as well as a colossal catastrophe.

This should suffice as background for my amusing anecdote.

Scheffler was for a while a visitor in the philosopher's celebrated Tuesday Afternoon Seminar. There was nothing unusual about this or about his participation in general — until he once asked a conspicuously simple-minded question. Taking him, as I say, to be simple-minded by choice, I rushed to respond to his question in a brief, simple-minded manner. I was out of order; in that seminar, however, this would not matter overmuch: what counted there was not whether one was in order but whether one was contributing to the ongoing discussion. The philosopher winced: he saw in the brief answer I had darted at the questioner an insult to him, a tacit declaration that he was, indeed, a simpleton. In an incredibly sophisticated and rapid maneuver the philosopher erected a terrific sketch of a set of background assumptions that rendered the question most sophisticated and subtle. Though I knew the philosopher was in error in his attribution to his young guest the question as it had newly emerged, I found the question not without interest and I listened attentively to the philosopher as he unfolded a pattern of ideas with which to respond to it. He got fired with enthusiasm; he got up from his chair and delivered the rest of his response in an animated fashion, walking up and down in the narrow passage between the wall and the chairs around the collection of tables that nearly filled the seminar-room, gesticulating with his whole body. Towards the end of his speech he said, what should I do with it, kick it? and he gave a small kick with his foot of an imaginary football representing the imaginary idea he was discussing. After that he slowed down and landed safely on his chair with a certain, rather justifiable satisfaction at his own performance. To my horror Scheffler expressed dissatisfaction with the answer; what is more, he did so by repeating his clearly simple-minded question. This time nobody responded. The philosopher finally did: he gave a repeat performance. The repetition was exact, including gesture and all. When he kicked the imaginary football yet again, it was my turn to wince. Fortunately, in the animated situation no one noticed. Things got even more ghastly: Scheffler repeated his question yet again. The repeat performance was inevitable — yes, the kick was there too. Fortunately, by then gentle Scheffler gave up. At the tea in the faculty club after the seminar was over, I asked Scheffler if he had found my answer satisfactory, or at least to the point. He said yes.

(My apologies for the omission of the report of the substance of the discussion. All I remember of it is that it had to do with the demarcation of science from metaphysics and the need — what need, I still cannot tell — to ensure that a metaphysical statement does not gate-crash into the exclusive club of science by clinging on to the coattails of some respectable member of the club; the gate-crasher won no sympathy from the philosopher but he refused to kick it. It really is all neither here nor there, even though Ernst Mach had made the discussion of such trivia respectable by declaring in a high-handed fashion that absolute space is a gatecrasher illegitimately clinging to Newtonian mechanics. Mach was in error, since Newton needed absolute space, or at least a part of it, in order to account for the effect of the rotation of a bucket on the surface of its water. Nevertheless, Mach's discussion was redeemed by his interest in space. The modern discussion of stinginess with granting entrance to the club was snobbish without discussion of why all metaphysics should stay out of the club, let alone have a bouncer see to it, regardless of whether the bouncer should kick gate-crashers like footballs or merely show them to the door.)

I have a point to make in this story, and I request your indulgence as I am now going to spell it out. The time was really bad. The philosopher could hardly have been born at a time worse for taking his opponents seriously: they were not serious and they made a virtue of their callousness. At that time Gilbert Ryle, a leading representative of Oxford philosophy, of a persuasion akin to that of Goodman and Scheffler, launched a complaint about those colleagues who habitually displayed philosophical virtue by failing to comprehend while debating philosophy the very same locutions that outside the philosophical seminar-room they would take in their stride. He thought they were thereby exhibiting bad faith. Possibly so, yet in accord with a philosophical dictum that was received on the authority of Ludwig Wittgenstein: the vernacular, Wittgenstein said, was perfectly in order as it stood. God forbid that we should refrain from using in the same vernacular expressions that in philosophy Wittgenstein found so very objectionable![33] Wittgenstein's friend and foremost colleague and even disciple, G. E. Moore, was enraged by McTaggart's declaration that time is unreal, regardless of the fact that it is ordinarily repeated not only in ordinary Japanese but also in ordinary English. Wittgenstein himself exhibited that mode of conduct — the refusal to comprehend as a philosopher what one can perfectly well comprehend as a normal speaker of normal language [Munz, 2004, 14] Thus, when Russell criticized him for his prohibition on the use of existential statements he refused to admit that he understood any. Russell helped free Wittgenstein from a prisoners of war camp and met him in Belgium. They discussed Wittgenstein's manuscript. Russell objected to Wittgenstein's refusal to admit existential statements as they stand. Russell said, there are at least three entities in the universe. No good. He said, there exist at least three churches. No good. Russell took offence and gave him up [Russell, 1968, 100-1]; [McGuinness, 1988, 291].[34] This was their last debate. Russell's obituary on Wittgenstein in Mind [Russell, 1951], the leading British philosophical periodical, ends with the time of that incident, even though thirty years had elapsed since. "Of the development of his opinions after 1919 I cannot speak", he concluded. That is to say, after that exchange he could not take seriously anything Wittgenstein was engaged in (though he continued to help him in any way he could).

(In my early days in Jerusalem, when I pored over the writings of Carnap uncomprehendingly, I met a long, complex sentence there, written in a formal language, which, its author said, means, "there exist at least three churches". I do not remember the text, but I remember that it made no sense to me to replace such a short and simple sentence with a barely readable long one. And so I remember well the three churches. They too puzzled me: why churches of all objects? When I learned about the Russell-Wittgenstein exchange retold in the last paragraph, it became clear to me. They were sitting in a restaurant in a square full of churches, and

---

33 Wittgenstein scarcely referred to religion in his writings. He first said [Wittgenstein, 1922, 6.432], "God does not reveal Himself in the world." He repeatedly declared that the name of the divinity has no referent, namely, that God does not exist: his mysticism was atheist. Later on he forbade even the heuristic use of the assumption that God exists [Wittgenstein, 1953, §346]. His kind permission to use the vernacular, then, is a permission to use metaphors. This looks reasonable, but it may amount to giving up his philosophy, as it comprises the demand to attain "complete exactness" (§91) and "complete clarity" (§133), and not as a mere ideal: "we are not striving after an ideal" (§98). Vague sentences too he viewed thus: "there must be perfect order even in the vaguest sentence" (loc. cit.). Does this vindicate or forbid the use of metaphors? This question is undecided. Hence, he was unclear about essentials.

(The book that contains these pearls of wisdom caused a sensation when it appeared in 1953, so that I skipped a few lunches to purchase a copy of it — with no idea as to what it was about.)

34 According to young Wittgenstein, the meaning of "some x is A", is "a is A or b is A or ..." Thus, it is unclear what in 1919 Wittgenstein could admit. Yet, to make any sense his replacement of any existential statement with a long disjunction needed a complete list of objects. As Russell observed on a later occasion, the claim that any list is complete is impermissible in Wittgenstein's language, as he deemed all informative statements atomic or composite [Agassi, 1994b]. Wittgenstein's *Tractatus* is shoddy book. As Russell said ("Faith and the Mountains"), even sillier ideas win popularity.

Russell was contesting Wittgenstein's claim that the term "there exists" means nothing: it signals an abbreviation, the exact meaning of which is context-dependent.[35] Hence, clarification by the way of rendering a statement exact is not always felicitous, though by the empiricist or the intellectualist standards, of the inductive and the deductive mode of discourse, it always is[36]; by the critical standard, which is dialectic, what would have helped me comprehend Carnap at the time was the claim that he was engaged then in taking a side in a dispute. Carnap was reluctant to speak openly of a dispute, especially when he was taking sides against Wittgenstein, even when siding with Russell, since, you should know by now, he advocated the thesis that there is no room for genuine philosophical disputes.[37])

At the time, Oxford philosophy led the way, and its practitioners discussed deep questions such as, when is it legitimate to say of a conversation that it is interesting?[38] At that time, the attraction of Popper's se-

---

35 This is odd, since Wittgenstein could accept Russell's challenge and read his sentence as generally understood (in accord with his permission to use the vernacular). Evidently, his disagreement with Russell went deeper: it concerned the ideal language. As the ideal language died, he let this drop out of sight.
36 For a recent example see James Conant, "Two Conceptions of Die Überwindung der Metaphysik: Carnap and Early Wittgenstein", in [McCarthy and Stidd, 2001, 13-61]. As it mentions no problem, it makes no sense to outsiders. The problem is this: what modification of Wittgenstein's view does Russell's just critique of it require? This is unsayable: it still is a skeleton in their cupboard.
37 My "The Secret of Carnap" [Agassi, 1988] is about my inability to elicit from fans any explanation for their admiration for him. I suppose he won it when he rescued Wittgenstein's early view after accepting Russell's criticism of it. He should have said that his major work [Carnap, 1937] was a response to that criticism. But then, the very admission of the possibility of a philosophical dialogue would have been self-defeating, as the thesis that Carnap took from Wittgenstein is that philosophy is only clarifications, not genuine dialogue. He deemed disagreements between reasonable people quickly resolvable and often due to misunderstanding [Friedman, 2000, 148-9]. Ayer seems to have differed: he said that the "Vienna Circle" had entertained disagreements [Ayer, 1959, 6]. Yet he said this in support of his dismissal of Popper's claim that he was in fundamental disagreement with them, admitting a small sin as a means for denying a bigger one. All the participants in this sordid disagreement about disagreements, from Schlick to Stadler, avoided Popper's claim that Edmonds and Eidinow recap: Popper's main disagreement with Wittgenstein is on the possibility of philosophical ideas and fruitful debates about them. On this, they conclude, Popper has won.
    Nothing can stop the "Vienna Circle" apologists endorsing the censure of Popper for his having declared his disagreement with them fundamental. Ayer expressed this censure in a sentence, Cartwright in a book [Cartwright, 1996], [Agassi, 1998], and Stadler in two [Stadler, 2004 and 2005].
    The most famous of the open disputes between members of the "Vienna Circle" was between Carnap and Neurath on the status of observation reports. The more important one is Carnap's rejection of Wittgenstein's rejection of the meta-language of and strictly universal statements [Carnap, 1937, 101, 186]. He returned surreptitiously to the view of universal statements as not statements: he dismissed them when they refused to allow him to raise their probability in the light of confirming evidence [Carnap, 1950, 572].
38 This sneer of mine is a bit unfair, as in it I overlook here the constraint under which Wittgenstein labored: he banned the idea that different informative statements may have different degrees of significance [Wittgenstein, 1922, 5.556, 5.5561] (as Francis Bacon did, and for the same reason). Those who took Wittgenstein seriously sought to overcome this in search for the more interesting. In this quest they used his technique: analyze meanings by viewing them as uses, and choose common examples. This technique always works, said Gellner, if for no other reason than that it condones procrastination. Gilbert Ryle, one of the less dogmatic Wittgensteinians, said, we should not talk about philosophy but do philosophy. This secures indefinite procrastination.

riousness was unbelievable. He was always extremely busy, extremely in a hurry and extremely available to anyone who was seriously interested in anything serious. He was always in a rush. He always entered the underground train carriage that was nearest to the exit in the station he was going to use. He took taxis whenever it was quicker to move by taxi than by public transport. There was always someone waiting at his office door when he arrived. He was always late going into his lectures and late leaving them. He always collapsed, exhausted, when he reached home.

This sounds too good. Though it is true, it sounds misleadingly very efficient. It was not. Example. Once I accompanied the philosopher to a physician. We entered in a rush Edgeware Road tube station aiming at High Street Kensington station. We faced a choice between two lines, the District Line and the Circle Line. These lines, I am told, are remnants of an era when each line was owned by a different enterprise. Perhaps so; and perhaps due to this and to some odd conservatism, in the late fifties the two lines still entered and left the station on different platforms, a couple of hundred feet away from each other. The philosopher found it exceptionally irksome as it is so obviously silly. We rushed to catch the first train. We missed. We rushed to the other platform, to catch the train of the other line just arriving. We missed again. I did not think we were in such a hurry: when we entered the station the huge clock on the wall opposite us showed 10:40; he was due at the clinic at 11:00; the trip should have lasted 10 to 15 minutes; but then I did not know how far the clinic was from High Street Kensington station, nor how punctual the appointment was supposed to be. We rushed to and fro; as you can imagine, I kept quiet, fearing to add to the philosopher's aggravation. We caught a train at 11:00. We reached the clinic in good time. All's well that ends well.

My memory of these days was of excess excitement, moving around on the great thinker's coattails as I did. There were things missed, no doubt. These were the days of the decipherment of the genetic code, and I was hearing about the promise and threat of genetic engineering from friends in biology, but I do not remember any echo of it in the philosopher's circle, even though William Grey Walter, the leading physiologist and artificial intelligence researcher, came to the seminar to discuss the mind-body problem with Popper and later Sir Peter Medawar began collaborating with him to some extent. (We discussed the genetic code while we were in the USA in 1956-7. I met Medawar in person only years later, when he visited Israel.) Yet there were the great debates about the philosophy of physics and the history of the natural sciences and about cybernetics and artificial intelligence and about the need to protect the environment and about the social sciences in general and about the welfare state in particular. (As economists recognized, at least in the London

School of Economics, the requirement to legislate means for the protection of the environment was possibly a deathblow to neoclassical economics.) And there were other exciting matters. The philosopher was the center of activity of the British Society for the Philosophy of Science where there were about eight exciting meetings a year, and each of them included exciting commentaries from Popper. The calendar was overstuffed with meetings — of that society, of other societies, in different colleges on different matters that he was unable to skip, including the meetings of the Aristotelian Society, lectures on logic by Paul Bernays[39] and by Van Quine, both of whom got converted to his philosophy of science there and then, and more. And when it was important enough, I was instructed to attend even if the philosopher could not. There were the meetings and seminars in the London School of Economics, and there was the philosopher's own seminar. I was invited, in the last years of my stay at the School, to speak about his work to different seminars in different departments, as I have already told you in the beginning of the previous chapter of this record. All this was a clear expression of interest in Popper's ideas, though it was also the outcome of the difficulty of getting him to appear in person — both as he was so much in demand and because many people felt his presence was too disruptive. My activities were unbelievably intense, and in the philosopher's shadow.

It did not look that way then. On the contrary; the philosopher regularly complained about the refusal to recognize his work. The department of sociology in the School was in a semi-official state of war with the philosopher and the anthropologists disliked him sufficiently to fail their students who were innocently cooperating with me, you may remember. The philosopher's regular complaints that he was regularly ignored and plagiarized made a lot of sense then. It was amazing to see, in particular, that Paul Feyerabend plagiarized from him as soon as he had reached the learned press. The philosopher took me aside for an excited walk in the corridor — he had no patience to wait until we met at his home a day later — and he told me he expected everyone, his former students in particular, to plagiarize from him. (Did this include me? The forecast did not look positive, he admitted.)[40] Some of his complaints

---

39 I do not know what Bernays and Popper shared, and whether it relates to their shared appreciation of Leonard Nelson. When they met in London, Bernays tacitly argued against natural deduction (that he had debated with Gentzen at length) and Popper tried to convert him to it.
40 Popper finally considered me a plagiarist. This should not surprise me. One of his very earliest remarks to me was that copyright laws apply to lectures. It sounded odd to me, since publication was not exactly on my agenda. Later on, I learned that he was not as fussy about other people's priorities as he was about his own. He often evaded acknowledgement of ideas to others by creating improved versions of them and skipping mention the ancestors of his versions. Yet he saw himself a paragon of a scholar careful to acknowledge all his debts. He told me a few times that he deemed this or that casual remark of mine worthy of acknowledgement and he promised to grant me one. He said he had chiseled (this is the word he used) this promise on his mind. I dismissed this. (I still do.) Once, after he broke such a

looked then much more convincing to me than they look now (I leave the details to the next and last two chapters of this record). When Robert K. Merton reported that Thomas S. Kuhn was — still is — most popular in the philosophy of science next to Popper, I was a bit puzzled, as I was not tuned to the idea of him being popular at all. But then I remembered the story that Frau Freud realized how important her husband was when the Vienna post office was kept open on the weekend of his seventieth birthday to receive all those greetings telegrams. The peculiarity involved is not in the blood; it is in the Viennese air.

Of course, the two stories diverge: both in Britain and in the United States there were a few philosophy professors far more popular than Karl Popper, except that their death has drastically lowered their reputations. (An example is Hans Reichenbach, who dominated every session of every annual meeting of the post-war Pacific Division of the American Philosophical Association until his death; the meeting after his death, I was told, had no mention of him — no mention at all.[41]) Still, there is no denying: my view of Popper's position was so lopsided that Merton's comparison of Kuhn with him was strange to me, even though I knew that Popper's reputation had sky-rocketed in the years since I left London. For, what was strange was Merton's evidence: Kuhn is cited in the learned periodicals and books over the years half as often as Popper. (I do not know how good the citation index is as an indicator; I supported the suggestion to give it a try when Derek J. de Solla Price announced it in the Toronto annual history of science conference in 1964; later I opposed it systematically as too establishment-oriented and he protested that there is nothing wrong in that; now that it is no longer established, we should recognize it as a reasonable though crude initial indicator.) I do not know of a better way to tell you of how distorted or disoriented my view of Popper's place in the general scheme of things was than by telling you that when I checked his citation index I was incredulous. The references spread over the years and the decades, and most of them meant little or nothing to me. Yet the frequency and length of period of his being cited by people who had no contact with him is evidence that he is an authority of sorts (of the

---

promise, I reminded him of it. He denied it outright. Once I drew his attention to an insignificant formula that he soon made great fuss over, as it helped prove Carnap (and Bar-Hillel) inconsistent. I put the new formula to him in its most general manner possible to prevent him from generalizing it and then bragging about his generalization. So he bragged about having narrowed it down. I am afraid I chuckled then. This made him insist at length that this trite change of a trite formula is very significant. As he continued obsessively, I said, have it your way. This understandably angered him.

41 This story about Reichenbach may indicate no more than that his colleagues were tired of him for a while. I may very well be in error about him, and I should repeat that Russell, for one, expressed respect for him "both morally and intellectually" [Russell, 1968, 216]. This is not the whole story of Russell's view of Reichenbach, however: on another occasion he complained of his plagiarism (of the formal theory of relations). The remark just quoted may be an act of kindness: there is a hint that he did not consider him a serious philosopher [Schilpp, 1944, 683]. Yet the support for him of both Russell and Einstein (and not only for his excellent popular science) is very impressive one way or another.

underground sort, of course). I could not believe that at the time.

I have mentioned that the philosopher was always surrounded by eager listeners — especially in his intense seminar. The seminar was Popper's own: with no department and scarcely any graduate students of its own. Yet it was always well attended by graduates and by guests, not to mention John Wisdom and soon also John Watkins. They, after all, were the first disciples in London and members of his growing department. I will not mention the august guests, except for Gombrich who was a kind of a member of the family.[42] The philosopher's graduate students were there regularly, including A. I. Sabra, the present Harvard University historian of science and Arabist, who read there from his dissertation on seventeenth century optics, Bartley who read his on rationality and Jarvie, who read his on Malinowski's revolution in anthropology. Lakatos, you remember, should have been there as a graduate too, but having reached England when the philosopher was in the United States for a year, he registered at Cambridge. I told you already of his visits and valuable contribution to the seminar. Jarvie's paper was one of the most remarkable, though he was then still an undergraduate. I cannot adequately report on it, as it became his doctoral dissertation and then his lovely The Revolution in Anthropology and I easily confuse the diverse versions. What I remember distinctly is the pleasure of that seminar session.

I asked my friends regularly to come and see the miracle of that seminar for themselves. One of them was Shmuel Ettinger, then a mere young historian struggling with his immensely rich material. He responded very adversely to the tense atmosphere. This was my first indication of a criticism. Oh, there were complaints on the seminar to hear everywhere, but these were of unappreciative audiences or else good humored (I like particularly Wisdom's reversal of the traditional rule: "thou shalt not speak while I am interrupting!"). Ettinger had no doubt about the high value of Popper's outlook, even though he was then a fideist and flirting with chauvinist conceptions that he later repudiated in the Israeli media as politically and culturally most dangerous. And having had a thorough Talmudic background, Ettinger loved disputation with all his heart, in and out of seminars; he detected an atmosphere of disingenuousness there, however, on which more in the next chapter, as it has to do with the philosopher's craving for recognition.

No. I do not mean in the least that there is anything wrong with the

---

42 To repeat, Gombrich responded to this report by denial: he never visited Popper's seminar, he said. (This may be due to Popper's absence from the seminar on that occasion.) I reminded him of his lecture and my debt to it. It was on art experts. His case study was the success of the famous master forger Han van Meegeren who fooled experts repeatedly. He succeeded because his forgeries confirmed a widespread theory (of Caravaggio's influence on Vermeer). In retrospect it is easy to see through his work. Hence, experts can be uncritical and reliance on them is no guarantee, although overall their performance is best.

190   Chapter Six

craving for recognition, even with the insatiable craving for it. My counter-example to the censure of the craving for recognition is the case of Bertrand Russell, the teacher of us all.[43] His craving was probably stronger than Popper's, and, of course, it was much less reasonable, considering that he was a legend in his own time, recognized as one of the greatest lights ever. Yet, barring his nearest friends and relations, he did not impose of his craving on others,[44] certainly not in the form of the demand for justice and historical truth and all that, and he made genuine and constant effort to prevent it from clouding his judgment. I will return to this later. I now wish to speak of the master of a workshop at his anvil.

I have mentioned in passing a colleague who was under stress and who had committed suicide (before my arrival) and the philosopher's involvement in his troubles till the very end. I have also mentioned George Zollschan whom the philosopher had appointed his first research assistant in order to help him out financially. It is hard to imagine the sacrifice this took: the philosopher had to speak to the Director of the School and, having over-prepared himself, he misfired. Desperate, he went to Lionel Robbins, the School's uncrowned king and one of his earliest real fans,

---

43 "The post-Russellians are all propter-Russellians" [Wood, 1959, 257].
    Quine saw Carnap as Russell's heir. He said that on the strength of Carnap's *Aufbau* — despite its being an utter failure, as he reluctantly acknowledged [Moulinas, 1991]: that work continued Russell's efforts to apply logic to philosophy. This way Quine was unfair to Wittgenstein, whom he always dismissed; he called him "the prophet".
    The leading post-Russellian logicians are Gödel, Tarski, Church, and Quine. They have all improved on Russell's logic and extended it. The leading post-Russellian philosophers are Quine, Popper, and Bunge (whose significant study of logic is in a different tradition). Popper added his extensive use of the logical asymmetry between contradiction and coherence, with his deductivist methodology of error-elimination by refutations, and his expansion on Russell's idea of the value of democracy as negative. Bunge's exact philosophy, especially his *Treatise*, is nearest to the application of logic to philosophy that Russell was first to attempt [Russell, 1944, 20] and that Quine admired in Carnap. He has written more widely than any of them. He also repeatedly expressed dissent from them, and his criticism of Quine holist view of meaning shows clearly the superiority of his systemism (that is a restricted version of holism) over traditional, radical holism. It is nevertheless not clear to me if, how, and to what extent, Bunge differs from Popper. His main expression of dissent here is his claim that negative arguments are insufficient, that positive ones are also important. Since he is a fallibilist, it seems to me that there is no disagreement here. (Remember the slogan, "pessimism of the intellect, optimism of the will" that Antonio Gramsci ascribed to Romain Rolland.) Likewise, they obviously agreed on systemism [Popper, 1957, §7]. Their disagreement on the mind-body problem is unclear: both rejected immortality and reductionism. So it is hard to see where they differed. I am not clear about views of emergence: with a law of emergence, all theories of emergence are deductive explanations; without it, they are not explanatory. Where do Bunge and Popper stand on this I do not know and I failed to engage them on this [Agassi, 1990b]
44 Russell's balanced view of himself finds expression in his self-obituary [Russell, 1950, final chapter]. The running theme of Monk's biography of him is his constant need for approval, the deep cause of it, and its significance for his whole life [Monk, 2000, 337]. In Monk's view, behind these stood a fear of insanity that was his dominating feeling and made him admire Joseph Conrad who had faced it valiantly. Russell knew other people dominated by this fear, Monk notices, but he could not appreciate their handling it with less than total honesty. Although a Wittgenstein fan, Monk ignores some central facts. Despite his great fear of insanity and tremendous mystic yearning, Russell stuck to commonsense and rejected off-hand Wittgenstein's mysticism — perhaps because he considered all mystic philosophy facile and uncritical (in contradistinction to the mystic yearning that he expressed forcefully, e.g., in the end of his *Problems of Philosophy*).

and complained to him most bitterly. Robbins rang the Director in amazement, only to learn that the Director could not understand what the request was. I have heard the narrative in much more amusing detail from Robbins himself: being the philosopher's close friend, he befriended me too, and though Kurt Klappholz and I criticized him openly [Klappholz and Agassi, 1959], he was magnanimity personified. As I have told you, but for the appointment of Zollschan, and for his having left the School the year after my arrival, and for Feyerabend's refusal to replace him, my life story would have been different. But I am now discussing the philosopher's personal attention to the personal needs of the individuals with whom he was interacting. I have told you that before I became his assistant he attempted to secure me some grant. (It was Robbins who advised him on whom to approach in the City.)[45] I will later tell you of his attention to my education above and beyond philosophy.

The choice of details obviously colors and biases the story of any series of events, no matter how true each of the details may be — not to mention the fact that memory misleads: the import of an event is often remembered better than its details; these are narrated not because they are remembered, but because they play a remembered role in the narrative: this is a matter of interpretation, as tacitly presumed by the narrator; and then, willy-nilly, narrative alters as the result of a change of narrator's opinions (unless the narrator boasts total recall). (This theory of memory is due to J. J. Gibson and his followers. It holds also for collective memory, as discovered by Collingwood, the great philosopher-archaeologist, as he explained in his Autobiography why we should not trust our best reading of archaeological remains and so we should not destroy the sites: we should continue to keep them as records in order to allow future generations to rectify our interpretations of them.)[46]

It is the initial impact of the event that creates the memory, not the later reporting of it. People often remind me of events of which I have no memory, because the events taught them something but not me, regardless of the question of whether I concur with their lesson. (Often the remembered result is informative to one party about the other's characteristics, and then only the one, not the other, will remember it.) Similarly in cases where an event is remembered differently by those who drew different morals from it at the time (whether or not they agree about its im-

---

[45] Popper tried hard to secure me a grant, but he was always at a loss when it came to practical matters. He once took me to listen to a lecture of Lord Herbert Samuel in the hope of introducing me to him and asking him to help. I did not know then what a sacrifice it was for Popper to go to a lecture. And, of course, he had no opportunity to meet the speaker for, as usual, after the lecture he was surrounded by many who wanted to catch his eye. It was my gain, though, as Samuel ended the lecture with the gem that I cite here as the motto of this book.

[46] A detailed record, photographic or holographic, will not suffice, as when we make the records we check their fidelity and in checking we pay attention to what we deem significant.

port). Let me elaborate with an example.

I heard a few details of a correspondence between the philosopher and an American colleague who was then of some renown. The only detail I remember of it is that he had addressed the philosopher as "Sir Popper" rather than the customary "Sir Karl" (it was a short time after he was knighted, in the early sixties). I do not know why the philosopher found it useful to correct that famous colleague's mistake and I take it to have been nothing but a friendly act. In any case, he did correct his famous colleague; the famous colleague and friends and admirers, I was given to understand, took it amiss. Despite the elaboration on the story kindly offered to me, I failed to learn whether the correction was considered a violation — whether of good taste or of morality — or a sign of a defective character or neither. Though the matter is of the minutest significance, I confess I found it deserving of some meditation. This, obviously, is a reflection on my own self — my poor character or my violation of good taste or of ethics or I know not what.

The reason for my concern may indeed be in my peculiar character. In my adolescence I was extremely lonely and unusually socially unfit. When I remember how I behaved in my late adolescence and years later, I blush to this day, and my present awareness of the very low level of manners generally acceptable in my country to date does not help. Let me admit, however, that I am at least proud that I managed to improve my manners somewhat, and largely due to my selection of good teachers in these matters. Before I looked for teachers in the matter of manners, I did what most people in such circumstances do: I tried to struggle alone. For example, I learned table-manners from books and from watching British soldiers eating in public places. (Looking at people hard is at times embarrassing, and it is a custom I regrettably have not yet learned to shed.) Similarly, I tried to behave in ways that are socially approved, but with no success: there are things one can only learn with the help of a coach, and even a poor coach is then better than none. Fortunately for me I found good coaches. I found that a joke is more readily enjoyed if it is told in the name of a popular, good joke-teller. I found that the same holds and even more emphatically for proper conduct, for political judgment and even for philosophical judgment. (The same holds even in science: I once made a point that Einstein had made previously, you may remember, regrettably making no acknowledgement, and I was told by competent physicists that thereby I proved myself an incompetent physicist. But in science this is least harmful, since science is still the most critically-oriented practice.) In brief, the significant fact is that people are afraid to express an independent opinion; yes, even on a joke.

The result of the lack of independence is that the praise people lavish on their friends and associates is highly misleading: regrettably rather than befriend people they approve of, they approve of people who happen to be their friends and they are often defensive about their weaknesses rather than accept them despite their weaknesses. For one, like my adolescent self, who attempts to take expressions of approval seriously, there is bound to be a great disappointment. The same happens in the social setting of science. Doing the right thing is often very important for one's career. Often students and even researchers surmise what is the right move in science the way I had tried to surmise what is the right move socially; I have heard many heart-breaking stories of this kind. When I tell individuals who admire members of famous academic institutions that they will be very frustrated if they try to emulate their distinguished peers they are often both amazed and hurt. Perhaps this is why I was so sensitive to the myths of science[47], including the myth that research begins with data collections that Popper always combated, always eliciting the response that he was merely knocking down straw men. Myths are, indeed, straw men; but when one is knocked down by a strawman one is bound to feel hurt and humiliated.

It will thus not surprise you to learn that I have come very late to the opinion that the philosopher's manners were not a model to emulate. Ettinger was the first to give me the hint that way, even though, you remember, I had regularly heard people complain about his manners before and after. I knew that from time to time he needlessly hurt people; I even attempted to debate this with him; I had no doubt that he had very strange views of some people; in particular, he repeatedly let me understand that Ernest Gellner could do nothing right: even when Gellner attacked the Oxford school of philosophy as phony, which was very brave of him, Popper refused to praise him. (Only late in life he publicly defended Gellner, who was attacked for his criticism of the popular pronouncement that Heidegger's allegiance to Nazi ideology was a philosophically irrelevant temporary misjudgment.) Yet for years I defended the philosopher's conduct as honest — that it doubtless always was — and as educationally unavoidable. On this latter point I have changed my view most radically. At the time I dismissed the endorsement by most leading adherents of the analytic school of the philosopher's judgment as but an expression of their lack of autonomy, of their inability to decide, their inability to see the great value of his philosophy. Admittedly (as my own case indicates), it is even hard to notice that the philosopher's manners were questionable, but as it was the widespread opinion, it was easy to endorse; at least it is easier to endorse the impression that he was rude than that he is a leading thinker — at least as long as this was contrary to popular philosophical lore. What is contrary to philosophical lore was for many hard to enter-

---

[47] I came to Popper very ill prepared; his welcoming me nonetheless was an act of admirable generosity for which I am constantly most grateful.

tain. It seems therefore a fair exchange to endorse an impression of his manners as the cost of permission to dismiss his philosophy. This last sentence, however, amounts to the idealized image of philosophy, to saying that poorly behaved individuals can hardly be good philosophers. On other occasions the same people dismiss the misconduct of famous individuals — even the cooperation of Heidegger and his likes with the devil — as irrelevant to the value of their ideas. As I say, approval comes first and reasons are then often tailored to support it. (The claim that this is always so is Bacon's doctrine of prejudice or Festinger's theory of cognitive dissonance that is fortunately false: the critically-minded refute it, as Whewell discovered nearly two centuries ago. One can even be a great thinker and still a Nazi sympathizer, such as Konrad Lorenz. Heisenberg was not a party member but)

A young colleague once pleaded with me not to discuss philosophy with him, since he feared that boosting his evaluation of Popper's ideas would hurt his career: he was under pressure to mute his praise for Popper. The pressure, incidentally, came from Oxford, from a philosophy professor there who had denied in a debate with Popper over the radio that he (Popper) was more-or-less ostracized there (in Oxford). He still is, by the way, although now with a bit more courtesy, as one whose ideas Lakatos has overthrown. (Of the output of Lakatos Oxford acknowledges diatribes against Popper, not great contributions to the philosophy of mathematics.)

I once heard someone praise a questionable scholar as at least very provocative. I was glad to hear someone hold the (false but unusual) view that provocative presentation is a virtue. I then asked whether the person praised for being provocative was better at it than Popper. I elicited no reply, merely displeasure.

One of the great attractions for me in philosophy in general is what I read in a preface by Kant[48] that the judgment we have on an opinion should be independent of the question as to who holds it[49]. I was then too young to comprehend the book, but this I did comprehend. And I know the point is deemed trite and the repetition of it is therefore likely to be deemed a sermonette. Perhaps I simply like sermonettes, perhaps even sermons. Perhaps I was drawn to the philosopher's seminars and lectures because they were full of this kind of sermonette. I think this kind of sermonette is still very much needed. True, the philosopher's sermonette was resented by most of his colleagues as redundant and so as an insult to

---

48 Kant did not live by his own standards [Kaufmann, 1980, i, 159]; [Agassi, 1975, 464-8]. This is the gist of Heine's jokes about him.
49 Kant viewed objectivity as inter-personality. Popper concurred [*LdF*, §8]. Good criticism, he added [*LdF*, Preface to the seventh edition], my translation), is "free of personal attacks, and independent of distortions." To this a sharp *aperçu* he adds: "Misunderstandings are at times fruitful: clearing them may lead to interesting results". Deplorably, he misapplied this to former students.

their intelligence. (That is to say, they seem to suggest that lectures with too much redundancy are unintentionally intimations that their intended audiences are rather ignorant or unintelligent, and thus they are insults. There is truth in this observation, even though every lecture must include a lot of redundancy or else it will not be comprehended even by the initiate.) I do not think philosophers are generally sensitive enough to poor lectures and redundant sermons: I observe that important people from all walks of life deliver regularly and frequently sermons and homilies and exhortations of all sorts and ceremoniously deliver lectures whose messages are commonplace; moreover, these are generally extremely popular, in and out of the academy; what is generally disliked, I think, are sermons and homilies and exhortations whose thrust is the advocacy of individual independence. I still observe my peers violating the tritest points of etiquette, and almost always on points of intellectual independence — unless these are the violations I am more prone to observe. In line with this observation, I should also report that I find that wherever the philosopher was admired, his manners were uncritically considered proper — but this is another matter.

When I came to London, my manners were much improved, but still atrocious — especially since there they had to be judged by the local standards that were incommensurably higher than anything I had encountered in my own country where very few individuals I met exhibited reasonably good manners and where the professors seemed to me particularly ill-mannered in the classical pompous German style. Oh, there were exceptions: I knew some rabbis whose conduct approached that of saints; much as I admired them, I realized early in the day that the manners of gurus of all sorts befit only gurus: it is easy to be well-mannered when withdrawn from worldly affairs. There were others, especially my uncle Yehuda Reichmann who had obtained a bachelor's degree from Manchester. (I do not think you can appreciate the efforts it must have taken an adolescent orthodox East European Jew who had grown up in a desolate corner of the Ottoman Empire to go to England and study for a degree there.) From him I heard for the first time in my life that one can be religiously observant — as he was — out of one's own judgment rather than on the authority of a rabbi. And there was then the already mentioned Professor Fraenkel, the world renowned mathematician and the only teacher in the Hebrew University who had helped me. He too was religiously observant, autonomous, extremely gallant, and very kind. Unfortunately they were both dismissed as cranky — my uncle Yehuda mainly by other members of my family and Professor Fraenkel by other teachers of mathematics whose classes I attended. There was also my already mentioned philosophy professor, Shmuel Hugo Bergman, the old-style European. Unfortunately, I did not think he was a model, as he lacked the courage of his convictions. (In my paper on his ideas [Agassi, 1985b] I said that only his timidity or lack of tenacity is to blame for his not being viewed as the originator of the great theory of science as levels of approximation to the truth.) By the time I had learned to admire the gentle manners of my uncle Yehuda and of Professor Fraenkel, and to understand that they were maligned only because in their very manners they made their company feel that they were demanding a little more independence than usual, and in no way because of anything else, I was on my way out of the country. (Let me narrate one incident, though. My uncle saw me once standing at a bus stop. He took the occasion of my waiting to start a friendly conversation. As was usual with him — he was a teacher — he spiced his conversation with delicious scholarly anecdotes. As is usual with my manners, I corrected him on a minor point of information, though it was of no import. He thanked me. He did not hesitate; he did not flinch; he did not fuss; he did not explain; he thanked me and continued with his narrative.)

And so I arrived in London with very bad manners. I was willing to learn, though, except that in England it is bad manners to comment on people's bad manners — even when they request it: even when corrections are requested, it is not easy to offer them without hurt, and even when one is not hurt, one may wish to put some arguments in favor of one's mode of conduct, and that is embarrassing too — especially in a country where debate is allowed only in certain well specified circumstances, akin to debating club arrangements. For a foreigner like me this fact was very odd: I found odd the rule that a debate should not begin without an explicit expression of agreement to play made jointly by the opposing parties. I learned this rule only later (from Plato's early dialogues, that in later years I regularly used as books on rules, in accord with Ryle's admirable "Dialectic in the Academy" [Ryle, 1990]) and I confused the debating club sessions with the political free-for-all that is so common in my country. (In the Israeli Parliament some rules of debate are obviously more strictly observed than anywhere else in the country; yet even there debates often turn into abusive shouting matches; an Israeli president with English background once commented on this in public.)

I have come to mention the oddest of my peculiarities now, I am afraid. I have the reputation as an advocate of a peculiar style of debate: cruel, dangerous, one-up, urban-guerrilla-type. (I have a letter of complaint from my friend Abner Shimony testifying to that.) Some of my best friends tell me that for my own good I must improve my debating style and follow the rules a bit more closely. I need not mention the opinion on my debating style of those who do not particularly like me[50]; they have

---

50 For example, R. N. D. Martin, in *Philosophy of Science*, 50, 1983, 346.

many ways of rubbing it in and they (rightly) suppose that they need not spare my feelings. My own self-image is definitely the opposite. Oh, I am hot-headed and may lose my temper in a debate, but this hardly worries anyone; rather, the story is that I lose my temper only after calculating that this will harm my opponent more than myself. But at least here I am on safe ground: my reputation as bloodthirsty is somewhat ameliorated by some evidence to the contrary: no one as far as I know has ever ascribed to me any personal vendetta. Still, the fact is that I see the role of philosophy largely as the study of the rules of the game, and my role as a philosophy professor mainly as that of a coach in the art of debating (orally and more so in writing: an art — or rather a craft — that I have acquired from the philosopher, of course). Naturally, students are more prone to declare dogmatic the dialectic method of teaching than that of boring lectures — but I will not dwell on this here.

In London, the seminar I most regularly and ardently participated in was, of course, Popper's. On top of his terrible manners as the leader of the seminar, the philosopher openly treated me as teacher's pet. This made me more intolerable than I had ever been. I can see now, in retrospect, that I greatly enjoyed the role: for years I had lived with almost no approval from teachers or coaches or other senior individuals or peers, and now I had my fill; but I had no idea about this then, except on a rare moment when the philosopher protected me against criticism the way he would occasionally defend himself: unfairly. (We are all unfair from time to time, though, so that this observation is neither here nor there.)

Finally a complaint was launched. I do not know who by and how, but it was. It was no doubt a well-conceived and well-executed move: usually it was not easy to have the philosopher's attention for a sufficiently long time on matters not engaging him intellectually at the time. In any case, he took the matter to heart and decided to educate me. (At long last, I have arrived at my story.)

It was a long walk in a country lane that lasted a whole afternoon. It was a propitious afternoon: the sun shone all the time and the philosopher was sweet reasonableness personified: it occurs to me now that he was then wearing a parental robe, but then I had no idea. I was more puzzled by the surroundings than by the rationale for them. Usually we could walk for hours in his garden — up and down the unused grassy tennis court or among the trees — quite oblivious of the surroundings. But he evidently wanted to control the situation better and far from Lady Popper's observant eye, so he chose a country walk. We argued about manners. His view of manners is not mine; his image of English manners was amazingly incorrect and unfair for the true Anglophile that he was and for the naturalized Briton that he was and for the observant, wise social and political commentator that he doubtless was and for the avid student of the writings of Jane Austen and of Samuel Butler and of Bernard Shaw that he definitely was, not to mention Elizabeth Gaskell and George Eliot and Anthony Trollop and other terrific writers who were observers and commentators on matters social, and whose works he was thoroughly familiar with.[51] Anyway, all this made the discussion rather heavy going, and myself rather defensive, I am afraid. He had an idea: he cut the discussion short: I was hurting people and so, right or wrong, I had to change my manners. I could experiment, I could try whatever seemed right and see if it hurt others less. I agreed. This was a turning point in my life.

At the time of that story I was the philosopher's research assistant. I do not know what the role of a research assistant is and it is usually left to master and assistant to work things out for themselves. The only limiting factor, as far as I know, is the working hours. But it was understood from the start that the philosopher had no sense of time and that I would work as much or as little as I would find fit. I was, of course, very lazy: from time to time I even went to the movies.[52] In any case, as a matter of course, the least I could do was to visit his lectures and take notes. He was bursting with ideas; his lectures, though set to a very strict plan, were full of diversions, many of them quite original, and he would constantly complain that he forgot these original ideas. The lecture notes, incidentally, were of no use: at best they convened a good outline of the content of the lectures in accord with the initial plan, never a report on the casual remarks and observations and unplanned *aperçus*. Yet there I was, taking notes and participating. The students disliked me intensely: I was a graduate and my participation was raising the level of technicality that was hard for many of them anyway. And I was foreign, my English was bad, my ability to express myself was poor and my manners were what they were. I had asked the philosopher if I should not stay silent in class as my participation was annoying to its members. He said, no. And in our conversation on manners he said I should find a way to talk there without

---

51 Popper read these English classic writers repeatedly for relaxation. He said he knew them all by heart. He found reading new material too exciting. (He also found movies too exciting. He watched a couple of movies based on plays by Shaw and they stirred him to excess. The Poppers went only once to see a movie with the three of us. It was Michael Anderson's 1956 *Around the World in 80 Days* displayed in the original impressive wide screen of Michael Todd. Following Jules Verne's sweeping depiction of his heroes as stereotypes, Popper responded to the movie with regret that it was David Niven who portrayed Phileas Fogg, as he fits the stereotype of the Englishman less than Leslie Howard, whose Professor Higgins Popper greatly admired.) He rightly found reading the current philosophical literature too tedious. To repeat, he preferred to skim through texts and then, if need be, to test his conjectures on them and study very carefully what he deemed key passages. He tried to read them more closely if he decided to publish criticisms of them. And he read carefully philosophy books that he found important, including everything by Russell.

52 Popper objected less to movies than to wasting time on them: he viewed as waste of time any attention to art, even the best. He found playing music relaxing, so at times he allowed himself to sit at his grand piano for a few minutes.

causing pain. I tried.

It was a funny moment; a mixture of solemn self-examination and self-mockery. I wanted to begin my intervention like a true Englishman, always lost for words and hoing and humming before starting. I had no idea how to do it or what is the secret charm of this affectation if any. But I was resolute in my attempt. I raised my hand and was recognized. I stood up slowly. Very slowly. I waited. Everybody looked at me in surprise. I then tilted my head a bit and said, Eh. I waited. Then I said again, Eh. And again. And then I went on with my business as usual. It worked. The students liked it. They were not fooled for a moment, of course. I supposed then, and I suppose so now, that they knew exactly what I was up to. But they liked it all the same: they were pleased to see me make an effort to reduce my cocky and vain and haughty conduct.

If the image emerging from these pages is incoherent, do blame me, by all means. The facts do not allow me to present a coherent picture. The excitement of cooperating in a workshop constantly mingled with poor manners and with complaints — the lesser lights have learned to emulate the big light and we began accusing each other of insufficient attention, recognition, whatnot. Into this atmosphere Imre Lakatos entered like royalty: he was a true intellectual, enjoying a clever idea and a sharp debate most, and he was a first-rate master of intrigue: he was educated in a Stalinist society where one paid for an error with a prison term if not with life. He had paid with a prison term: he had spent four years[53] in jail, one of these in solitary confinement. He found mere relaxation in the intrigues he concocted, and the arm bending that he practiced, first in the philosopher's back-yard and then in the international philosophical community and beyond. He had great plans and he went on executing them relentlessly. There is no telling what he might have achieved had he not died so young.

Perhaps he was attracted to me because of sibling rivalry of sorts; perhaps he decided that he could easily steal my thoughts (except that they were no more my own than the Brooklyn Bridge; I simply used them as everyone else could); perhaps he was taken by the challenge of his inability to impress me with his intrigues or to put pressure on me, emotional or social. He followed me to Boston and he followed me to Israel and he used his ever growing power in attempts to constrain my steps, except that we moved in different territories. When he visited the territory I usually inhabit, I welcomed him and we had lovely discussions; he liked them and came for more when he could take leave of his urgent business in the territory where he spent increasingly longer time. It is hard to blame him. Who would have believed that he could throw a wedge be-

---

53 The Lakatos WebPages report that he had spent three year in jail and say nothing about solitary confinement. Perhaps he exaggerated to me, and perhaps he later preferred to conceal some information.

tween every pair of friends and colleagues in the philosopher's entourage and beyond? Who could believe that he could create the international Popper school in one swoop and make Bartley the Judas of the new movement and all that?[54] He managed to make Watkins say to me that I was the intrigue monger who had incited Bartley to rebel against the philosopher, and also that as long as I was friends with Bartley he would not be friends with me. Watkins has since regretted this remark, I know, but the memory lasts.

Yes, I do think it is a reflection on the philosopher — in part at least. It is not only an idiosyncrasy of his, though no one who knew him would deny that he was prone to fall for an intrigue any time. Rather, it is his view of the political role of leadership, even of politics as a whole, that is too naïve, indeed not sociological enough. This will be the closing discourse (Chapter 8) of this record; before that I should speak (in Chapter 7) more of intrigue: I wish to expand on the following view that I advocate: rather than condemn intrigues, it is better to increasingly restrain them, as can best be done indirectly — by promoting the open society.

P. S. There is no bigger offence to Popper than to employ the essentialist mode of thought when discussing anything even remotely related to him, yet this I do now.[55]

Essentialism is the idea that certain characteristics make a thing what we know it to be. It is a tremendously popular idea and even thinkers who oppose it are often enough prey to it. For a conspicuous example, logical positivists endorse a refined, seemingly positivist version of essentialism, known as the theory of explication (of Rudolf Carnap): like all essentialists, explicationists take it for granted that some known concepts, that are vague, invite close examination so as to render their meaning exact. Even Popper, whose attacks on essentialism are so very influential ever since Feyerabend has succeeded to stake a claim for them for Wittgenstein,[56] even he often falls prey to essentialism, as, for example, when

---

54 Lakatos called himself an "inquisitor" [Motterlini, 1999, 150] — half-jokingly as usual. Feyerabend called him *"Intrigant"* [Duerr, 1995, 155] and *"Parteisekretär"* [Baum, 1997, 36].

55 Robert Boyle said, users of water suppose that ice is essentially water while users of ice suppose the opposite. Essentialism is thus anthropomorphic. In its traditional, verbal version, it impregnates words with magic. As to abstract essences, Frege asserted their existence; Russell denied it; Popper left the question open, except for the identification of the essence of artifacts (including art and science) with the intention behind their production. This he fully admitted as commonsense.

56 The popular story that Wittgenstein opposed essentialism is largely due to a misreading of the famous review [Feyerabend, 1955] of [Wittgenstein, 1953] that includes a surmise: Wittgenstein advocated an instrumentalist view of language. This misreading is hilarious, since a huge gap lies between the essence of language and essences of sticks and stones. Wittgenstein's first book assumes the existence of the essence of language (the ideal language) [Wittgenstein, 1922, §5.532]. The demise of this assumption entails the instrumentalist view of language. Wittgenstein never discussed this (as he was reluctant to admit his errors and his loss of interest in logic). He was a methodological nominalist and a metaphysical essentialist (e.g., [Wittgenstein, 1953, §§ 89, 92, 97, 113, 116, 217, 239, 371, 547). Some of his fans seem to disagree about this matter, e.g., [Pitcher, 1964] versus [Robinson, 1998, 113-14], as they confuse metaphysics and method despite Popper's clarification of this point [*OS*, Ch.11].

he declares us all essentially tribalists and essentialists, so that day and night we must fight these evil tendencies in us. Essentialism is, in effect, a poor thinker's form of metaphysical engagement. The opposite of essentialism is, traditionally, nominalism, the theory that there are no essences of things, no substance, that what there is is what one sees, that the world is flat. This is positivist metaphysics, of course, a contradiction in terms, no less. It brings the positivists repeatedly to metaphysics: metaphysics, said (sexist) Immanuel Kant, is the mistress with whom we forever quarrel only to return to her (*Critique*, A850; B878; see also there A753; B781 and A756; B787).

A number of writers have suggested what Kurt Vonnegut has presented with good crass humor in his *Cat's Cradle*: the essence of the family is a special sense of family affiliation. We are ready to consider essential to the family too many inessential characteristics, he suggested, including blood ties. The essence of a family, however, is a sense — a strong sense of sharing something very important, of caring for it and thereby for each other, of working to further the cause of what we deem important. (This very idea is present in a more sober manner in the writings of Wilfred Bion, the student of group dynamics, who viewed the more stable and serious groups as task-oriented.)

In this sense, the philosopher and his close associates were, still are, a proper family. We are scattered. We are not all alive. We have other concerns. We have a traitor or two to the cause, and we do not know who they are. We have newcomers and we do not know all of them. But we share the concern and dedication. This makes us one family Vonnegut-style. All this sounds pompous. It is. How is the pomp to be removed and the essence of the concern retained? This is a big question.

For my part, I deem significant the support that family members need and may expect from each other. How should we offer it without regress into tribalism? I do not know; I do remember, however, the philosopher's caution early in the day about being a closed society: we should not worry too much, show too much concern for security against it: fear and suspicion weaken the open society.

I once came across a paper [Campbell & Hutchinson, 1978] on the interaction of Louis Agassiz with his disciples. I only glanced at it as I never appreciated him: he was much too pompous and conservative for my taste. It drew my attention anyway as it concerned his quarrels with his disciples. He quarreled with them hard. They had to leave and strike out on their own. The story somehow had a happy ending, though, and this appealed to me tremendously: it was at the time when I was striking out on my own and still hoping for a *rapprochement*.

I was very impressed with John Schlesinger's movie *Madame Sousatzka* (1988) for the same reason — perhaps due to the sensitivity of author Bernice Rubens for the problems created by pressures exerted to achieve and due to the inimitable touch of screenwriter Ruth Prawer Jhabvala, but evidently also due to my strong identification with Sousatzka's pupil, whom she trains to be a star performer, who performs publicly too early in defiance of her advice and pressure, fails predictably, and escapes back to her unqualified support. Oh, the story is much more complicated, and is a tear jerker anyway. I thought I would tell you of my reaction to it all the same.

To balance the picture, may I mention "Laurence Olivier: A Life", a TV mini-series on him (1982) that I saw once, with its hero as chief commentator and interviewee. He projected a terrific image, of a person as self-critical as responsible, as benevolent as humble, honest and inundated with a sense of proportion. The personality that he projected is, in brief, one that emerges from the pages of some of Popper's writings (I trust on this the testimony of a number of his readers). The literature [Davies, 1982], [Lewis, 1996], present the real person as different from the one projected in that series. This is greatly disappointing, of course, but also thought-provoking. It shows that there was a whole network, in the sense of a web of affiliation, of a club Simmel-style, advocating a new ethos, the ethos of the open society, one doubtless propelled best by Einstein, but expressed best by Shaw, say, in his famous letter to his biographer Archibald Henderson of June 30th, 1904, but elsewhere also more generally[57] and placed in a proper philosophy by Russell[58] and much more profoundly by Popper.[59]

This invites much commentary that I cannot venture in this record. Nor does it matter much here whether Shaw's self-portrait is correct; for all I know Russell portrays him differently and more correctly [Russell, 1956].[60] And my mention of Russell and Popper here in one breath, as philosophers best expressing the new non-justificationist philosophy,[61] as William Bartley has christened it, is due to the idea of Jagdish Hattianga-

---

57 [Shaw, 1965, 801]:
"...the way to get at the merit of a case is not to listen to the fool who imagines ... himself impartial, but to get it argued with reckless bias for and against. To understand a saint, you must hear the devil's advocate".

58 [Russell, 1944, 707]:
"Throughout these pages, I am endeavouring even more to explain what my opinions have been than to defend them; for I consider that some of them have value as hypotheses even if they are [false]."

[Russell, 1968, 223]:
"I regard it as mere humbug to pretend to lack of bias."

59 Beware of people who say they have no prejudice, said Popper in his classes, as they obviously have lots and lots of them. For, being complacent, they do not take precaution against them.

60 Russell said, Shaw's major asset was his fights against Victorian humbug [Russell, 1956, 77]. His greatest weakness, he added, was his power worship. The whole Fabian Society, Russell noted, the Webbs included, were "fundamentally undemocratic" (109).

61 [Wood, 1957, final page]: Russell "showed how an agnostic can be unafraid. While cynical scepticism is sterile, a passionate sceptic can live a life of courage and achievement."

di [Hattiangadi, 1985] and John Wettersten [Wettersten, 1985], of Popper's philosophy of science as an ironing out of the aberration in Russell's philosophy that is due to his inability to break away from his peculiar empiricist tradition that is alien to the spirit that he expressed in his magnificent "Free Man's Worship". Feyerabend has a point, then, when complaining that Popper is wrong to suggest that he was without antecedents, though, of course, Feyerabend intentionally overlooked the fact that the new faith in the open society raised certain problems that were too difficult even for Russell and that urgently demanded solutions, some of which Popper was the first to offer. This will be seen clearly when a comprehensive and intelligent study of the ethos is made — or even of a part of it, such as a comparison of the completion by Roman Jakobson of the theory of language of Bühler with that made by Popper.[62] So far, to the best of my knowledge, Popper seldom referred to the theory of his teacher, and his reference was not taken up by any leading thinker except Noam Chomsky, whose high-handed dismissal of Popper's completion renders his discussion trite.

P. P. S. One does not need a moral principle to oppose the doctrine of hard work: some sense of proportion will do. I am indebted for this, and very profoundly, to Ronnie Edwards, the economist. As I was sitting once in the London School of Economics Senior Common Room he joined me and offered me the privilege of a brief conversation. Look around, he said, and observe all these people here; they all share one chief characteristic: they are all very busy. And look at me, he added; I am a professor here, a full-time government employee and the president of my own company; and I have nothing better to do right now than to chat with you. Ever since then, I am never busy except when I have an appointment, or, at times, some urgent assignment. Whenever I am asked when I am free, I try to say, now, or, when this meeting is over. Most people are puzzled and ask how I manage to be so free. It is a state of mind, not a matter of agenda; and the state of mind is that of unseriousness.

Perhaps since art is playful, one might expect it to accommodate this quality – unseriousness – more often than science does. Alas, this is seldom the case. It was my fortune to bump into the great pianist György Sándor just when I needed to be told to beware of excess seriousness and excess investment of energy. He helped me more than he knew, in the very light touch with which he conveyed to me this great lesson. I find it very hard to resist the temptation to improve and rework, but I now do so when there is a distinct reason for it, any reason at all, but not otherwise.

---

62 This is neither to endorse nor to reject Popper's presentation of Bühler's theory: his intent was to offer an improved version of it. He did not mention any problem that either version came to solve. And, after all, he would be the first to admit that any division to levels is acceptable as they all are arbitrary although they can be more adequate or less adequate for the tasks for which they are designed to perform.

In this I am the very opposite of my master, as I have managed well enough to free myself of his pedantry.

When once the philosopher complained to me of how busy he was in the last stages of preparing his *The Logic of Scientific Discovery* for the press, I said, the history of humanity would not radically change if he submitted the book a few months later; he did not comment on this, but he was evidently greatly displeased. And rightly so, of course — my remark was no doubt rude and uncalled for and I regretted it at once.

Still, I was greatly puzzled. Why did he, the greatest philosopher of the sense of proportion ever, exercise this wonderful quality so rarely? I explain this puzzle by the hypothesis that his acceptance of the moral duty to work hard as the highest, his acceptance of the idea of moral duty as serious to the utmost, limited his exercise of his sense of proportion. The Talmud advocates such an attitude: behave as if the next act of yours will decide whether the balance of your conduct will then be in the black or in the red; nay, as if the balance of the whole world's conduct will.[63] Max Weber said, a scholar should write as if the balance of his scholarly worth depends on the correctness of his next conjecture [Velody et al., 1988], [Agassi, 1981, 436-43, 438] [Agassi, 1991]. Martin Buber's last word is the story of a Hasidic rabbi's reply to the question, what was his most important deed? He replied, unhesitatingly, what I am doing right now — meaning, in principle so [Schilpp and Friedman, 1967, 736]. I dissent. Who can live by such a high standard? Why should a sense of proportion apply everywhere except in matters moral?

One final coda. My reports about complaints as to the philosopher's isolation are not meant to offer a report on the matter of his current reputation — nor even on his reputation then. Reputation is a tricky thing anyway. Article "Popper, Sir Karl Raimund" in the popular *Chambers Biographical Dictionary*, 1984 edition, manages (despite its brevity) to be more correct and lucid than the professional reference books: it declares Popper's classic *The Logic of Scientific Discovery* "the greatest modern work in scientific methodology". So there.[64]

---

63 *Talmud Babli, Kiddushin*, 40a; Maimonides, *Codex*, Rules of Repentance, Chapter 3.
64 Steven Weinberg crowned Popper "the dean of modern philosophers of science" [Weinberg, 1992, 165]. Don A. Howard's "Einstein's Philosophy of Science" (*Stanford Encyclopaedia of Philosophy*, 2004) refers to Carnap repeatedly and ignores Popper, in disregard for Einstein's dismissal of Carnap [Schilpp, 1959-60, 491] and support of Popper [Popper, 1959, 461].

CHAPTER SEVEN: THE OPEN SOCIETY, ITS MENTALITY AND STYLE

The belief that science leads to wisdom — to an actually rational self-cognition and cognition of the world and God, and, by means of such cognition, to a life ... truly worth living ... this great belief, once the substitute for religious belief, has ... lost its force.
(Edmund Husserl)[1]

For two different reasons I see no way to escape from some straight philosophy at this juncture of this melancholy account. For one thing, if I am to offer more than tidbits of what my apprenticeship was, I should convey some of the excitement that some philosophy can still generate — and some of the difficulty to do so that was current at the time when I was an apprentice, at the time when the faith in higher values was promulgated by hypocrites and obscurantists, at the time when in the name of the faith in reason and in clarity all values were forsaken.

For the other thing, if I am to report the termination of my apprenticeship, insofar as it ever ended, I should report the progress of my own philosophy in any sense in which I have a distinct philosophy of my own. I do not speak of my contributions to the fields, whatever these are; I speak of my distinctive sense, not of any specific item that is or is not original in this or that sense. And I do not know what my distinctive philosophy is, if I have one in the first place.

The point of distinctiveness is baffling, not only on the individual level. There are distinctive schools of thought, and their distinctiveness is seldom a superficial matter of mere style, and often easier recognized than characterized. From the very start I took it that the distinctive style of Popper's school was the encouragement of dissent, of striking out on one's own — in matters of choice of interest and otherwise.

This is not only a matter of philosophy. Popper has characterized schools in general by dogmatism, by their sticking to a dogma, to a practice, or to any other characteristic, and then freezing it, or, much worse, changing it surreptitiously. Thus, Freud's disciples had to emulate him or pretend to.[2] Not all teachers want their students to emulate them. Roger Sessions was the leading *avant-garde* composer of his day. I met him seldom, but he left a strong impression on me. He said to me, when we discussed the matter at hand, he did not see the point; he found it beneath his

---

1 [Husserl, (1929) 1969, 5].
2 Freud wrote under his portrait by Robert Kastor, 1925, intended for inclusion in a book on the greats of the world (!), "There is no medicine against death, and against error no rule has been found."

dignity to try to bring a student of his to emulate him; he found no temptation in that and he failed to see how and why others were so tempted. I do not know if I ever wanted to tell you — or anyone else — about anything that I consider my credo, if I have such a thing, but if I have one, it is the one I just attributed to Sessions. He was by far not the first to hold this attitude; I found it in a number of thinkers before him, and I daresay I held it consciously ever since I left London, even though Hong Kong was the hardest place to practice it, but the simple prosaic presentation of it that made a great impression on me I heard from Sessions. Perhaps this links to his somewhat peculiar role, as a leading *avant-garde* artist, scholar and academic.

But I am speaking of the philosopher now. I discussed this matter with him in Hong Kong, long before I heard of Sessions, and I was not as articulate as Sessions, but I said it anyway. I do not remember what disagreement we were airing, and I remember I was wary of debating anything with him then because a few days earlier I had annoyed him so much that he called me a liar, and I did not wish to worsen matters. At his insistence, we discussed my wish to publish my disagreement with him. He was wondering about the wisdom of the project I was describing. I said then, we of the critical school have to stress diversity most and recognize as a merit any attempt at a criticism and make a song and dance about any criticism of each other that we manage to present to the world. He found this amusing and he sniggered. I do not think I ever saw him snigger except then. He said, yes, he would gladly recognize my merit. I was taken aback finding this response an insult (I still do), but I said nothing. I was in serious error. I should have responded, and even as forcefully as I could. I should have said, Karl, you are not true to yourself now and your conduct right now is plainly embarrassing. But I did not: I was his host and I found playing host to the great thinker a great privilege and above all I wanted no more discord.

That should do; I should begin with the heavy stuff.

Let me begin by making some very general observations about Popper's major works, apologizing to you for their superficiality in case you are already familiar with the material. This is not a philosophical text, allow me to repeat.

His first published book, his first vintage, is also his *magnum opus*, *Logik der Forschung* (= The Logic of Research) of 1935. The book appeared in Vienna, his hometown. The philosophically inclined public there was then fond of the "Vienna Circle" and showed interest in the mock-logical pseudo-scientific outlook known as logical positivism (which was a pretentious version of anti-religious propaganda presented as scientism allegedly buttressed by logic but with no shred of anything like logic). The politically inclined public there was busy with frantic preparation to welcome or to resist the Nazi takeover planned about that time (1935) and executed soon (1938). The book was lost between a hostile reception by the "Vienna Circle" and the Nazi takeover and the ensuing political turmoil, which had forced the whole of the "Vienna Circle" into exile, Jew and Gentile alike.

It has been recorded a few times that these exiles took it for granted that the Nazis hated them for their sponsorship of clarity. It is too extravagant to assume that the Nazi authorities noticed them one way or another. It is compliment enough that these spiteful hoodlums in authority disliked them as intellectuals. What can philosophers qua philosophers do against the forces of evil? Short-range or practicable interests can make them join existing organizations designed to fight evil, and long-range interests or ideals and aspirations can make them participate in the slow process of education for higher values. These are not distinctively philosophical. What philosophers qua philosophers can do against the forces of evil belongs to the middle-range, where ends and means interact powerfully, since the discovery of practical methods of implementation of aspirations may render them possible candidates for the task of some short-range or immediate goals. Unfortunately, this field is almost empty: social and political philosophers and scientists sorely neglect it.[3] So please allow me here one paragraph for the role philosophers qua philosophers can play as educators of the public on matters of values and aspirations.

Did the "logical" positivists educate the public for higher values? Did they preach higher values? My long study of their literature does not offer a judicious answer. I presume nevertheless that they preached logic and clarity in the hope that this is the best way to boost values, including social and political progress. This they did not say, much less discuss critically, though they should have, at the very least because they disagreed among themselves on what counts as progress.

Popper's *Logik der Forschung* was virtually unknown until its (ex-

---

3 [Pinder & Moore, 1980] concern the middle range of generality, not goals. Middle range political plans or forecasts or projections are scarce, and they rest on useless historicist gut-feelings. Economists have still not responded to Keynes' response: in the long run we are all dead. Yet long-range discount rates or economic plans or forecasts are of the middle range. Investments take place in accord with middle-range considerations, some good, some not. We may deem them exogenous, namely, out of the range of economic theory, yet strategic investment theories do exist (regarding investment banks, securities, and such, as well as economic class action and such). These consider factors in the middle-range. There is room here for discussions of attitudes of governments to these matters, beginning with the interests of politicians. These are short-range and too narrow. Conscientious politicians may care for long-range interests, but these should not be utopian or historicist. As interests of the next generation are getting increasingly urgent, they demand more study — also from the economic viewpoint, since changes of government economic policies influence markets most. Some economists hope that globalization will prevent the development and implementation of ever newer national economic policies, as globalization requires world peace and relatively open migration policies everywhere. Economists who support it hope that it will support the cause of peace; but they are more concerned with local markets as the mainstay of globalization. So they hope that globalization will constrain implementations of large-scale economic reforms. Regrettably, this is the full extent of the middle-range plans of many politicians.

tended) English translation appeared a quarter of a century later, in 1959, as *The Logic of Scientific Discovery*. (The title of the translation regrettably differs from that of the original.[4]) Subsequently, many similarly updated German-language editions have appeared as well. Meanwhile, after the defeat of the Nazi regime, the power worshipers, its initial supporters, had to consider it faulty. Doing so, they faced a choice between two powers, East and West; the professional philosophers among them split much the same way; they split into the Critical or Frankfurt School that was indulgent to the East despite their laudable severe criticism of Stalinism, and the Criticalist or Popperian school, which opted for the West. As to those indifferent to power, they were a small minority: to a large extent power dictates popularity.

This helps explain why popularity is so shifty: the moment a popular doctrine begins to wane, its former supporters will betray it fast and conceal their betrayal by vociferous expressions of timeless contempt for it (like the ex-spouse who wipes out all memory of any past agreeable moment in the now-lost union). Criticisms of popular doctrines that meet with success and deserve recognition as slayers of dragons, regularly receive instead sneers from official quarters at the formerly admired dragons as obvious paper tigers with no teeth ever. "No-one I know", leaders protest, "has ever held that view." The sneer comes to convey that successful criticism, being successful, proves that the doctrine it demolishes is minor, one that was never of any significance, certainly not one ever seriously entertained by those who now sneer at it; sneers, thus, often have the role of unspoken, outright deceit.

The prevalence of this attitude made the philosopher feel like a laborer cheated of his wages. His regrettable tendency to express contempt for opponents, even if by mere implication, assisted the official speakers for philosophy in treating him unjustly — except when they declared that he was one of their own. As the philosopher had something to offer to established power worshipers, he soon became quite popular. He thus became the most popular defender of science in England and the most popular defender of Western democracy on the Continent of Europe; this way his reputation grew despite the public pretense of the leaders of his profession that he did not count. The work of his that should have been most popular on the Continent, therefore, is his classic political work, *The*

---

4 Popper explained to me the change of the title of the book in translation, but he refused to discuss it. He said, he wanted to stress the realist aspect of his views. The original version presents (ingeniously) methodology as neutral to the dispute between realism and idealism. His later advocacy of scientific realism is very significant and not limited to discovery: it extends to all theories: we should take them literally as true or false — usually as putatively true, if not now then when they were young. (There are exceptions: classical metaphysical theories often display the constraints laid down in the light of the severely restricting demand for proof. Russell, the last of this line, explicitly presented his view as a possible reconstruction, and decidedly not as idealist metaphysics.) Tradition allowed viewing theories as true or as neither-true-nor-false (meaningless). Popper's innovation is thus far-reaching.

Open Society and Its Enemies of 1945 (translated into German by Paul Feyerabend, who at that time was still an acknowledged disciple of the philosopher). Yet the messages in that book, the call for individual autonomy and for democratic control of the authorities and for the injection of a hefty dose of ethics into politics, these messages could not make the book popular where philosophical discourse is still mostly collectivist. Rather, his *Logik der Forschung* is more popular with the German speaking philosophical public than his *The Open Society and Its Enemies*, and other of his writings beat that.

Contrary to Merton's myth of instant recognition, as long as the author of *Logik der Forschung* was obscure and lacking a power-base so-called, the book had no popular success. It could not make a dent in the commonwealth of learning either, let alone revolutionize the whole subfield of the philosophy of science. It had little chance anyway, flying in the face of the well-organized, well-publicized monopoly group in the same city as it did, a group that he refused to join, feeling as he did that accepting the conditions under which it would welcome him would violate his intellectual integrity.

The assumption that a power-base is important is well illustrated by the attitude of a leading German sociologist, Max Weber, who died soon after World War I. He penned two immensely popular academic sermons, "politics as a vocation" and "scholarship as a vocation"; the latter presents as self-evident the idea that scholarship is only for the very few because very few professorships were then available and scholars had to make a decent living as professors. How narrow. The very concept of a vocation or a calling was invented to distinguish between it and a profession or occupation; yet the leading sociologist (con)fused the two in his major works on the matter.[5] Nor was the distinction purely abstract: it had living examples: a rich traditional culture, now utterly extinct, then flourished just to the east of Germany and there scholarship was common as a vocation or calling yet rare as a profession or occupation: the Jewish Pale of Settlement promulgated generations of admirably dedicated scholars who put to shame the ideals of dedication to scholarship as Weber earnestly prescribed them in high words: the standard that he described as ideal was below the one practiced daily by Jewish scholars — and they were as proud of their poverty as Socrates of old, and most of them had no power-base at all, even those who reluctantly served some rabbinical function (at extremely low pay). By contrast, the modern commonwealth of learning, originally a charming amateur gentlemen's club, abandoned private scholarship (in the 1830s) as soon as their brand of learning had won income, recognition and prestige.

---

5 The same error occurs in Weber's other sermon: he ignored the novelty of the modern method of payment for politicians. In his indifference to democracy he spoke of bureaucracy, not of government.

The demand for a power-base is, perhaps, the most important obstacle to the growth of public understanding of *The Logic of Scientific Discovery*. For, the challenge to the reader that the book's technical difficulties pose fades in comparison to that of its central thesis. The public views the need for a careful demarcation of science in order to insure that we recognize the authority of a theory if and only if it is scientific. Popper demarcates science because he recommends the view of scientific research as autonomous. This way he challenges public opinion that accepts the right authority, questioning not the authority of science, but the scientific character of this or that theory. This is particularly transparent in the obscurantist derision of science for having undergone revolutions that constitute radical doctrinal changes. Thus it came to pass (as late as 1970) that Popper conferred on Thomas S. Kuhn, of all people, the title of his best critic, although Kuhn viewed as autonomous only the leadership of science; the "normal" scientists (most of them rank-and-file employees in science-based industries) he deemed intellectual journeymen. Kuhn solves the problem of demarcation of science: he admits to science whatever the normal scientists admit on the authority of the established scientific leaders of the day — those who mightily hold court in their academic power-bases. He demanded of non-scientists too submission to the authority of established science of the day: although they hardly comprehend current doctrine, they should endorse it, submissively and universally, until their fearless leaders issue a signal for change. This signaling he called a revolution. Under his impact, a problem-shift took place. The urgent question ceased to be, what is science? It is now, who is the rightful scientific leader? Kuhn did not ask that question: he took it as self-evident that the established leading university is the home of the leading scientists. According to his book on the quantum revolution in physics, however, transmission of leadership is through cooperation: induction into the élite research group brings about general respect and recognition by the ageless method of induction into the council of elders.

In his autobiography of 1974, Popper explained his reiterated declaration of his never having been an affiliate of the "Vienna Circle"; he did so by the use of a humble disclaimer. The "Vienna Circle" was a private seminar of Moritz Schlick, then the professor of philosophy [and of physics too] in the august University of Vienna. As it happens, Schlick never invited him to that private seminar; had Schlick invited him, he added, he would have gladly accepted the invitation. This is intriguing and revealing, yet it is puzzling: why should Popper have denied that he was invited? Why did he do that repeatedly and insistently? His denial accords too well for comfort with the establishment's view of itself. The puzzlement will disappear, however, upon the addition of some relevant information. Popper spoke of the "Vienna Circle" as no more than a private seminar, and for a good reason: there is a sharp contrast here between membership in a movement and in a seminar. The one relates to accord on a doctrine, and its end is the dissemination and implementation of that doctrine. The other is a matter of readiness to partake in critical debates. Thus, the discussion of his affiliation or its absence is related to the question of whether or to what extent his theory of science is a variant of (any of) the doctrine(s) advocated by (members of) the "Vienna Circle". Attendance in a seminar is a personal affair and its end is intellectual. So Popper's story reads thus: were Schlick a serious intellectual rather than a power broker, he would have invited Popper to his private seminar. And then, of course, the Cambridge Moral Sciences Club Incident (a.k.a. the Popper Poker, a.k.a. Wittgenstein's Poker) would have taken place a couple of decades earlier.

Schlick and his crew had managed — only God knows how — to convince the intellectual world that following up the revolution in logic they were making a revolution in philosophy (and in politics?). This may easily convey the impression that Popper meant something else when he said he had received no invitation to the private seminar: the impression conveyed might be the suggestion that it is regrettable that Schlick did not pass on to him the mantle (or the leopard skin). This impression finds its refutation in the solemn invitation to join, one that Carnap issued, with Schlick's full authorization, but on condition. That condition he found too compromising and so he did not accept the invitation.

Popper's judgment of the offer is too conscientious. He had the right to reject it, of course, but not the duty: he could speak in favor of Wittgenstein in a manner comfortable to both his conscience and the need that the "Circle" had to save face. He could say, for example, that Wittgenstein has priority concerning the attempt to apply Russell's achievements in logic to philosophy at large. No doubt, this would have sufficed for the "Vienna Circle" as its members were eager for him to join them. Otto Neurath, who was Schlick's successor as the leopard-skin chief although he was not a member of a university and had hardly any power base, referred to Popper as the official opposition of the "Vienna Circle"[6] (the title comes from a Polish review of *Logik der Forschung*), and in his famous post-war report on its members' whereabouts he mentions Popper too. (He put all his hostility aside as he tried to impress the world with a long list of distinguished refugees who had participated in his movement; this is not too clever, as the participation of one person of the stature of Kurt Gödel in the "Vienna Circle" should count more than extended lists; for, Neurath included Gödel in the list too, ignoring his open endorsement of Leibnizian metaphysics in preference to the hostility to metaphysics

---

6 [Popper, 1963, 269n44]; [Schilpp, 1974, 970]: "(Neurath used to call me 'the official opposition' of the Circle, although I was never so fortunate as to belong to it.)"

that was the trade-mark of the "Vienna Circle".) Had Popper presented himself as one of the "Vienna Circle", they would not have contested it. Hempel told me that Popper's refusal to appear as one of them is disgraceful. Hempel could never suggest that Popper should have pretended that he had been a member; his idea is simply that Popper should not have denied the ascription made by others, at least not as vehemently and repeatedly as he did. Hempel himself, incidentally, did just that: he emphatically denied that he had been a member, saying he was too young to be one. (And here considerations of invitations to Schlick's seminar or their absence suddenly lose their relevance.)

Please indulge me two paragraphs on the contents of *Logik der Forschung*, since it is such a philosophical landmark. It raises two questions, discusses the two traditional answers to each, and presents alternatives to them. The questions: One: David Hume's problem of induction: how does theoretical learning from experience take place? Two: Immanuel Kant's problem of the demarcation of science: what theory is scientific? The inductivist-empiricist answers: One: theoretical learning from experience occurs when experience backs a theory. Two: a theory is scientific when experience backs it. (No one could say exactly how experience backs a theory, and debates about this matter still rage.) Much more sophisticated are the conventionalist-instrumentalist answers: One: theoretical learning from experience occurs when theories fit existing experience. Two: theories are scientific when they fit given experiences. (No one can say exactly what this fit is, except that it is of a formula and of given cases that serve as examples for it, and the fit renders the formula a mere tool; each tool is in good shape, but the whole tool-kit may be wanting.) The new criticalist answers: One: theoretical learning from experience occurs when researchers manage to refute theories by new experiences. Two: theories are scientific when researchers can submit them to empirical tests that may lead to their refutations in case they are false. By the inductivist-empiricist view, science is the set of all presently empirically-backed theories plus all known empirical information; by the conventionalist-instrumentalist view, science is all extant mathematical formulas, of all known empirical information, and of diverse known ways of fitting some of the formulas with some of the information; by the criticalist view, science is all known refutable hypotheses and all known possible empirical refutations to them, potential and actual, true (whose truth refutes some hypotheses) and false (whose falsity corroborates some hypotheses).

What is dated in this exciting book is its sustained critique of "logical" positivism, so-called, namely, the conceit — now no longer advocated but then the official creed of the fashionable "Vienna Circle" — that philosophy is all metaphysics and metaphysics as a whole and every portion of it is condemned by the mere rules of grammar. The members of the "Vienna Circle" had little knowledge of the grammar described at the time by innovative linguists; they talked about finality in the natural sciences as if Einstein had never existed (Schlick, remember, was a physicist who received a fan letter from Einstein for his defense of realism [Howard, 1984][7]). Moreover, they made unbelievably extravagant assertions, particularly about logic and about mathematics — all in the effort to give their sentiments (especially their hostility to religion) the air of obvious and incontestable truths. They had to, since their principle was that only incontestable scientific truths count, since any sentence outside the domains of logic, mathematics and science, is of necessity (which necessity?) pseudo-scientific and pseudo-grammatical in one go.

(Their attack was not on pseudo-science proper; they showed no hostility to magic and sorcery and witchcraft, to astrology and alchemy and homeopathy, to chiromancy and phrenology and graphology, and not even to the social Darwinism and racism that they naturally abhorred; their daily bread was hostility to traditional, established theology and to other religious doctrines. In the early days of the official Soviet antireligious propaganda, that ran contrary to the traditional proposal of the Enlightenment Movement that Marx and many others had endorsed, to let religion alone, Neurath, the second leader of the "Vienna Circle", hoped that their arguments from logic against theology would prove useful to the Soviet authorities and gain their recognition and support. This is exonerated to some degree: at the time, the Catholic Church (as practically all other religious establishments) played a particularly nauseating role both politically and educationally. The place of religion — in philosophy, society and politics — still requires much discussion; the "Vienna Circle" policy of high-handed dismissal of all religious doctrine as muddles and of backing this by obscure references to modern logic only blocks intelligent discussions of this grave matter. Back to *Logik der Forschung*.)

I hope I am allowed the (admittedly slightly smug) observation, that the obvious textual points made here on Popper's classic book can be found almost nowhere[8] in the literature, other than Popper's writings and in mine; not even in books dedicated to Popper's ideas; not even in books on them written (singly or collectively) by self-styled disciples; and respect for his teachings is expressed only in the works of a handful of his closest disciples. It is on the strength of this observation that I count myself a foremost exponent and one of the very few genuinely active fol-

---

7 For the temporary realism of the early philosophy of Schlick and of Reichenbach, see Appendix below.
8 Anthony O'Hear rightly reports Popper's equation of learning from experience with refutations [O'Hear 1995, 2]. Hempel says [Hempel, 2000, 203], Popper "sees scientific progress in the transition to ever more highly corroborated and ever better testable theories" in crude, total disregard for Popper's praise for refutations of theories as theoretically informative. Believe it or not, Hans Jürgen Wendel discusses Popper's philosophy as concerned with cognitive validity [Stadler, 2003, 79-94].

lowers of Popper's teachings — always allowing that I may be judged a follower at all, given my comprehensive and severe criticisms (not to mention my having been ostracized by the Master himself). Still, it is interesting that occasionally some critics present the teachings more correctly than most disciples do, not being averse to presenting aspects of the master's view that they find objectionable: they are in error, but their critical attitude allows a better presentation than the defensiveness of so many disciples. The criticism in question is that Popper is a conventionalist in a new guise. The exhibit displayed by critics and concealed by servile disciples is the conventional element in science à la Popper, which item relates to the tentative nature of refutations: since any refuting evidence may itself undergo refutation, its endorsement is due to decision, and thus it is partly conventional. Now the exhibit is neither here nor there, as the conventional element in science is entrenched in a much wider territory: Popper establishes with ease the existence of a conventional element in science, boldly in 1935 but to everybody's satisfaction these days, on the conventionalist view of language: science uses language and language is partly conventional. The alternative, the naturalist view of language, is erroneous. (It was then respected on the authority of Russell and of Wittgenstein and by now only on that of Chomsky.) Yet, whereas the conventionalists propose defensively that the place of the conventional element of science is mainly in its theoretical part, Popper proposes critically that it is better (since less defensive) to place it mainly in its observational part, where it is open to risk. All this is incidental to the observed fact that disciples are all too often paralyzed by the fear of criticism.

(Did you notice that the previous paragraph twice poses claims on the strength of observations? Did you ask, is it consistent to make any definite claims, especially as empirically founded, while advocating non-justificationism (whether Popper's or Bartley's)? This is a good question. The answer to it is in the affirmative: consistency merely requires of non-justificationists that they should offer their standards of evidence tentatively. More generally, non-justificationism is consistent with the reasonable, tentative recognition and grading the value of some standards and of some authorities. This point, however, is problematic; it is central and deserves careful examination.)

My claim that the book's attack on "logical" positivism is dated is not in disagreement with its author's claim: in a provocative, well-publicized statement in his autobiography, he claims that the doctrine of this movement is dead and takes credit for its slaying. Nor can the book be abbreviated by the omission of the dated attack: it comes whole. And this is very fortunate, as there is some value to the dated attack: it was extremely brave of the philosopher to stand up against the mainstream, and in this respect his work of that period still commands respect and admiration. But the intellectual interest of the criticism is gone: unlike many errors in the history of ideas, this one never had any worth: it has left no sedimentation; with the sole exception of Russell's criticism of Wittgenstein's first book, no criticism of it has enduring interest. Moreover, brave as the philosopher's stand was, he was not entirely free of contamination by the very doctrine he was contesting. While rejecting "logical" positivism, he advocates a more traditional brand of positivism, the view that metaphysics and pseudo-science are identical; it is weaker than (= logically implied by) "logical" positivism, the latter being the conceit that the former is logically demonstrated. (The claim that a doctrine is demonstrated should not be contestable. After all, this is the whole point of the demand for demonstration.) His very prolonged critical engagement with "logical" positivism shows, by his own lights, that he appreciated it more than other contemporary doctrines. Otherwise, he would resent the failure of the "logical" positivists to appreciate him or his ideas or his criticism no more than the same failure in others. (He says this in the 1959 preface to his *Logic of Scientific Discovery*.) And had he not been busy combating the poor doctrine of "logical" positivism, he would have criticized traditional positivism to his own satisfaction earlier in his career: though much the superior view, traditional positivism too is myopic. (Why myopic? you ask. Well, because metaphysics is ubiquitous whereas science, whether genuine or fake, is typically Western; only in Western medieval culture was metaphysics pseudo-scientifically presented and possibly even there not systematically so. Moreover, though to date many metaphysicians deem their doctrines more scientific than those of natural science, not all of them do: some, perhaps most of them, do not pretend that their output is scientific, so that their doctrines are not pseudo-scientific.)

My biggest complaint regarding the philosopher's public conduct is that his withdrawal from traditional positivism (which is not systematic but fairly frequent) is not sufficiently open even if not quite surreptitious. In 1935 he said his intent was to abstain from all metaphysical controversy. Accordingly, he declared his realism a private affair [*LScD*, end of §27],[9] and he later became a terrific champion of realism, without having clearly revoked his earlier attitude; though in 1935 he clearly disapproved of employing of metaphysical ideas in scientific research (end of his

---

9 This view of realism as private is Kant's [Critique, A782; B810] and Mach's. Popper's *Logik der Forschung* follows Mach here in an ingenious though erroneous way that he regrettably played down. So does his *The Open Society and Its Enemies*, even as it rightly opposed campaigns against metaphysics and recommends commonsense metaphysics. He later criticized Mach and took side in the dispute between realism and idealism. The new view fell into place when Wisdom invited him to write an essay to *The British Journal for the Philosophy of Science* issue that (as editor) he devoted to Berkeley. Popper's brand of metaphysical realism says, scientific theories are true or false. (The idea that a scientific theory can be false is more radical than the idea that acid can be sweet.)

Logik der Forschung), he later explicitly appreciated it, but he could have revoked it more clearly. His admissions of changes of opinion seem to me sparse and forced, as if they were admissions of some guilt. And this despite his view that all change under the force of criticism is progress and that surreptitious change is among the worst intellectual crimes.

The clearest example is his attitude to Tarski's theory of truth. He admitted his error in having avoided discussing truth in his first masterpiece, and he (rightly) protested that he always was a realist in inclination but feared the ambiguous character of the realist ("correspondence") theory of truth on account of the ambiguities it contained prior to Tarski's treatment. This is a lamentable use of important truths as a cover-up. As to the inclination to realism, it is neither here nor there, as we all possess it (as he rightly stressed on quite a few occasions). As to his reason for his early attitude, it is clearly stated in his classic book and is an integral part of the book's general policy: in that book he stays aloof from all metaphysical controversy. The book notes that even though there is controversy over the nature of truth, the inadmissibility of contradictions is generally endorsed and should suffice for methodology. This is a lovely point that even the metaphysically inclined can appreciate, and Popper's suppression of it by telling a different story is regrettable: however problematic the concept of truth was (it still very much is), the most problematic aspect of it is the questionable idea that there can be a final (finite) true proposition that describes correctly the whole universe. (Russell says [Russell, 1917, 32], obviously this idea is false.) Popper's early theory sidesteps it elegantly and successfully, even though his later and considerably better theory (1956) reinstates it as a (metaphysical) guiding principle (regulative idea).

Rudolf Carnap, to return to my narrative, was decent and paid friendly attention to the new book despite its author's lack of lustre. Not so Moritz Schlick, the Vienna University professor who had a flair for public relations and whose star-studded private seminar won fame all over the philosophical world and beyond with the aid of a flashy label and by claiming a monopoly over scientific philosophy and by other ploys. Schlick had a different attitude to the new book, the philosopher told me: he called it masochist[10], yet, as the publisher had accepted it for publication, he added it to his own series.[11] Neurath, the next and last head of the "Vienna Circle" (after Schlick's tragic death from the bullet of a student), was a free-lancer soon to become a refugee; his power base was its philosophical periodical, *Erkenntnis*. He was more hostile, but with reason:

---

10 Behind most of the criticism of Popper's ideas stands not so much the dislike for refutation, as the contempt for error: Popper said, refuted errors is valuable, and that refuting them is progress, so that proposing good new explanations is always progress and refuting them does not call for contempt.
11 Malachi Hacohen mentions evidence that throws doubt on this story [Hacohen, 2000, 210-13].

he had a strange theory (a vision, rather; as a theory it was stillborn, and discussed of late only by myself) of meaning that conflicted with everybody's views. Yet he too was not straight: public-relations considerations motivated his attitude to disagreements: he emphasized those he had with Popper and played down those he had with Carnap [Agassi, 1987].

The existence of doctrinal disputes within the "Vienna Circle" led to the publication in *Erkenntnis* of three reviews of *Logik der Forschung* by three stars: Carnap, Reichenbach and Neurath [O'Hear, 2004, 1]. One was appreciative if qualified, and two were dismissive and frankly hostile all the way. All three used the occasion to blow their own horn. (Reichenbach used his review to square the theory of probability as frequency with the theory of the acceptability of a hypothesis as its probability; he said he took the occasion to do that there for want of an opportunity to have said it before. This was funny, as he had just published a thick book on probability. He simply had not thought of this earlier: the book he was dismissing prompted him to think of it, but could not make proper acknowledgement without softening his harsh dismissal [*LScD*, 260n].) Even on the wild presumption that their criticism is valid, their dismissal of this obviously remarkable book is a shame. It betrays their lack of disinterestedness and/or a sense of proportion that stifled their intellectual leadership. (Max Black, a life-long supporter of the doctrines of the "Vienna Circle", showed better sense: being in England, away from the local politicking of its leadership, he presented in the leading British philosophical journal, *Mind*, a review that was both favorable and critical [O'Hear, 2004, 1, 222-4].) Admittedly, the loss of responsibility by the intellectual leadership on a large scale was most marked in the post-World War II era, yet the present information concerns the good old days. The ability to force leaders to behave themselves supports the saying that people receive the leadership they deserve. This is particularly so in the international intellectual world, as cheap leadership in it is more often a symptom than in the case of national political leadership: it is a symptom of the leaders' assessment of the level of education of their intended audience, much like the relatively cheap propaganda of some governments which is clearly due to their assessments of the level of education of their own people as poor. The case of leadership will take up the next chapter of this record, since the philosopher's greatest fault is his refusal to view himself as a leader.

The publication of the book increased Popper's reputation somewhat, even though it was at the expense he won it as a member of the "Vienna Circle". His disclaimers were useless before he gained his great reputation and they lost most of theirs.

Soon after the storm in a teacup over the publication of the book (including Schlick's summons to his lair in order to give the audacious author a good talking-to), its members fled for their lives and the book left the public stage in silence. But the mock-verdict which the grand old man of Viennese philosophy had pronounced before his tragic death was carried by his crew to the four corners of the earth as they became refugees: they claimed repeatedly that Popper had exaggerated his differences with them (and, they added, out of personal ambition). This is senseless in the absence of a measure of intellectual similarity. Friendly Carnap repeated Schlick's mock-verdict to his dying day, thus transmitting it to his heirs and successors: Wolfgang Stegmüller has repeated it in his obituary on Carnap and elsewhere [Hintikka, 1975, LXV]. So did Hempel. [Jeffrey, 2000, 269]. I am afraid it still is a part of the lore of the flotsam and jetsam of the "Vienna Circle" [Ayer, 1959, 6]; [Stadler, 2004, 176, 448].

The book's all-round ill-luck contributed to its author's switch from the philosophy of the natural sciences to other fields of philosophical study and probably also to his exile to a remote corner (Christchurch, New Zealand; there was a minor opening in England, but he would not compete with another refugee). He published a few very important papers, one set of which, *The Poverty of Historicism*, appeared in the war years and reappeared in 1957 as a book [Passmore, 1975]. Historicism is the name he gave to the doctrine of historical inevitability or of destiny. His *The Open Society and Its Enemies* of 1945 concerns that doctrine too. The doctrine is now much less popular (though its influence is, alas, still strong and pernicious) but the book has not lost its freshness and great significance because it is an original, forceful, and passionate defense of democracy and because the doctrine of historical inevitability still is very important as its stamp is visible everywhere. Although the democracy that Popper defended in *The Open Society and Its Enemies* is not the pluralist participatory democracy of these days, nevertheless, the paucity of philosophy texts defending democracy (without stumbling into populism)[12] makes it a rare treasure. In addition, the book bursts to the brim with a tremendous array of contribution to a great variety of topics, mainly to theories that are significant for democracy, such as the idea that individual responsibility is not a part of individual autonomy and that democracy is not a matter of government by the people but of their democratic control; it also includes many ideas concerning different fields, from classical Greek philosophy and philology and the history of Greek mathematics to contemporary welfare economics and international relations, from social philosophy and political science to a dazzling analysis of Wittgenstein's major work.

(Oddly, this analysis led the logician Bogusław Wolniewicz to conclude that the object of that forceful analysis must be important, and,

---

12 Amusingly, in the TV series *The West Wing*, President Josiah "Jed" Bartlett refers to this difference as he overrides populist sentiment.

being isolated in post-war Poland, he attempted to reconstruct it [Wolniewicz, 1969]. The result was of some interest; regrettably it is now forgotten. I mention it because Wolniewicz got it right: severe criticism is a mark of appreciation, regardless of its author's intent.)

The Open Society and Its Enemies had a different fate altogether from *Logik der Forschung*, and a very odd one. Having enjoyed for over four decades the status of a political best-seller it has broken a record; it is also assiduously ignored by the (pseudo-)Marxist literature and it still regularly receives very hostile appraisals from colleagues, publicly and (much more so) by word of mouth — which often brings more notoriety and thus publicity. It is a scholarly eye-opener in ever so many respects (as its author was at home in many fields, including the classics and natural science); and, above all, it is profoundly humane, politically and philosophically. It has, you will see soon, changed the scene. Its major thrust, however, was political; it achieved its chief end that is directly political and concerns political public opinion, namely, the public discrediting of the theory of historical inevitability or destiny.

In this respect, the book is outdated, but only in this respect, and only to some degree, as the doctrine's overtones still prevail. Some established professors of political philosophy and political science in the United States concluded very early that on this issue Popper's campaign was going to make it, perhaps because it suits their positivist temperament. So they jumped the bandwagon and outdid him (so as not to have to refer to him too much) and declared the post-war period the era of "the end of all ideology". The celebrated snobbish "end-of-ideology" dispute was as poor as the very theory of the end of ideology, which, incidentally, tacitly assumes the belief in historical inevitability or destiny to be the only possible ideology (so that the end of one belief is the end of them all) and it tacitly reeks of the faith in historical inevitability (the faith in the historical inevitability of the end of the faith in historical inevitability) — as do all who take intellectual fashions more as fashionable intellectual matters than as matters of fashion among intellectuals. (It is a great compliment to our culture, though, that the culturally and intellectually fashionable are often in possession of cultural and intellectual value, even though it is unreasonable to expect excellence of them.) The debate on "the end of ideology" is obviously anti-intellectual, and of the pragmatist sort: former adherents to the doctrine of historical inevitability or destiny often view the ban on it as the harmful rejection of all theory, concern with theory, with human history, etc. The absurdity of this attitude acted like a boomerang: the last chapter of *The Open Society and Its Enemies*, called, "Has History any Meaning?", consequently received more exposure than the rest of the book; possibly it was the first open recognition of the book's significance in the United States — the country that can boast the world's leadership in intellectual fashions (including the fashion of anti-Americanism, if this fashion deserves to be called intellectual). And if it is true that the last chapter of *The Open Society and Its Enemies* is its author's most popular piece, then it is so by right (despite its regrettable positivist denial of genuine disagreements about the broad outline of history): of all of his output, this chapter is the nearest to a manifesto: history will have the meaning that we give to it, if we are human enough to try. The idea and sentiment is not new, I daresay, but its presentation is breathtaking in its freshness and monumental presentation.[13]

The book's first volume presents Plato as the first great (meaning profound, not necessarily influential) anti-democrat, and as one of the most influential thinkers ever, especially through his influence on educators through the ages. It was Plato who presented as historically inevitable the trend from aristocracy to democracy and then through anarchy to disorder (anarchy, lack of rule, is still taken to mean disorder) and thus to tyranny. But Plato did not fully accept the doctrine of historical inevitability: the almost inevitable decline can be prevented: a philosopher-king can restore perfection and arrest it. The discussion of Plato presented in the book is marvelous; it is profound and highly scholarly and yet accessible to the general reader and even in an exciting manner. In particular, it is the first rounded portrait of Plato in the whole vast literature on him. (Many a commentator has rhapsodized about the impossibility of achieving one, since Plato hid self-effacing behind his characters.)

The narrowness of such commentary as there is on the book's achievement is remarkable. (Only Russell said, Plato will never be the same again. Ryle, doyen of Oxford philosophy, wrote a favorable review [O'Hear, 2004, 1, 257-63] that did not impress me, especially since he had written earlier a glowing review of Heidegger's major output.) Popper's description of Plato's philosophy was declared a travesty and an insult, an accusation that Plato was a totalitarian guilty of having paved the way for the Nazi war crimes. (The book does contain a discussion of the Nazi war crimes and how to handle them after the war, should the Allies win, but with no inkling of the enormity of these crimes. I know of no printed discussion of the Shoah by Popper or any member of his close circle; what I, and more so Judith, wrote on the Holocaust deviate from his views.) His description of Plato's traditionalism and élitism must be

---

13 Popper mentioned some *bona fide* anti-historicist precursors, who wished to avoid both historicism and the view of history as "one damned thing after another". (His avoidance of the latter view won him the ascription of a "whiff of Hegelianism", quite wrongly in my view.) He mentioned the famous liberal historians H. L. A. Fisher and Lord Acton, ignoring Shaw, who charmingly dismissed Marx's historicism as "liberal fatalism" [Shaw, 1965, 814] and argued against it in his *Fabian Essays* (in the conclusions of his prefaces for the 1908 and the 1930 editions). The same goes for Russell. This is not to belittle Popper's arguments; Russell rightly if generously appreciated them and their novelty despite their resemblance to the terrific anti-historicist conclusion of his *History of Western Philosophy*.

familiar to all students of medieval philosophy and anyway so obvious as to be scarcely questionable (it won an explanation even in a feature film[14]). Yet some leading philosophical authorities repeatedly dismissed even this as lacking all textual evidence. Many commentators utterly ignorant of Greek repeatedly flatly denied the philosopher's proficiency in Greek (instead of admitting that they rely on other people's judgments). Ignorant camp followers still dismiss this magnificent kaleidoscopic book with flimsy excuses of this kind.[15]

No matter how obvious a thesis is, if sufficiently clever people put their minds to it and manage to make it questionable, then this should count as progress. But the more obvious a thesis, any thesis whatsoever, the harder it is to make it questionable, and the success of doing so is more significant. Only a little sense of proportion is required to make this point obvious — and if someone does make it questionable, I would be very happy to know how and to declare it very significant. In the meanwhile, let me observe, with few exceptions (see below), all those who detracted from this the best modern[16] presentation of Plato's philosophy, did not offer any explanation, or any indication, as to what in it exactly is objectionable or at least questionable. Were they to do so, then their detraction would be a significant contribution to an ongoing debate. As things stand, their detraction is high-handed, self-righteous, and impeding progress.

This is not to endorse the book's presentation of Plato's doctrine of historical inevitability or destiny to the full; I will soon explain. Also, its presentation of the politics of Socrates (as Richard Kraut has noted in an

---

14 Popper's view of Plato as anti-democrat raised vociferous hostility even though it is common and often deemed unproblematic. In evidence, here are some items.
   The 1949 movie, *Mr. Belvedere Goes to College* (of Elliot Nugent), presents a lecture in which the professor in a small college cites Plato's chief argument against democracy, his famous likening of wise rulers to a ship's captain who will not consult inexpert passengers.
   Russell, Crossman and Popper read Plato the way George Grote did [*Autobiography*, §24]. The same goes for Field (see Appendix). The same goes even for the historicist Toynbee. Russell dismissed Plato's politics in a few sentences [Russell, 1945, 95] that call him a liar [Russell, 1968, 249]; [Russell, 1950, 1-20]:
      "I disapprove of Plato because he wanted to prohibit all music except Rule Britannia and The British Grenadiers. Moreover, he invented the Pecksniffian style of the Times leading articles."
15 The flimsy excuse for the dismissal of Popper's reading of Plato is that his translations from the Greek are inaccurate. This hurt him deeply. He repeatedly referred to the authoritative judgment of Oxford classicist Richard Robinson [Robinson, 1951]. He agreed with everyone involved that final judgment here depends chiefly on the decision as to whether Popper's translations are less partial than the received ones. And he sided with Popper's translations, although, he added, they are not always the very best available. High and mighty Richard Kraut nevertheless declared Ronald Levinson's *In Defense of Plato* "the fullest defense" of Plato against Popper [Kraut, 1992, 489] – in insulting oversight of Popper's devastating response to Levinson [*OS*, i, Appendix].
16 The Mediaeval readings of Plato as well as of Aristotle, from Al Farabi on, were very far from the modern liberal ones. Both groups agree with these texts as they read them. The mark of the new, *avant-garde*, mainly twentieth-century, style is that of allowing for criticism and even censure of thinkers while expressing sincere admiration for them. Already Grote showed this trait. Popper is the first to ground it in a commonsense (fallibilist) philosophy and employ it systematically.

otherwise bizarre collection of essays allegedly on Popper and the social sciences [Currie and Musgrave, 1985]) is somewhat idealized — in the direction opposite to Plato's idealization. That Plato's Socrates is misleadingly idealized as an anti-democrat is no reason to make him more of a democrat than the reading of Plato warrants; moreover, the exaggeration makes one miss the enormous democratizing influence that Plato's early works had on Machiavelli (of the *Discourses*), Hobbes and Spinoza[17], and thus on Enlightenment political philosophy in general (as expressed by David Hume[18] and by Benjamin Constant, for example). They all shared the view that to some extent all regimes are conventional — meaning governed through a consensus of sorts — so that they are all democratic in that sense and to that extent. This oversight is strange, as Popper endorses the Enlightenment political doctrine. He conjectured that Lycophron said that all regimes rest on some consensus and so that they all exhibit some democratic characteristics. I disagree, but my want of Greek prevents me from discussing this point. Still, it seems to me obvious that his conjecture that Lycophron advocated this view lends plausibility to the conjecture that Socrates, too, advocated it [Kraut, 1985, 1992]. Since this view stresses a factor common to all regimes, the ascription of it to Socrates would facilitate the understanding of the shift in Plato's political principles, from the ones he must have held in his youth, that were not hostile to democracy, to the ones he advocated in his maturity. Still, the first portrait of Plato ever and the fresh interpretations of Plato's theories, especially his political theories that were given to the learned world in *The Open Society and Its Enemies*, are still the very best around and they still go with insufficient acknowledgement.

The literature on that book is generally of very little value, and most of it is painfully unscholarly. The following are the exceptions I know of: one rather grudgingly favorable review of the first volume by Richard Robinson; a book by Ronald Bartlett Levinson, one of the few books on it, all of them very hostile (and with this one exception all are by now justly forgotten); one neglected, interesting, and appreciative-and-critical review by Alfred Cobban of London [Cobban, 1954]; and a few

---

17 Spinoza understood Machiavelli to have supported the political use of greed and lust as the most reliable motives while discouraging the trust in princes (*Psalms* 146:3). He interpreted this as a liberal and as democratic. He recommended checks and balances (*Political Treatise*, 8:20) and legislation with the intent to remunerate lawful conduct.
18 Hume said ("On The First Principles of Government", opening),
"Nothing appears more surprising to those who consider human affairs with a philosophical eye, than the easiness with which the many are governed by the few; and the implicit submission, with which men resign their own sentiments and passions to those of their rulers. When we inquire by what means this wonder is effected, we shall find, that, as force is always on the side of the governed, the governors have nothing to support them but opinion."
Similarly, Rousseau said (*The Social Contract*, Bk. I, Ch. 3),
"The strongest is never strong enough to be always the master unless he transfers his strength into right, and obedience into duty."

interesting minor comments from some Indian socialists — all of them on the second volume and none ever commented on in the West as far as I know. Having read various reviews in the learned press while a graduate student, I noticed their poverty. The one in the *British Journal of Sociology* is from the pen of the famous Oxford historian of political thought John Plamenatz. (Sir Isaiah Berlin's essay on him is a warm and friendly and moving obituary cum portrait [Berlin, 1980, 116-23].) His review of the book exhibited ignorance of it. (The book is full of italicized passages, and, contrary to tradition, some of them express important opinions that the author expressed appreciation for before attacking them. This misled the reviewer seriously, I suppose, because he took for granted the traditional uncritical opinion, forcefully expressed by Sir Francis Bacon, "it is hardly possible at once to admire an author and to go beyond him" [Bacon, 1620, Introduction to The Great Instauration], so that the appreciation of a view amounts to its endorsement.) Even to the best of us may make a faux pas and publish a sub-standard essay. The less said of such cases the better, especially when they involve worthy people. But just such cases may cause pain that would be all the more strongly felt. Gombrich came to the rescue. The exchange [Bambrough, 1967] is not enjoyable to read. Plamenatz himself was personally not half as hostile as were others who used his authority; his review thus looks suspiciously like a put-up job. Let me just speak of one other instance [Grant, 1954], a very hostile review (in a respected learned periodical) that astonished me: toward the end of the review its author said he had got the drift of the book only in reading its last chapter, Chapter Twenty-five that, he says, he quite liked, since there its author finally comes to the point and discusses "the dualism of facts and decisions". My puzzlement is due to the fact that Chapter Five of the book is on that matter. These examples should do: there is no longer any point in correcting the follies of respected intellectual leaders of yesteryear: now that the book has proven a hardy perennial (as it won an award of the American Political Science Association as the best work in political philosophy still relevant 25 years after its first publication), the damage they have incurred is a part of our past. Rather than correct old forgotten errors, we better block their present-day repetitions by instituting some democratic safeguard and thereby alter the tradition that allows the use of silly errors by the intellectual leadership as excuses for postponing the recognition of new important studies, be their conclusions true or false, as this recognition would facilitate public access to them and encourage public debate as to their value. The way to this reform should be the demand that reviewers should distinguish between two questions: is a book important? and, is its message true? Moreover, a reviewer who judges a book unimportant should explain why it merits a review nonetheless.

I was foolishly pleased to credit the Subject Index to the great work that I had compiled with the later reduction in the number of unacceptable errors in discussions of its contents. I have evidence that I was in error: the reduction of the unacceptable errors was only temporary. The logic of the situation of the rate of errors is simple: when the reputation of a work increases, the public may expect it to receive better treatment for a while — until it becomes the fashion to discuss it and then everybody chips in. The level of discussion thus is first low, unless the book is by an established author; it then rises when it wins general recognition, and it then falls to the general level of competence. If the book itself remains on the shelf for a longer period, then the level of commentaries on it may rise again as commentators on it may be people who have learned about it when they were still in school, before they lost the habit of reading; but even then, the book may become a classic and then the standard of discussion of it will soon again fall to reach the average level of competence. Incidentally, in professions where literacy is generally expected even after having left school, the process is a little different and the time lag between the publication of an idea and its appreciation is shorter. (See my *Science and Society*, chapters on cultural lag in science.)

Such is the way of learned commentary, and that on Popper's work is no exception. One should not expect too much, he used to say. But he did. He said so to me many a time and he wrote in his autobiography that he was the happiest philosopher he had known. Since he knew so few, this remark may be not much of an exaggeration. Still, it does look like one.

Once he showed a kind interest in whether I was happy. It was in my car as I was driving him to Brandeis University where he was a visiting professor soon after his retirement from London. Sitting in a car driven on an open freeway for some time, he was relatively relaxed. Was I happy? he asked me in a friendly way. As Solon said to Croesus[19], one cannot say yes till one's dying day, since Lady Luck is so fickle; but barring that, I added, the answer is very much in the affirmative: I had been ambitious in my adolescence and the sheer distance between prospects and expectations made me break down; I had recovered only after I had relinquished all ambition other than the normal and the average; after graduation I was accordingly ready to become a humble high-school teacher in Israel. I had, thanks largely to his help, achieved more than I had dared hope for in my wildest adolescent dreams. I fell silent. Just then, we approached the turnpike, slowed down considerably and then stopped behind a long line of cars waiting to go through the barrier. He said, that is very nice, that is very nice; yet his whole body tensed and his

---

19 Herodotus, *Histories*, 1.29-.33

hands moved with excess nervous energy. But do you not have people misinterpret your writings? he painfully[20] asked. I chuckled: of course, I said; I can anticipate the ways they will do so and it is their liberty. Very nice; very nice.

Just so you have some idea of how little the run-of-the-mill average erroneous responses to the philosopher's works failed to disturb his happiness, I shall now speak of Levinson's *In Defense of Plato* — the one book against *The Open Society and Its Enemies* that I found an exception and even liked. Its author used a heavy scholarly apparatus, organized it very clumsily and filled it with lots of complicated, barely relevant scholarly material; what is worse, he attacked the integrity of anyone who dared doubt that Plato was a liberal; yet he argued his case so very passionately that it read surprisingly well (and went to a reprint [Levinson, 1953, 1970, 1987]) despite its obvious faults, so its hostility was easier to take than usual. (Walter Kaufmann surprised me by making some of these points [Kaufmann, 1955] while noticing that Levinson conceded to Popper and hid his concession behind a verbal torrent and while complementing Popper enormously though very briefly.) I told the philosopher on some occasion that I liked the book. He was stunned; perhaps he heard me say that I agreed with the author, or he (unbelievably) confused my expression of appreciation with an expression of agreement, or he oddly thought that I was victim to this kind of confusion; whatever it was, his reaction upset me somewhat. So I insisted on conveying at unnecessarily great length that I liked the book but did not endorse its views. Indeed, rightly or not, I thought it was singularly in error. I could not judge, for want of any Greek, but, for better or worse, that was my impression. The philosopher always suspected foul play when his impression was corrected, you may remember. But now, perhaps due to my excessive emphasis, he was ready to change his mind: so write a paper against it, he proposed. I suppose it was a test, as it was in the days of my graduate studies, when I devoted to the writing of my dissertation every spare moment I had after discharging my duties as his research assistant. He once persuaded me to write a letter to The Times of London after a terrible

---

20 Popper viewed optimism as the proper approach not as a personal report. His mood was constant melancholy, relieved only by hard work. See the odd 1986 "Postscript" to his autobiography. That he was not recognized is no sufficient reason for his feeling of hopelessness: Faraday was in a similar situation and never lost hope. Faraday too was depressive, and even to the point of mental paralysis. But to the extent that his environment was to blame, it was not the disregard for him but the violation of etiquette that this disregard was and his inability to recognize that researchers violate etiquette regularly. As to Popper, it was less the disregard for him and more his inability to communicate with his peers that refuted his view of communication. He said, good intentions overcome all communication barriers, and he said, his colleagues were very nice and intelligent. ("I see Victor Kraft, and also Schlick, Carnap and Feigl as philosophers of outstanding achievement" [Schilpp, 1974, 976].) Hence, he should have seen this as a refutation of his view on communication. He could not. He strangely viewed the intentions of his peers as good and those of his students selfish and mean. Perhaps he changed his mind about this: there is evidence that he did, but from an obviously garbled-up record [Popper, 1997, 54].

railway accident, comparing standards of safety on the British Railways and the Swiss. He evidently felt very strongly about it, so I did it — to be greatly relieved when my letter was rejected. But to write about *In Defense of Plato* would have really been way out. I said, Karl, I have no Greek, no Greek at all! He said, this is no trouble: I will teach you Greek in no time. He had helped his girl friend, later his wife, as well as other of his youthful companions, to pass their high-school matriculation Greek exam, he explained. He knew what he was saying, I know; but he deliberately ignored his daily time-budget — yes, you are right, he had 24 hours a day like everyone else — and he was simply stuffing his agenda with one more absurd item. I said, all right, when you think I am ready I will. I forgot the incident pronto as I forgot many other similar ones. I was reminded of it with a jolt when I read in Hong Kong in the latest edition of *The Open Society and Its Enemies* a couple of appendices, one against that book and the other on truth, I do not know apropos of what, but Bartley took it to be against him, even though it contains an acknowledgement to him. I found both new appendices unpleasant to read on account of the hostility of their polemics. I was distressed. I remembered that the philosopher had confessed to me he was sorry that the style of his attack on the German philosopher Hegel in the second volume of *The Open Society and Its Enemies* was so acrimonious: it was written in the typically German style and had he written it in England he would have written it the English way: he sincerely admired the cool English style of polemics. And here was a new appendix, written many years later, in as German a style as they come: the acrid tone of plaintiffs who take each statement they disagree with as profound personal injustice to them. The worst of it is, of course, that the idea that criticism is a friendly act does not square with this passion to get a strong hold of the opposite party and run them down to the ground and put one's foot on their chest and let go with a tremendous yell of victory and momentarily feel great. I was not surprised, as Popper's polemic style had been similarly unreasonably hostile in his attack on the authors of the standard text on the Presocratics in his "Back to the Presocratics"; but there it was merely the wrong color. The appendix on Levinson was of necessity poorer than his "Back to the Presocratics": it was not embedded in as exciting a context, as it was an appendix devoted to mere responses to a book not nearly as exiting as the standard text on the Presocratic speculations. Also, I felt, perhaps quite mistakenly, its polemic style was more acrid and hostile, presumably because Popper had an amply justifiable grudge against Levinson but not against the authors of the standard text on the Presocratics. Consequently, his appendix did not even present the book or the book's thesis. He did express there appreciation of both Levinson's liberal creed and of his understandable hurt at the unmasking of Plato as anti-liberal; he complained repeatedly, though quite correctly, of much misreading and injustice; he picked on far too many errors, while hardly bothering to attend to relevance and proportion. In sum, his rebuttal was not very happy, or so I felt at the time. (I may very well have over-reacted, of course, especially since the appendix includes lavish praise of Levinson's liberalism.) Regrettably and unusually it fell short of the Popper's standard of lucidity: it was over-detailed, with no overview of the work it discussed; yes, almost like the present discourse. Except that this one is meant to amuse you, not to complain about injustice; the complaint comes next.

In Hong Kong, the philosopher asked me specifically for my opinion of his appendix on Levinson. I could have said that I endorsed all of its points, which I did, but I dismissed this option as cowardly. I had the option to dodge the question, or to postpone answering it. Instead, I was gauche, and said bluntly that I did not like it. Invited to explain, I said the style of his rebuttal was too hostile. He had argued that all translation is interpretation and that Richard Robinson, the celebrated Oxford classicist, [grudgingly] admitted that the unorthodox translations of crucial passages of Plato in *The Open Society and Its Enemies* are both correct and crucial to the case against Plato. I felt uneasy at the rebuttal's claim both that all translation is interpretation and that Levinson's is dead wrong; both stressed claims are valid, but I was troubled by his coupling of the two with no comment though they are seemingly contrary: putting them together is a cause for unease, yet there was no unease in the appendix: it was all self-righteousness on account of the conspicuous injustice of Levinson's attack. It would have been, I suggested, a much better policy to accept first, for the sake of the argument, all of Levinson's criticisms of the unorthodox translations without reservation, and then show that he still fails to make a case for his view that Plato was a liberal and to answer Popper's charges against Plato. There is no disagreement here: in a sense this is what the appendix does; yet not as a matter of a policy of answering Levinson's criticism. I remember giving as an example Levinson's reading of the Greek word *"douleia"*: it means, he has protested, servitude, not slavishness. Whereas, to repeat, translation is not a hard-and-fast matter, accepting Levinson's translation here for the purpose of the debate, his defense of Plato as a liberal is still decisively refuted. (The debate is significant, since one of the important new points made very forcefully in *The Open Society and Its Enemies*, is that Plato's expressed hostility to the new liberal attitude to slaves indicates that at the time a movement for the abolition of slavery was developing in Athens. If true, this would invalidate the excuse that Levinson and earlier defensive commentators made for Plato: they all said that what is illiberal in Plato was standard in Athens at the time (not a pleasant excuse for the greatest philosopher in the western tradition, though). These days, incidentally, scholars tend to follow Sir Kenneth Dover's lead and agree with Popper — albeit tacitly, need I say. When historians of ancient Greece and classical scholars learn to do justice to *The Open Society and Its Enemies*, they will acknowledge Popper's discussion of Greek attitudes to slavery as a significant contribution on many counts.)

So much for my dislike of the appendix against Levinson. I have no wish to insist; I do not report that I expressed my response to the full while driving and touring the lovely island of Hong Kong with the philosopher, but this and more is what I had successfully conveyed; let me repeat, however, that in conversation with the philosopher his speed and impatience left neither the need nor the occasion for detailed explanation, so that, as always, there was ample room for communication failure.

His response came at once: I was shifting my ground surreptitiously: I was a liar: first I had thought Levinson was right; when the appendix had convinced me to the contrary, I picked on its style instead of admitting defeat. How nice of you to remember: yes, this was the sad incident mentioned in the Prologue to this account, in which he called me a liar apropos of a small error. No, I am not going to let go — I am just as embarrassed as you are, but I am going to speak some more of surreptitious shifting of grounds before I reach the title of the present chapter. It is "The Open Society, Its Mentality and Style", in case you think I forget.

A shift of scene to the time when the philosopher was not on speaking terms with Bartley (nor with me for that matter). Bartley had the temerity to accuse the philosopher then of having surreptitiously changed his ground. Naturally, the philosopher heard about it. Naturally, he was furious. He demanded proof. (Interesting how skepticism works in daily life.) Bartley provided some. Exhibit A was the classic, lovely, most impressive "What Is Dialectic?" (1940), published between the philosopher's first two great books, and republished in his best-selling collection, *Conjectures and Refutations* (1963). The concluding assertion in the original essay is positivistic (it opposes speculative philosophy and says that the task of philosophy is to study the methods of science: the philosopher's most Carnapian statement ever) and it is appropriately altered as reissued in the *Conjectures and Refutations* collection (where the study of scientific method is no longer declared to be the sole task of philosophy: the statement is altered to read, the study of scientific method is one of the tasks of philosophy). (How is it possible to understand adults spending time on such tiny matters, such as the change of two or three words in a text, I for one cannot say, except that acrimony does tend to render petty even noble souls, even able people, even close friends and collaborators who have much better things to exchange than petty resentment.) The book's preface contains a blanket mention of some minor changes (my review of it in *The Jewish Journal of Sociology* [Agassi,

1988, 277-9] takes them to be minor; I stand by this judgment. I find the end of "What Is Dialectic?" still too positivistic and at variance with the philosopher later condoning of speculative philosophy and even participating in it). To say explicitly what changes the book contains would have been pedantic and the philosopher's disposition to be a pedant is at times in check: it could not be too pleasant for him to remember and to own up that he had been off guard and had capitulated in a bleak moment of lonely exile. (In his contribution to *The Philosophy of Rudolf Carnap*, also reprinted in his *Conjectures and Refutations*, he does confess: Carnap had put the fear of God in his heart.) The philosopher was told about Bartley's response; he was not lost for words, oh, no. He drew out his evidence and showed the emissaries his own copy of the reprint of the original paper. You see, he used to keep at least one reprint copy of each publication of his, clearly marked with large letters with the inscription "My copy" or "M. C."; this was very important as he would give away copies of his reprints in the excitement of discussion and then be left with no copy to correct the misprints of, which was intolerable, of course. Anyway, he produced his own copy of "What is Dialectic?" and it was found corrected by a fountain pen, clearly not by a ballpoint pen — thus proving it had been done very early, before Bartley and his likes appeared on the scene. Why does the age of the change prove it had not been surreptitious? Because there is a statute of limitation on surreptitious change, perhaps? But this is not the end of it. The story crossed the Atlantic and reached Boston, where I was teaching then, and landed right on my doorstep. It rang a bell: I had seen the pages of the collection once on the philosopher's table, resting there and awaiting last inspection on their way to the printer. (I was not closely cooperating with the philosopher by then; I chanced upon the matter because Ernst Gombrich's son Richard, now a famous Oxford Indologist and then a young student, was helping the philosopher with the manuscript and I told him I thought the philosopher was overdoing the preparation of the work. He did not feel comfortable squeezed between the philosopher and myself, and arranged for us two to meet.) I had chanced then upon the original ending of "What Is Dialectic?" and drew the philosopher's attention to it. To my surprise, he saw nothing wrong with it. I directed his attention then to its positivistic character. He looked around for a pen. This was a contingency I was very well used to since my long days as his assistant; I automatically offered him mine. He refused it. With some effort he finally found and used a familiar old fountain pen, the one with the thick nub that writes so very beautifully. He staunchly refused to tell me why he chose it in preference to a simple ballpoint pen. I have an odd habit: since learning from experience is by refutations and since refutations are puzzling, a puzzling event may very well be enlightening; so I try to remember puzzling events until they cease to puzzle. This one, his staunch refusal to explain, still does not cease. No, this is still not the end: you may skip to the next paragraph if you wish, and leave me alone to run this gauntlet to the end. I had later an occasion to mention to the philosopher this, the whole fountain-pen story: I put it in a footnote of a manuscript of a paper he was supposed to respond to (in *The Philosophy of Karl Popper*, a book on his philosophy that, in accord with the formula of the series it belongs to, contains his autobiography, critical essays on his philosophy and his replies). We were then on reasonably good terms — this was during his stay in Brandeis University. So I was happy that he consented to read it even before it was submitted. He said I should not publish the story as I had access to the information in my capacity as his research assistant (this is true only in a loose sense) and thus subject to the rules binding a confidential secretary (I do not know what these are, but I doubt they hold for assistants). I respected his wish for decades. We all have small faults, and some of us have big ones too and most of us have nothing particularly redeeming; the philosopher's small faults would not be worth mentioning but for his greatness and for the immense emotional load that went with them and for the amplification they suffer. And so, at last, I am ever so slowly approaching the title of the present chapter: "The Open Society, Its Mentality and Style". You did not think I forget, did you?

I do not know what the right mentality of the open society is. I would like to take as my models the mentalities of some real individuals who have excelled in their attitudes and who can be role models for us. Though history is full with noble examples that can serve as role models of one sort or another, they may be so remote from the ideals and needs of contemporary society as to disqualify. This includes, in particular, all the religious role models, no matter how noble, from Gautama the Buddha to my idol Pope John the 23rd. During my apprenticeship, I was encouraged in many small ways to take Socrates for an adequate role model. This is difficult, as the character of Socrates is controversial and as the problems of the modern world are rather different from those of ancient Athens, particularly as I see Socrates as a religious reformer of sorts. For these reasons, even Spinoza, the modern individual who came nearest to a secular saint, even he disqualifies. My own choice is Albert Einstein. And his verdict on matters of gossip seems to contradict my decision to tell you of all the little gossipy snippets that seem to belittle the great philosopher, my master and mentor and benefactor. Einstein wrote once to a girl who had written to him of her own complaints concerning her own education; he advised her to keep the record of her complaints to show to her children in due time. And he found his advice useful enough to have his letter published[21]. Perhaps his advice is valid for my case too.

Some unpleasant information needs airing; to make the matter free of any suspicion of animosity, the usual form for such information is fiction or semi-fiction. The paradigm is Samuel Butler's terrific classic, *The Way of All Flesh*. Although it is semi-autobiographical, it is quite accurate. (The book went to press posthumously, yet relatives tried to suppress publication to protect his parents' reputation.) The suspicion that the author uses the story to express animosity is extremely cleverly allayed there, as a major point of the story concerns the way the hero learns to overcome his animosity (to his old teacher). Let me take the opportunity to report that I would have probably remained ignorant of Butler to this day, despite his tremendous importance as a philosopher and as a writer, but for the fact that Popper was a great fan of Butler and of his greatest disciple, Bernard Shaw. He felt very strongly about literature. One of the minute details that had made a difference, as it had brought us together early in the day, was our common admiration for Selma Lagerlöf, who was then even less known in England than now, despite her being a Nobel laureate. (He wrote later on about her influence on him [*Autobiography*, §3]; [Popper, 1994b, 99].)

I have no evidence that this account is no expression of animosity. Should I then avoid publishing the personal detail, in accord with the advice of Sir Ernst Gombrich (Appendix to Prologue)? Perhaps. It is, in any case, useless to protest that my aim is not to be vindictive and not to betray trust and all that. For, objective facts are not decided by intent, and the question is, do I not harm the philosopher's reputation by telling of his small failings? Are my stories not useless anyway since we know *a priori* that we are all fallible and the details for this or that individual are insignificant? Are my stories not an impediment to learning as deflecting from the philosopher's important teachings and as lending arguments to the arsenal of his detractors?

In the passage of time, differences between facts and semi-fiction diminish. If readers of some near or distant future will glance at this gloomy record, they may find some historical interest in these stories; today's readers may prefer to view my narrative as amusing semi-fiction, especially if they do not care too much about the philosophy it involves. I have no objection to this. I do not wish to discuss the status of this work as semi-fiction, though, as sufficiently many friends, including ones who

---

21 Albert Einstein, "A Letter to a Young Girl" [Einstein, 1949b, 21-2]; [Einstein, 1954, 56-7]:
"Dear Miss —,
I have read about sixteen pages of your manuscript ... I suffered exactly the same treatment at the hands of my teachers who disliked me for my independence and passed over me when they wanted assistants ... keep your manuscript for your sons and daughters, in order that they may derive consolation from it and not give a damn for what their teachers tell them or think of them. ... There is too much education altogether. ..."

have advised me not to publish, have shown more interest in this work than in anything else I have ever written: gossip has its own fascination and it is much less taxing than philosophy proper. But as gossip about a great philosopher, is the present account wrong? Does it show myself as petty and vindictive and treacherous, as repaying the philosopher's immense kindness to me with petty and malicious gossip about him? Will I be (misunderstood and) taken to be vindictive?

I sincerely hope not: I never wished him harm. But pious expressions are less interesting than discussion. It is much more important to take the point at issue in all its generality rather than center on the question of the motive behind this specific, account. The question then is, since we all know that we are all fallible, why, oh why is it always deemed harmful to tell stories of unpleasant events? Why do we find the biblical story of King David exciting and that of King Solomon dull, yet wish to be described as the latter, not, heaven forbid, as the former? No, the comparison of the present stories concerning a modern struggling philosopher with the biblical stories concerning an ancient legendary king is definitely not ridiculous; why should I not take the Bible as a model, however inimitable, for this account? I criticized once the philosopher's view of the open society as purely Greek in origin, and as an oversight of the story of King David and his extraordinary acceptance of censure from an unbelievably brave court-prophet. He found this amusing, but quaint. As to Einstein, I do not think he would have found it quaint; I suppose that he recommended keeping in the drawer an educational autobiographic story simply because he found it not sufficiently interesting for publication; this one I hope is.

The second volume of *The Open Society and Its Enemies* has Marx as its center and the progenitors and progeny of Marx in the outfield. One chapter demolishes Aristotle, and the next is a bagatelle on Hegel. The former includes the classic attack on the classic Aristotelian theory of definitions (labeled there "methodological essentialism"). Thanks to some misunderstanding and to the strong need Wittgenstein's devotees felt to ascribe to him some intellectual idea, Popper's critique of essentialism is now regularly ascribed to Wittgenstein. As he had enough original ideas, he should have not minded.

Of course, like all wrong attributions, this one, too, must misfire: it is a tacit admission that Wittgenstein's devotees have nothing to ascribe to him. In addition, this specific wrong ascription is a much worse admission: all declarations to the contrary notwithstanding, the admission is of the correctness of Popper's reading of Wittgenstein's anti-metaphysical first book as representing a metaphysical view akin to that of David Hume and Ernst Mach, according to which there is nothing in the universe except our perceptions of it, a view of the world as "flat", to use the term invented by Bernard Shaw and used by him disapprovingly (a flat world, he said, would be a real hell) and by Neurath approvingly, since, he taught, this metaphysics is the only one that is anti-metaphysical, meaning, anti-religious. (He used a different terminology, of course, and many words, to say this brief, simple message, but he did say, "everything is on the 'surface'." [Neurath, 1973, 326]. Malcolm said it in the title of his *Wittgenstein: Nothing is Hidden* [Malcolm, 1986]; see also [Gellner 1958, Chapter IX.4].)

The chapter on Hegel in *The Open Society and Its Enemies*, the bagatelle on Hegel[22], has aroused most of the hostility to the book. Walter Kaufmann's famous "The Hegel Myth and Its Method" is the *locus classicus* [Kaufmann, 1951]; [Kaufmann, 1959]; [MacIntyre, 1972, 21-60]. Despite my want of German at the time, I accepted the philosopher's challenge and wrote, while his research associate in Stanford, a detailed critique of Kaufmann's attack that was very much to my liking and even the philosopher was agreeable to it. For years I tried to get it published and failed; this is another example of the openness of the commonwealth of learning to critical exchanges. Kaufmann is right in denouncing Popper's method of composite quotations and in noticing, apropos of hardly anything, that not Hegel[23] but Jakob Friedrich Fries, the hero of Leonard Nelson whom young Popper greatly appreciates, invented the Final Solution to the Jewish Problem [even in detail: offer them the opportunity to leave and then kill all those who stay]. Kaufmann nevertheless failed in his self-appointed task of defending Hegel against Popper's[24] severe charges. The funniest example is his complaint that Popper quotes from Hegel's famous book on physics to illustrate Hegel's ignorance of what he was talking about. Kaufmann complains about the quote. Is it a piece of inadmissible evidence? Why? According to Popper, Hegel was plainly

---

22 Surprisingly, John Findlay defended Hegel's philosophy of nature [O'Malley et al., 1973, 72-89] in disregard for Popper's ridicule [*OS*, ii, 28]. Frank Collingwood responded sharply: "I do not see any grounds upon which Hegel's adoption of this thesis about the four elements is defensible" [*op. cit.*, 90].
23 Popper rightly insisted that Hegel influenced Nazi thinking significantly, especially about national destiny and glory. Kaufmann objected: the Nazi leaders were ignorant and that the idea of a Final Solution belongs to Fries. True but irrelevant: Hegel popularized and lent respectability, especially in Germany, to chitchat about destiny and glory. This chitchat was the daily bread of the Nazi leaders.
24 Popper noted that the view of Hegel as a windbag is traditional; he ascribed priority here to Schopenhauer. Heine, who was Hegel's former student, called him a snake (Preface to second German edition of *Religion and Philosophy in Germany*) on account of his use of nouns and descriptive phrases that designate the divinity to refer to humanity, to the German people, to Prussia, to himself, and to his own writings — as his musings moved him. Whewell said, he could not take the German philosophers of his day seriously because they valued Hegel. William James found his metaphysics confused and quietist [James, 1909, 114, 116]. Russell said, his reading of Hegel's pseudo-mathematical "muddle-headed nonsense" had cured him of his erstwhile Hegelianism [Russell, 1944, 11]. And he deemed Hegel's worship of the state astonishing even by comparison to other power worshippers [Russell, 1946, 709-10]. He also said [Russell, 1950], "To anyone who still cherishes the hope that man is a more or less rational animal, the success of this farrago of nonsense must be astonishing."

a charlatan who polluted the German language[25] and intellectual atmosphere, thus making intellectual irresponsibility respectable. This way, Popper declared, Hegel had prepared the ground for the rise of Nazism with no suspicion of things to come. This charge Kaufmann ignored, perhaps because he had a strong admiration for those who could put the German language to good use.[26]

The significance of the matter goes much deeper: the failure of the French Revolution (the failure to achieve the ideals of the Revolution, the subsequent terror, the Napoleonic wars and the crowning of Napoleon and of his kin) invited the Reaction and Hegel was its great representative. (Alfred Cobban,[27] mentioned above, noted Popper's oversight of that important historical defeat and also of the unpleasant aspects of Kant's version of the Enlightenment.) Hegel's idea of history as the proper substitute for any social science was thus reactionary and thus quite anti-scientific: it opposed scientific method and scientific prediction. It was Hegel's scientifically minded disciples, Saint-Simon and Marx included, who (putting Hegel on his head) said, historical inevitability or destiny gives scope to scientific historical predictions.[28] Pseudo-science, much more than mere historical inevitability or even mere destiny, was the ingredient in Nazi ideology that Popper noted too, speaking of both historicism and social Darwinism and scientific eugenics. He was very insightful here, as he could not then know about the Holocaust, much less aware of the fact that, as Lucy Dawidowicz has since empha-

---

[25] In addition to calling his own philosophy *The World Spirit* and *the Spirit of the Age*, Hegel used repeatedly "subjectivity" and "objectivity" to denote individualism and collectivism. For this, and for his diatribe against the Enlightenment Movement, see the opening of the last chapter of his *The Philosophy of History*. He also called "science" and "the science of logic" any idea that he promoted.

[26] Kaufmann was very critical of Buber but admired his translation of the Hebrew Bible for its clarity. (Wittgenstein disliked it for the same reason.) He could not praise Hegel's language. This however is not to endorse Popper's assessment of Hegel, even though his ridicule is more than just. Hegel's very worst, his philosophy of nature, had a liberating effect on a few thinkers (as Owsei Temkin and others have argued): Hegel's dialectics that is a silly defense of inconsistency was for them the lovely idea that intellectual growth involves the corrections of past errors, a sentiment that Popper managed to express superbly ("What is Dialectic?"). As to Hegel's criticism of the radicalism of the Enlightenment movement, it has a point. Popper says Hegel took that point from Edmund Burke. It is a pity that he did not distance himself from the doctrine of the Enlightenment, although it is understandable: he rightly viewed himself as a follower of their heritage. See next note.

[27] Popper never responded to Alfred Cobban's criticism of his great book [Cobban, 1954]. I do not know why. It is intelligent, friendly, enlightening and full of important criticism. (My tendency is to endorse Cobban's criticism of Popper's idealization of the Enlightenment Movement, especially of Kant. But this is debatable. What is baffling is the absence of such a debate.)

[28] Marx was obviously friendly to science. He spoke of relative truth, and regrettably this led him or his followers to relativism and thus to irrationalism. The recognition of relative truths is consistent with absolutism, as absolutism allows for both relative and absolute truth (truth by nature and by convention, to use ancient Greek idiom); relativism dismisses the absolute truth. Hence, though relativism looks weaker than absolutism and thus less dogmatic, it is stronger and hopelessly dogmatic. This wipes out Richard Rorty's effort to link Popper with the American pragmatists [Rorty, 1985, 17n]:
"The attitude toward truth, in which the consensus of a community rather than a relation to a nonhuman reality is taken as central, is associated not only with the American pragmatic tradition but with the work of Popper and Habermas."

sized [Dawidowicz, 1975, 131], it began with the destruction of handicapped non-Jews. (George Santayana had noticed that Nietzsche's aestheticism constitutes a danger for a neighbor who is not particularly beautiful [Santayana, 1968].[29] *The Open Society and Its Enemies* discusses beautifully the link between aestheticism and historicism.)

And yet the view of the doctrine of historical inevitability, of destiny, as pseudo-science is proper as it is a response to the claim that it is scientific. This claim is not Hegelian; it is post-Hegelian. Hence, Kaufmann is right on this. But he is in error when he presents Hegel as not having endorsed this doctrine: it was an error of both author and critic to fuse the pseudo-scientific and the anti-scientific variants of the doctrine of historical destiny. This is odd, since it was Popper who first distinguished the two versions of historicism and dealt with them separately — in his own terrific *The Poverty of Historicism* [Jarvie, 1982, 85].

What Hegel's devotees find annoying is hardly Popper's criticism; it is his off-hand, wholesale repudiation and dismissal of Hegel. They could accept most of the arguments, as long as these need not saddle on them the conclusion that Hegel does not count. My evidence is the work of Walter Kaufmann himself, especially his review of John Findlay's book on Hegel [Kaufmann, 1961], but also his own posthumous work [Kaufmann, 1990] in which he tacitly endorses most of Popper's criticism, but not off-hand. (His ascription to Hegel of holism there is pathetic, since that doctrine is ancient and ubiquitous: clearly, he had nothing to ascribe to Hegel, try hard though he did.)[30] As for myself, although Popper's contempt for Hegel is just, I do not share it. I have expressed appreciation of the phony and verbose Hegel as a perceptive critic of radicalism, of scientism, and of mechanism — in the volume dedicated to Findlay. He did not like it. Do I fall between stools or am I balanced between extremes? It does not matter.

The study of Marx in *The Open Society and Its Enemies* suffices for a whole lecture-course. The final chapters concern rationality and the

---

29 Santayana's critique of Nietzsche's aestheticism as politically dangerous is powerful: Nietzsche's vision is of a beautiful world that has room only for supermen, not for ordinary people. Nietzsche's regard for autonomy looks similar to that of Spinoza (whom he admired), yet it went with contempt for common people. The philosophy of Spinoza has no room for contempt.
30 Russell dismissed Hegel's philosophy [Russell, 1912, 143]; [Russell, 1967, 17].
He said [Russell, 1934, Chapter 18 (2)],
"Hegel, as everyone knows, concluded his dialectical account of history with the Prussian State, which, according to him, was the perfect embodiment of the Absolute Idea"
and [Russell, 1956, 185],
"The word 'liberty' ... in the hands of Hegel ... came to be 'true liberty', which amounts to little more than gracious permission to obey the police."
Popper's dismissal of Hegel, of his pretense at learning, and of his method of twisting words in the service of the enemies of freedom was not original. He was original in his detailed, sweeping, severe, and just appraisal as well as in his claim that Hegel lowered standards and vitiated the German tongue. (Georg Cantor blamed Kant for this. See his letter to Russell of September 19, 1911.)

meaning of history. They are magnificent despite their intolerable caricature of religion and haughty dismissal of metaphysics.

The chapters on Marx are still shunned by the (pseudo-)Marxist fraternity that is ever growing in philosophy and the different social sciences — including economics, to everybody's surprise. Popper's critique of Marx is the subject of one early critical book by a party hack, happily now forgotten, and a lovely essay by the leading Cambridge economist Joan Robinson, regrettably now utterly forgotten.[31] She observed that Popper was unfair to Marx by his own standards: he required refutability and here he was ridiculing a theory on account of its having been refuted. This criticism is as legitimate as possible; it even happens to be valid. Popper's response that Marx has followers who defend his doctrines come what may is neither here nor there. Proof: Newton had such followers too. The claim that Marx had encouraged fanatic discipleship is also neither here nor there, and for the same reason. The same holds for the view that science is irrefutable and for the haughty conviction that one had delivered the goods: these views were shared by Newton and by Marx and by most modern thinkers, as Popper has noticed repeatedly.

Popper's criticism of Marx is a paragon of fairness, the validity of Joan Robinson's counter-criticism notwithstanding. In any case, however astute and relatively accurate Marx's arguments were, they are by far too vague for current tastes: the program in *The Open Society and Its Enemies* was to show that these arguments are full of holes — Marx repeatedly overlooked options — and the exercise is as fair as it is generally enlightening. To carry out this program one should oppose Marx's inference from his own premises to his own conclusions, leaving the question open as to the truth or falsity of either the premises or the conclusions. Regrettably our present-day education is so dogmatic that even some leading and famous scholars (even leading American philosophers of science, see the next paragraph) exhibit a remarkable inability to understand the distinction between truth and validity, much less between truth and scientific character.[32]

This point is significant and merits a comment. The reluctance to

---

31 Joan Robinson's economic philosophy [Robinson, 1962] is close to Popper's views (even where she expresses disagreement with him). See Chapter Ten.
32 The theory of induction as probability does not help here, unless one holds that the probability of a hypothesis is its degree of truth as a proper fraction. Reichenbach toyed with this idea but only a few people have advocated it, and for good reasons. It grants the same truth-value to known informative statements as to a tautology. It renders probability no more open to dissent than formal logic. There is no logic of degrees of truth, since a valid inference transmits to its conclusion the truth of its premises but not its partial truth. The classical idea of inductive inference is that some (unspecified) invalid inferences convert the truth of their premises (observed facts) to a probability of their conclusions (generalizations). In this case, a probable hypothesis is one that is probably true, whatever this may mean. There is no discussion of inferences with probable premises and conclusions whose probability is equal-to-or-less than the probability of its premise.
   Discussions of partial truth ignore induction and probability [Hajek, 1998].

speak against the great defunct scientific theories of the past, due to the Baconian equation of criticism with contempt or of error with sin, has led to the locution "invalid", meaning false-but-not-condemnable. This is vaguely a euphemism that many take as precise jargon or even as clear-cut terminology proper. Consequently, they lose the most basic aspect of logic: falsehood and invalidity are distinct in every recognizable sense of these words. And this is confusion that spills over to other areas. Adolf Grünbaum, a leading figure on the American philosophy of science scene, was dumbfounded when he heard me say that in Popper's view some of Freud's ideas are true and even significant, but none of them is scientific.

Yet Popper's description of the doctrine of historical inevitability or destiny is inadequate as he makes it too unconvincing. He notices that the strong points of Plato's theory have to do with his analyses of the logic of situations of individuals; he notices the same about Marx; he compliments each of them for that; but he examines these analyses as independent of each other, and declares them both inconsistent with the doctrine of historical inevitability. He views Plato's (historicist) claim that regimes must decay as inconsistent with Plato's claim that it is within the ability of the philosopher-king to turn the wheel of history backwards, or full circle, that the ideal republic is thus accessible and durable. This is disturbing. If we see the decay of every regime as rooted in certain imperfections, and in its turn as rooted in the logic of ordinary situations but avoidable only in the ideal state, then Plato's doctrine begins to look much tighter. This is conspicuous in the case of Marx, who saw change as possible, not as necessary. He did not claim that progress is necessary, only that if and when significant change occurs, then it is the progressive, revolutionary outcome of individuals improving practiced methods of production. He viewed capitalism as an exception, though: it is so shaky that, of all regimes, it alone will have to give way to the next stage, to the socialist regime, and even that it will do so fairly soon (most likely in his own lifetime), simply because the workers will not miss the opportunity of a successful revolt: here the idea that there is no time-table to history, only a general direction, is well-squared with the claim that the demise of capitalism comes nearest to having a time-table.

Popper commended (and criticized) the situational analyses of Plato and of Marx. He deemed them as contrary, or at least incidental, to their authors' overall views on destiny — on history as inevitably marching along a predetermined route. This will not do, since both Plato's route and his situational analysis go the same way (both bring deterioration), and the same holds for Marx (both bring improvement). This is no accident: the theory of history of each of the two is the integration of individual situations into one global historical scheme (degenerative or progressive as the case may be). This facet of Plato's view is what Hegel called the Cunning of Reason and others called the unintended consequences of individual actions. This is the way Hans Albert reads Plato's and of Marx's social philosophies. He calls this reading "methodological historicism". (He was unfortunately too wary of openly expressing disagreement with the philosopher. He is an original thinker, yet he prefers to appear a mere popularizer — and indeed he is the best-known popularizer of Popper's ideas [Agassi, 1988, 451-68], [Morgenstern and Zimmer, 2005, 18].) Anyway, Plato and Marx sustain the view that individual and social ends are in harmony, a view that Simmel and Popper have rightly rejected [Agassi, 1977]. Though all the known theories that link individual rationality and historical trends are false and much too naïve, the question remains: in what way should we explain trends, such as the classic process of industrialization or the rise of golden ages [Agassi, 1981, 421-35]?

All this shows, I hope, how different were the social philosophies of the historicist giants whom Popper criticized from the religious eschatological revelations (= apocalypses) that he derided. It shows, in particular, that the problem he ascribed to historicism, which has to do with despair in matters social and political, even if it is true of eschatologists, is not true of these philosophers, as they solved an intellectual problem: how do individuals acting in their own interests coalesce into social and political trends? Moreover, this question rests on the assumption that trends are given. While concentrating on it, one may understandably ignore human actions not contributory to trends. They may then clinch their oversight by endorsing the view that all an individual can ever contribute to the public sphere is either aid or abet trends. These are the major errors of the doctrine of historical inevitability that Popper and other critics of this doctrine expose.

I hope you see the value of Popper's treatment in his *The Poverty of Historicism* of the pseudo-scientific and the anti-scientific versions of historicism as separate and how regrettable it is that he did not follow it up in his *The Open Society and Its Enemies*. His fusion of the two variants in his chapter on Hegel may be overlooked as a mere result of brevity; except that this fusion fed his funny urge to present the doctrine as Jewish in origin. He thus fuses not only anti-scientific and pseudo-scientific doctrines, but also eschatology (*The Book of Daniel*), that is neither scientific nor pseudo-scientific but pre-scientific. Admittedly, in the sense in which Sir James Frazer had demarcated magic as pseudo-science, we may view pre-science as pseudo-science. Yet this view of magic as pseudo-science is criticized by hordes of anthropologists. In its defense it can be noted that it squared well with the staunch faith in the empirical justification of science. This defense cannot possibly apply to Popper, of course.

This is the place to notice that Popper fused eschatology and historicism (pseudo-scientific and anti-scientific) with utopianism too, though historicism and utopianism are logically independent, so that utopianism can go with critical rationalism too, as Jarvie has noted [Agassi and Jarvie, 1987, 227-43]. A few modifications to Popper's philosophy are required here, so please be patient.

Popper's classic "Back to the Presocratics" presents the rise of Western philosophy as a totally internal Greek affair (except for the clash of cultures), totally ignoring the Hebrew contribution. This has to do with the popular theory that the Greek philosophers have discovered the idea of the cosmos, which is evidently false, as this idea is utterly prevalent, and as it includes the idea of the harmony that covers the laws of nature and of humans (truth and justice) intertwined and undistinguished. This idea (that Popper calls "natural monism") still occurs in the biblical *Book of Judges*. Long before the rise of Greek philosophy the Hebrew prophets discovered that justice does not prevail. This discovery must have been terrible, and it was the attempt to restore the cosmos that is common to the Hebrews and the Greeks. (Obviously some ideas were independently discovered and some traveled; current research of these matters is still unsatisfactory.) Attempts to recreate the cosmos received three different expressions: the proposal to struggle for social justice, the claim that justice will prevail in the end of days (or in the world to come), and utopian dreams. None of these has the slightest connection to the view that human history moves from one stage to another with inexorable necessity. All that one can find as a connection is the fact that some (not all) eschatological texts describe horrible revelations (apocalypse) of series of dramatic events preceding and heralding the envisioned end of the world. (*Ta eschata* are the last things: death, the Day of Judgment, the world to come.) True, some historicists have interpreted these series of events as historical stages, but these interpretations are no more than frivolous analogies of the kind Popper had little patience for.

This is not all that renders the Hebrew Bible relevant to the question of the origin of the critical tradition. A significant part of it is devoted to the removal of magic rites. Surely, the view that the rise of science is the abolition of magic (Weber, Frazer,[33] Gellner) invites a new look at the Bible that offers the earliest anti-magical documents (though it still is

---

33 Frazer advocated the historicist view of intellectual progress as the move from magic through religion to science. Religion that allows for miracles but not for magic (à la Frazer) is conceivable, but there is no decent distinction between the magical and the miraculous. Such a distinction would smuggle the view that all religion is monotheist and naturalist and all magic is polytheist and mythical. But a religion that is free of magic — whether it allows or forbids miracles — rests on the naturalism that characterizes pure science. A rare exception is Maimonides. He allowed for miracles as incomprehensible suspensions of the laws of nature but not for magic: he used naturalist arguments against it. Modern scientific method, incidentally, suspends judgment on everything supernatural, magic and miracles alike, as they are unrepeatable. Scientific metaphysics precludes magical and miraculous causality not as spiritual but as either amply refuted or unrepeatable-in-principle.

steeped in magic, of course). Popper's idea of the evolution of science from myth and his idea of the position of pseudo-science in the face of science, then, invite radical reform. ([Flusser, 1988, 610] may serve as a start.) I will not now discuss the critical attitude that appears sporadically in the Bible, nor contrast it with the pre-critical attitude that generally pervades there. Suffice it to notice that there are clear traces of the invention of criticism in biblical, pre-Hellenic times, such as the discussion (*Deuteronomy* 13:1) of the demarcation between true and false prophecy. This is but an early stage of criticism, as it presents the false prophecy not as the soothsaying of the period of the late prophets but as the preaching of worship of other gods, a preaching punishment is death, in accord with the oldest tribal custom. Still, here is the oldest text extant that demands individual autonomy in a clearly fashion.[34] This is not to belittle the Greek achievement, nor to claim any part of it to the Hebrews. Rather, it is to say that the collapse of the concept of the cosmos and the resultant distinction between human and natural law (not found in Homer), and the resultant abolition of (idolatry and) magic rites, had to precede the contributions of the *physiologoi*, the Greek naturalists engaged in dialectics, who shunned discussing human laws. And then the sophists arrived on the scene, and applied dialectic to morality — prior to the development of social science: the discovery of the individual contribution to political processes was, I conjecture, the discovery of social science and of historicism in one and the same step.

Allow me to remind you that I have no Greek and I am not a student of the classics. My venture to the origins of our civilization is prompted by my attempt at a critical reading of Popper's view and at relating it to others'. I now must leave it at that and return to my narrative.

As I had helped the philosopher with his manuscripts, I naturally wished him to expurgate the next edition of his great book of what I considered somewhat excessive errors and was greatly moved to receive an acknowledgement in the new preface. He also generously helped me with my work, a bit too generously. Once, after I had graduated and after I had ceased to be his apprentice, I wanted to alter in proofs a book-review he had helped me to write and he lost his temper with me. He then received an angry letter from his friend Herbert Feigl, the editor of the book I had reviewed, and for a very disturbing moment he sided with him against me: he forgot his own involvement.[35] He was equally involved when he

---

34 The demand that one should trust one's own judgment is not systematic in the Bible, but then few texts are so utterly free of appeal to authority. It is here that Kant excels even more than, say, the great Galileo, yet the first who were utterly clean of all appeal to authority were Heine, Russell and Einstein.
35 My review of *Minnesota Studies in the Philosophy of Science*, Volume II, has caused me an ostracism that lasted throughout my career. Popper should have thought about this but it did not occur to him. Or perhaps he concluded that this was my fate anyway. If so, then I support his view, but not his conduct: he should have openly shared it with me. It seems he used me to publish in *Mind* a review that he could

corrected manuscripts of my earliest publications — in *The British Journal for the Philosophy of Science*, whose editor, the admirable John Wisdom, was generously trying to help me build my career. I would not have noticed[36] the philosopher's pedantic hand, but for a funny slip: I somewhere spoke of cabbages and kings and the philosopher changed this to cabbages and diamonds. He was unfamiliar with Lewis Carroll whom he disliked — possibly on account of his (Carroll's) popularity with the devoted, Wittgenstein-style-ordinary-language-worshipping philosophy professors of the day, and possibly because he was contented enough with Hugh Lofting and his Dr Doolittle whom he had even most enviably met in person (he met the great Lofting, not alas the lovely miraculous Doolittle; it is beyond me how he could have omitted from his intellectual autobiography this exciting episode, as he was instrumental in arranging Lofting's Viennese visit as he told me upon discovering that I too admired him when I once spoke with him of the pushmepullyou) and he felt that in my insensitivity to the English and their national quirks I could all too easily give them offence. I reinstated the original text, of course, and remarked to the philosopher that it was not customary to correct other people's manuscripts uninvited. He blushed and I let it go. Oh, I blushed in his company more often, but this is neither here nor there. I hope you will understand from this long digression why I felt free to write as I please only after arriving in Hong Kong.

As we landed in Hong Kong before the semester had started I responded to a review in The Australasian Journal of Philosophy of the newly published *The Logic of Scientific Discovery*. The reviewer — David Stove it was — took the book to be one more inductivist failure to solve the problem of induction, taking its last chapter on empirical support or corroboration as its center. (Later the reviewer realized that Popper never

---

fully identify with. His refusal to allow me to alter my review when I proofread it in galleys decided my determination to leave London come what may, simply because his conduct was a violation of my liberty. For my part, when former students of mine try to pick a row with a big target and consult me about such a move, my recommendation is that we have a thorough discussion of the possible adverse consequences of that choice for their careers. Such discussions should make this clear: wonderful as the academy is, it is no utopia.

36 Usually I hardly notice when editors alter my texts without permission. A few cases stand out. An editor of a German journal, Hans Peter Duerr, added to an essay of mine reference to a celebrated Marxist-existentialist windbag and refused to publish retraction. Some editors changed my dating of Popper's *Logik der Forschung*. (The date on its cover is 1935; he insisted that it appeared in 1934; true but irrelevant.) Bartley copy-edited my *Towards an Historiography of Science* radically and sent it to the printer. My style was of juxtaposing short and long sentences. His was measured sentences. It annoyed me, though his publication of the work was the greatest boost to my career. I protested, and he apologized.

Schrödinger, who was a great stylist, told me that copy editors constantly annoyed him with alterations in accord with their house styles. Gertrude Stein tells a most amusing anecdote in her *The Autobiography of Alice B. Toklas*: the first American publisher to accept a manuscript of hers sent a copy-editor to her from New York to Paris to help her with her English, on the assumption that she was a native French speaker who violated English grammar out of ignorance.

equated science with corroboration; he then wrote a book saying that this makes him an enemy of science and of scientific progress. Despite all my reservations concerning that book, I have great sympathy with its distinction between Popper's theory of learning from experience and his theory of corroboration.[37]) I discussed the role of corroboration in science with the philosopher as my last conversation with him before departing for Hong Kong; he had acknowledged in a footnote in *Conjectures and Refutations* that I was possibly right, that possibly his demand that science have not only conjectures and refutations but also, from time to time, some corroborations, constituted a whiff of inductivism. (Yes, I did mention this in the Prologue above; thanks.) That footnote is a red herring: I do not know if corroborations are essential to science or not and I am more than willing to examine either possibility; but the following is crystal-clear and as yet quite beyond dispute. Popper's stand in that chapter signals a change of view, though he presented it as a mere extension or working out of the implications of that view. Perhaps he confused two versions of his view and perhaps he had surreptitiously altered his view from one version to the other. He did promulgate two different versions of the refutability criterion of demarcations of science: according to both versions science is a corpus of theories plus the attempted refutations of some of them, successful and unsuccessful, as they happen to be, namely their actual refutations and actual corroborations; as to the theoretical corpus of science, according to one of these versions, it is the corpus of all refutable theories unqualified, and according to the second, it is the corpus of only those theories that were corroborated, even if they were later refuted. Alternatively, what matters is that some scientific theories are corroborated now and then. (The picture seems especially vague since now the view in question no longer concerning the question, when is a theory scientific? but on an unspecified set of theories spread over an unspecified length of time: each of these has to be refutable and some of them require corroboration. Will one case of corroboration in the lifetime of the whole of science do? If so, then the requirement is very weak. Or does every theory that claims scientific status have to be corroborated before it be refuted? If so, then the requirement is very strong and runs contrary to known historical records of important scientific theories that never had any corroboration[38]. How much corroboration is required?) Perhaps I am in great error; possibly a third demarcation of science is pre-

---

37 David Stove's flippant style conceals his seriousness — in contrast to the common, heavy style that conceals insignificance. His *Popper and After: Four Modern Irrationalists*, 1982, admits that the problem of induction is insoluble, yet it praises philosophers who struggle with it and denounces other philosophers as irrationalists, as he identified rationality with induction [Watkins, 1985].

38 Descartes' theory of planetary motions never underwent any test, and so it could not have any corroboration. Popper deemed it significant. The same goes for the famous Bohr-Kramers-Slater view of the law of conservation of energy as statistical: its first test was its refutation.

sent in Popper's writings: science potential is the corpus of all known conjectures and all their conceivable refutations (given the state of the art), and science actual is the corpus of those refutable theories that have been tested and not (or not yet) refuted. Which of the three (or four) readings of Popper's texts is correct? It was his turn to enlighten and say what exactly he had in mind. He could report what he had originally meant; perhaps he would report that his original meaning was unclear on the issues that his commentators have in mind, that while writing his classic masterpiece he had in mind a different concern altogether; and he could tell us what his thoughts were at the eve of his life. It is his silence that I do not pretend to comprehend. And now it is up to the next generation to advance matters further.

I wrote for the same Australasian Journal a presentation of Popper's view of corroboration as I understood it, explaining its role in his early philosophy (without the famous whiff of his later philosophy) yet while explaining why it is important (even though not essential) — contrasting my exposition with that of the reviewer. When I visited England about two years after I had written that paper, on my way to the newly founded University of York for a job interview, I visited the philosopher twice, and in both cases he commented on it. All interviews I had for a job in England were arranged only pro forma: by the time I came to each interview its outcome was already secretly decided. In this case the position was already allotted to a senior person who soon came from Australia to occupy it.

One of the two meetings I had then with the philosopher was in the presence of summoned witnesses — John Watkins and Bill Bartley. This was ominous, and the two witnesses were more nervous than the accused. It was a kangaroo court worse than we had feared. The session began almost at once, with the philosopher's declaration that my critique was so perfunctory, it could only be understood as an effort to attract attention. Now that attention had been granted as a mention of my name in a footnote (in which the possibility is admitted that there is something to it, the possibility that Popper is guilty of a whiff of inductivism), I was obliged to admit the senselessness of my critique and desist. I was not surprised: the philosopher had made this point in a letter to me, to which I had replied that if his explanation of his having written the footnote is true, then he should not have written it in the first place. My response to his opening speech was the same. The response that followed was unbelievably acrimonious. Watkins cut in. He said, philosophically he was on the philosopher's side, but he found the denunciation of my conduct irksome. Bartley had stayed up the previous night in order to be too tired to participate. A prearranged emergency phone call rescued Watkins; Bartley joined him since we had all come in one car; I would return by train. The philosopher was stunned: it was one of the rare occasions for which dinner had been prepared for the guests and they had taken a powder. Whenever surprised he was very adult. He let me talk at last: he asked me to explain my thoughts. I then gave him a brief private presentation — the only one ever — of my criticism of his theories of corroboration and of the empirical basis as too empiricist. I also said, his excessive empiricism drove him naturally to positivism and thus to blindness to the role of metaphysics as providing interpretations of old scientific theories and thus as guiding scientific research (which role was the point of my doctoral dissertation, on the function of interpretations in physics). I later published in *Mind* the point about the empirical basis and attempted to simplify Popper's ideas by using his own general view of science. He declared my authorship of that paper an act of treason.[39] The paper was in galleys for four long years during which Lakatos advised him of newer and better ways of putting pressure on me to withdraw it. Fully four years after I had read the proofs, I wrote to the editor, Gilbert Ryle, to say that even if he delayed publication indefinitely I would not withdraw it. He apologized and it appeared soon afterwards. The philosopher went on resenting that paper enormously, but now his view of it as a betrayal can be checked; while it remained unpublished reservation was unavoidable.[40] I take it that the philosopher deemed scandalous the publication of the paper, not the views in it: but that evening, after the kangaroo court was truncated, after we were left alone by our friends, the philosopher did not object to my criticism, and even expressed appreciation. My discussion of the idea of the role of metaphysics in generating scientific problems appeared in the volume in Popper's honor on the occasion of his sixtieth birthday and though it reaped great success I was grieved that he was not pleased with it. I can only comfort myself with the memory that after the kangaroo-court session was truncated he was impressed. He said then, it was my fault that I had not presented to him earlier on all of my criticisms together, as together they make better sense than separately and he could then see clearly that I was going further with the help of his ideas — not circumventing them but going through them, as the jargon expression goes. He then recognized for a moment that I had a distinctive angle, and I was happy. (Do you notice how a whole critical and commonsense view of science can be expressed and understood in one jargon expression apropos of some insignificant, personal-vanity matter, but seem masochistically absurd when carefully and forcefully expounded? Intuition misleads.) The evening that started disastrously and got ever so much worse, ended on a very cheerful note. I was greatly moved as the philosopher

---

39 Popper never read my "Sensationalism"; Lakatos misinformed him about it.
40 Popper's report on my "Sensationalism" surprised Hans Albert; he later read it and was more surprised to find it inoffensive [Morgenstern and Zimmer, 2005, 76, 104].

told me he hoped very much for my sake and his that I would get the position at University of York; he too had not heard that the position was already assigned to some Australian.

As fate would have it, the philosopher soon spent a semester in Australia. His last appearance there was a public lecture with all the expected pomp and circumstance to be followed with a grand reception. After the lecture — I see you can hear the anticipation in the background music — the Australian-on-his-way-to-York just had to be the first to ask a question from the floor. The philosopher's response, I am told[41], was the peak of explosive rudeness; he left the lecture hall in a storm and there was no reception. When told of this I felt an extreme conflict: I sympathized with the fellow yet I was complimented by the philosopher's rudeness to him more than by all of his friendly gestures put together.

I have promised to talk in this chapter on the open society and its mentality and style. I regret I presented more anecdotes than discussion, but then this is no philosophical treatise. So let me end with one more anecdote and a little moral, perhaps. Before Feyerabend became an enemy of the open society and declared he had never been a friend, he used to say to his students — I have it from the horse's mouth — join the open society, but leave me behind! Query: can this be done? Can one preach any gospel and yet not join? Of course, one may be a liberal and not preach liberalism or even think it out yet; but can one be illiberal and preach liberalism? T. S. Eliot, that first-rate poet and substandard reactionary essayist, declared this the rule: in his fine "Journey of the Magi" he says, the childhood of rebels keeps their hearts hankering backwards even as their reason looks forwards. In Popper's view, it seems, one can be taught liberalism, if one is not a liberal already, but not the liberal way: teachers should coax the not-yet-liberals, as they would not voluntarily listen to the liberal gospel [*OS*, ii, 276]. This is like saying, only dirty people bathe, not clean ones; so I do not wish to ascribe this idea to him except in the case of children. Are children so special? Or is Eliot right in considering us all children at heart?

What is at issue may be a different matter altogether: some people can admire liberalism only from afar. This is not very pleasant. I admire the honesty about this that Eliot has displayed in his grand poem, and that Feyerabend has exhibited as long as he could take the pain it undoubtedly involves. The culmination of each volume of *The Open Society and Its Enemies* is the thesis that we all suffer the burden of civilization, that civilization is a cross we all must bear. This, I have explained above, is

---

[41] Berl Gross told me this. He was about Popper's age, an amateur with deep concern for political philosophy. His posthumous *Before Democracy*, 1992, is excellent on a few counts, even though possibly outdated by now. I met him in Popper's classes. At the time of the story he lived in Sydney and frequented Popper's seminar in Canberra by special permission.

too Freudian for my taste, and too revealing of the puritanism hidden in the message of that great masterpiece. (My *Towards a Rational Philosophical Anthropology* [Agassi, 1977] is in part an effort to modify the image of humanity it presents.) I was unjust, all my intent to the contrary notwithstanding; perhaps I even misread the book: I ignored the fact that some do greatly suffer the pain he had called the strain of civilization, and possibly he meant to draw attention to that pain, not to endorse the puritan view of it as unavoidable. I regret even more that I did not take the pain into account when I repeatedly and severely and uncompromisingly criticized and censured Feyerabend — publicly and more so in private. I still cannot praise his mode of conduct,[42] but to the end I had hopes for an opportunity to be more compassionate with him.

I do not know if I have a distinctive philosophy, and if so, what it is; but I do hope I am relatively free of the stern poise that the philosopher has inherited from his predecessors and that Feyerabend rightly and painfully rebelled against, even if a bit childishly. I do hope that my philosophy is distinctive as one that expresses a somewhat considerate and friendly feeling for the unnecessary pain that humanity still inflicts on itself: there is too much suffering anyway and we should deeply regret any case in which we add to its stock. This vision is not mine; I learned it from my master, and if it has an originator, it is Democritus of Abdera, whose immense sensitivity to suffering shines through the fragments of his writings that are extant, and who was known in Antiquity as the laughing philosopher.

---

[42] I objected to the contradictions between Feyerabend's private and public assertions, not to his (public) assertion that I praise science because the capitalists pay me. After all, they paid him more. Putnam said this about me (in a public debate) too, and he did not raise any resentment on my part.

Incidentally, Feyerabend contributed a beautiful essay to the volumes of Jarvie and Laor in my honor. It appeared after his death. I regret I could not express to him my gratitude and appreciation.

CHAPTER EIGHT: THE MATTER OF INTELLECTUAL LEADERSHIP

Democracy is aristocracy without snobbery
(Russell)[1]
and though only few can originate policy, we can all judge it.
(Pericles)[2]

One of the earliest philosophical texts I ever read was a preface by Kant in which he declared his aim not to convince but to help his readers to think for themselves: it matters little who is the originator of an idea, he added, in comparison to its merit. I was deeply impressed, not knowing that this was the sentiment commonly expressed in the Age of Reason: the Royal Society of London had as its motto a sentence from Horace, saying he swore no allegiance (yes, it was used as a motto to Chapter Two above). I did not know that the sentiment was not commonly practiced, much less that even Kant himself violated it when he dismissed the ideas of his best critic, Solomon Maimon (branding him a parasite)3. I found the idea very powerful and I still do. My claim that I follow Kant on this matter and do not try to convince anybody is often dismissed. Perhaps this is why I find it so disturbing that Popper often sounded as if he was fighting, not to say pleading, for the convictions of his readers4, especially his opponents who were established in the seats of learning,5 when he repeatedly sought ever more powerful and ingenious arguments against the already amply refuted but still popular image of science as resting on some foundations, on inductive proof, that allegedly philoso-

---

1 In this passage, I forget where from, Russell echoes Spinoza, *Political Treatise*, 8:11 and 11:2.
2 Pericles, Funeral Oration, Thucydides, *The History of the Peloponnesian War*, II, 37-41. Popper has it as a motto for his *The Open Society and Its Enemies*.
3 Solomon Maimon endorsed Kant's epistemology as a hypothesis. This deprives it of certitude and of the status as logic of sorts. Thomas Young soon had similar ideas. George Boole even wanted Kant's theory empirically tested! (Jakob von Uexküll and Konrad Lorenz proposed empirical versions of Kant's ideas [Popper, 1963, 377-84].) Kant alluded to this option repeatedly, but he could not overcome his contempt for hypotheses. Popper made an excuse for him: the corroborations to Newton's theory of gravity were overwhelming [Popper, 1963, 180]. Yet Maimon viewed Newtonian mechanics as a hypothesis. Kant viewed skepticism as destructive as he was unable to entertain consideration of verisimilitude. He insisted that even the slightest imperfection of a theory makes it a hypothesis and all hypotheses are "counterfeit" (*Critique*, Axv). This polarization between the truth and all else is still rampant. Thus, Leo Strauss wavered between considering the Hebrew Bible perfect and declaring it [Strauss, 1997, Preface]. Buber had said earlier: obviously, both views are exaggerated; the text is a distorted memory that we may try to reconstruct (Preface to *Moses*). Verisimilitude is commonsensical.
4 I confess at times I find Popper's admirably clear and direct writings slightly disturbing as his tool to fight for readers' assent. (This is delightfully absent from his posthumous *The World of Parmenides*.)
5 Where exactly are the seats of the commonwealth of learning? Popper said, not universities and not research institutes, as these are dogmatic [Bartley, 1990]. His alternative is the California-style intellectual fringe. This is flimsy. Popper's view of science as a game suggests that it needs no structure other than its rules. To this Jarvie responded in detail [Jarvie, 2001], observing that games require structured social settings. Wittgenstein caused much confusion here when, instead of withdrawing his hostility to metaphysics he stealthily reinstated it as a game and referred to it as rules rather than as social settings (speaking vaguely of forms of life).

phers ought to examine. The right attitude is that of Faraday: he, too, found it undignified to try to convince; he was eager, however, to put on the public agenda his thoughts and his criticisms of the established doctrines — so that all concerned could hear his arguments and judge for themselves [Agassi, 1971, 27, 31-3, 134, 1136, 151, 286]. And he nearly failed, since what determines the public agenda, as all other public intellectual activities, are institutional arrangements, not intrinsic merit. Any attempt to change the public agenda has to reach the agenda, and this is a catch: unlike parliaments, the commonwealth of learning has no steering committee. Its establishment controls its agenda informally.

To create controls over power we need more study and better education for autonomy. Our current scientific education is still wanting here. The revolution in physics made it noticeably easier[6] to put on the public agenda thoughts opposed to established doctrines in physics, but not enough. To date new thoughts enter the agenda at once only if they have famous origins or hit the public eye. Examples to the contrary still save the day. Most conspicuous among them is young Albert Einstein; he could not secure for himself a small position as an assistant to a physicist he appreciated (his letter of request for a humble position was accompanied by a return post-card; the post-card was not returned), yet he was greeted at once by Max Planck, the editor of the established journal to which Einstein had sent some papers, even though these papers were so far out that for years established physicists judged them cranky and unfit for publication. Planck was exceptional, though according to the creed of the commonwealth of learning he was normal. That is, according to the myth of the scientific community that is popular in it, his conduct is proper, not exceptional.

At stake is the rule of impartiality, such as it is. The regular violation of this rule has been repeatedly observed. Michael Polanyi has justified this by reference to intuitions of the leadership that, in his opinion, may overrule all protocol (as quite possibly Planck did). The leading sociologist Robert K. Merton christened one violation of the rule, the favoring of the established over the newcomer, by the title "the Matthew effect" [Merton, 1968]. (He did not say that the very existence of the effect is a violation according to his own view of the rules, since he viewed science as democratic, yet "the Matthew effect"[7] is clearly a description

---

6 It still is hard to finance research projects that promulgate heresies [Bronowski, 1971, 16]:
"The shortcomings of present international organizations are trenchantly described by one of their most distinguished servants, Gunnar Myrdal, in his Clark Memorial Lecture given at Toronto in 1969. Myrdal calls them inter-governmental organizations because, he says, they are only 'an agreed matrix for the multilateral pursuit of national policies'. His account of the national pressures on the secretariat is a sad but salutary catalogue of intrigue, deceit, spying, and open threats."
7 Merton explained: his label "the Matthew effect" alludes to the Gospel verse that promises to enrich the rich. Merton recognized here discrimination in favor of the incumbent, contrary to his view of research as socialist and freely competitive. By neo-classical economics, competition excludes discrimination. It

of the meritocracy[8] that is now so rampant in the scientific community. He must have noticed this point: later on, he came ever closer to Polanyi's view that the intellectual élite governs science.[9])

This is the case in the circles of the philosophy of science or of scientific philosophy, where established professors and their associates are busy defending science and its practices against nobody in particular by techniques alien to the spirit of science as they themselves present this spirit [Agassi, 2003, 152-63]. The conspicuous example of the use of this technique is the way the philosophical establishment still treats one of the greatest philosophers of all times, Karl Popper, who was recognized first by the scientific community and only later, to some extent, by philosophers of science: he is the first philosopher to have been elected (in the mid-seventies) as a philosopher to be fellow of the Royal Society of London[10]: traditionally, its positivism was so strong that Russell was elected fellow (early in the twentieth century) only as a mathematician. The election of Sir Karl Popper was explicitly not by the positivist book: it was the Society's way of surreptitiously relinquishing its official positivism. Established leaders in the philosophy of science will not dream of attempting to criticize the august mother scientific society; and they will not reconsider positivism: they may give it up at a drop of a hat of any fellow of the Royal Society of London, but they will not rethink. There are exceptions, of course, but I am speaking of the established leaders in

---

does not. This refutes the theory. An excuse for this is the view of discrimination as the economically preferable default option, as one that will undergo correction later if need be (statistical discrimination). The infrequency of this correction refutes this excuse [Buber Agassi, 1985].

8 The Matthew effect is rational: it rests on the supposition that a published author is more likely to produce a publishable paper than a novice. Editors and publishers favor papers that attain praise and shun those that attain blame. By the incentive system for editors and publishers, fear of blame is a stronger motive. Consequently, doubt prompts rejection: editors cannot grant authors the benefit of the doubt that they owe them. So most of what the learned press publishes is now *a priori* marginal. (An admirable exception was Gilbert Ryle, editor of the prestigious *Mind*, who discriminated in favor of novices. *Mind* remained boring, but for different reasons.) The prestige of a journal then does not reflect the value of its contents. The Matthew effect is thus refuted whenever prestige counts more than publication records. An examination of publication indices will show with ease how common this is. (Prestigious periodicals and interesting ones overlap but do not merge.) My experience refutes the Matthew Effect: though I am very well published, it is still hard for me to publish my work [Agassi, 1990c]. Of course, possibly my published papers have some merit but not my rejected ones. Yet they often are accepted for publication after they are repeatedly rejected. The Matthew effect is indifferent to quality anyway: it says that regardless of quality my works should be increasingly welcome. Perhaps it is merely statistical. It should then undergo statistical tests.

9 Young unknown rebels or dissenters find it too hard to publish. Merton does not say that the worthiest will persevere rather than drop out of the game, but he suggests this. He is thus complacent about the wisdom of the system (as was Polanyi) instead of suggesting ways to improve it. Perhaps this is not urgent: the information highways can alleviate much of the frustration of young unknown rebels or dissenters. The information highways are terrific despite their immense duplication of rubbish and despite the inaccuracy of most of the information that travels there and despite of its bringing no prestige. So, as ever, the task is to try to improve the system.

10 This is not literally true. Due to its positivist ancestry, the Society has no rubric for philosophers. Russell entered it as a mathematician. Popper entered it in the category of odds-and-ends. This grieved him, but he was placated as he heard that the same held for Churchill [Watkins, 1997].

the philosophy of science. Were there among them leaders like in science[11], then they would have put an end to this charade. As to the underground, they were traditionally anti-positivist to this or that extent. Even Kant, arch-positivist that he was, said, "we shall always return to metaphysics as a beloved one with whom we have had a quarrel" (*Critique of Pure Reason*, A850; B878). But the point is, leadership is always a complex matter.

The idea that there is an established leadership whose task is social and an *avant-garde* leadership that forges new ideas is the idea of a two-tier leadership in contrast to the Polanyi-Kuhn single-leadership theory.

In my book on Faraday I have marshaled interesting and moving material: he complained clearly enough about being stonewalled, and clearly he was greatly disturbed by the silent censorship of his concepts, which were highly metaphysical [Agassi, 1971, 151]. The material I reproduced there clearly is in clear conflict with the established pen portrait of him by L. Pearce Williams as a sedate member of the scientific establishment [Williams, 1965]. A reviewer in *Science*, the official organ of the American Association for the Advancement of Science, reported [Finn, 1972] that my book is worthless as it is out-of-date as it merely replicates that of Williams. Science does not often display such conduct. Lakatos invited Williams to write a review of my book in *The British Journal for the Philosophy of Science*. It was titled, "Should Philosophers Be Allowed to Write Histories of Science?", and became an instant classic (as those who refuse to refer to my book but wish to prevent the charge that they are ignorant of it refer to this title; even my former student, Alan Musgrave[12]). The answer Williams gave to the question of this

---

11 Published information on scientific authorities is too scant. Polanyi uses it as means to reestablish scientific consensus in the absence of induction. So he only insisted that each domain has a leadership that curbs dissent. Kuhn went further and declared it a necessary condition for ascribing scientific character to a domain, on the understanding that leaders proscribe all dissent. He then reneged and allowed for bi-paradigms and multi-paradigm in science. Both Polanyi and Kuhn claimed approvingly that scientific leaders decree changes of orthodoxy. This is a post World War II phenomenon.

Until the French revolution all universities were officially authoritarian with rationalist pretensions. Mediaeval and Renaissance seekers saved the scanty freedom of thought, and they were mystic loners and itinerants who occasionally landed in courts (Harry Wolfson). These seekers lived in intellectual fog in which they kept curiosity as yearning for salvation and as keeping awake the critical spirit in opposition to the smugness of the universities. The Copernican revolution was primarily critical: Copernicus opened his discourse saying, as the Greeks disagreed among themselves, they are no authorities. Then Francisco Buonamico recognized the contradiction between Aristotle and Archimedes. His student Galileo then rebelled against the Aristotelian tradition (and became a courtier). He admitted in the opening of his first great dialogue that the Copernicans-Pythagoreans were unclear, and he promised to change that. Although some leaders of the revolution were courtiers — Kepler, Galileo, Gilbert, Harvey, even Bacon and Descartes — Copernicus was neither; Buonamico was a university professor. The rise of the Royal Society of London changed that structure: research was then the domain of the amateur scientific societies. Universities began to change after the American and French Revolutions.

12 This Musgrave has corrected. His [Musgrave, 1999, 257] refers to Williams correctly, with proper mention of my part. He legitimately ignores there Williams' specific censure, speaking only generally against censorship. Most references to Williams' essay still play the role mentioned in the text here.

title was "a resounding No!" [13] Philosophers, he recommends, should not be allowed
Should not be allowed by whom? By the established leadership, of course. The once so very famous member of the "Vienna Circle" Hans Reichenbach wrote a review of Popper's *Logik der Forschung*, you remember, saying similar things about it; I am in good company. I once asked established Harvard professor, Hilary Putnam, what in his view was the contribution of Reichenbach to philosophy? Joe (this is how most of my American friends address me), he responded with no hesitation, Popper is not the only philosopher around. I found this answer disturbing and began writing a paper about the situation, as I usually do when I encounter anything noticeable. I gave up the project as useless. Was I right? How should one respond to such a situation? Should I have attended to it? More importantly, how should the underground face it?

After all is said about the irresponsibility of today's established leadership, Putnam and all, one may observe that most underground leaders ignore such matters; the rest do all they can and regularly bump into frustration. That most underground leaders prefer to avoid clashes with established leadership is understandable: they prefer to concentrate on their intellectual work and are content to be left alone, as Caleb Gattegno[14] has observed so philosophically. He was a giant educator and grand underground leader.[15] The interesting and useful question is, how should underground leaders behave? This is problematic, since underground leaders do not stand for election, so that no one solicits their consent, and so they have no executive duties and, as their abilities are limited, so are their moral obligations.

A general theory of leadership should serve as a framework here. The Enlightened philosophers of the Age of Reason declared each individual autonomous and so unable to follow a leader (the autonomous endorsement of suggestions that some other individual makes, be the other individuals leaders or not, counts as an autonomous act, not as being led); hence, the so-called leaders are merely the individual citizen's representatives or delegates (and so they are still called in the United States, the child of the Enlightenment Movement). The Romantic Reaction to the Enlightenment declared common citizens heteronomous, in need of guidance (from autonomous leaders who are themselves guided by strong pas-

---

13 Williams mentioned errors in my book on Faraday as proof that I am no historian. So he has no choice but to conceal my criticism of the errors in his book on Faraday, even though he knows that my criticism is valid and is more significant than his, as it concerns a central question: did the Establishment recognize Faraday as a thinker?
14 Gattegno considered [Gattegno, 1970, 101] progress that
"... the Establishment burnt some of the seers, jailed others, exiled others, prevented some from making a living. Today we may be less ready to burn but as ready to ignore."
15 Gattegno worked at the University of London in mathematics education in 1946-57. I met him in his office there in 1954 as I accompanied Fraenkel when he paid him a visit.

sions, not by reason). The Enlightenment theory is too good to be true and the Romantic theory is too true to be good.[16] It is not realist but barbarian. We need a theory whose description is no adulations and whose application may help raise the general level of autonomy. (This situation is not peculiar to the theory of leadership.[17] It should apply likewise, for example, to the two polar theories of the role of *avant-garde* artists, the one that presents it as that of servants of the public and other that presents it as that of people free of all concern for the public. These theories should give way to the balanced theory that recognizes that role as educational in some sense.)

The theory of leadership presented in Popper's *The Open Society and Its Enemies* breaks new ground. It is the very best there is, yet it is incomplete; viewing it as complete leads to an indifference to the choice of leaders. It may be less responsible to ignore the leaders than to discuss the question at hand, namely, what leaders we want and who do we want to lead us, and, more significantly, where to should they lead us? Briefly, this is traditionally the central question regarding leaders: what do/may/should we expect of them. Popper criticized this tradition. Consider the positive and the negative questions: "who should lead?" and "who should not lead?" Since we cannot as easily agree about the positive question as about the negative one, it is better to give up the traditional positive question and return to the Socratic negative one. This renders quite central the important practical question: what can be done if we have the leaders we do not want? And indeed this important question should stay topmost on the political agenda of every nation. Democracy is repeatedly defined in *The Open Society and Its Enemies* as any regime in which those who are governed (the people) are able to overthrow the ones who govern (the leadership) without bloodshed or revolution.

This is very important. I have formulated it here in a way that raises a few bothersome questions. In the way I have formulated it, I have identified the governed with the people and the governors with the leaders. Both identifications are questionable, to say the least. (The first identification overlooks the difference between self-selected human groups, like voluntary organizations, and given ones, like tribes. The second identification overlooks the very prevalence of two-tier leadership, discussed above, even though, notoriously, when legitimate rulers, for example, the

---

16 [Gombrich, 1974, 948] argues against the polar views and speaks of "social testing" of aesthetic judgments, drawing attention to the growth of taste that is at times counter-intended.

17 [*OS*, i, 120-1] rightly criticizes Plato's question and replaces it by a better one: rather than ask, who should rule, we should ask, what institutions prevent tyranny? But, Popper wisely adds, after that we can return to Plato's question, though of necessity it will be in some institutional garb [*OS*, i, 122, 137, Conclusion to Chapter 7]. Thus, it is not that Popper overlooked the need to take his view a step further, but that he left it incomplete. In this respect Russell went further. He said earlier, "The merits of democracy are negative" [Russell, 1938, "The taming of power"] and he then tried to go further. As he did not put all this within a negative philosophy, he had much less impact than Popper.

hereditary ruler of the day, prove inept in moments of crisis, then it is the chosen adjutant who may — or may not — save the day.)

Many great critics of democracy, from Plato to Shaw, declared that decisions by popular votes do not bring about the best option available; that, in particular, the popular choice of political leaders does not secure the best candidate available. Some democratic thinkers admit this and say, democracy is not perfection.[18] Others, indeed the most popular, spoke differently: *vox populi, vox dei*, they used to say (or, as it was popular to say in the United States of America, you can't fool all of the people all the time [McClure, 1904, 184]): the standard defense of democracy was the dubious claim that decisions democratically arrived at are the best. We may agree that democracy is the best regime available without agreeing that the popular vote is a supreme court. This is all too obvious, since every one of us disagrees with the majority from time to time. This is the measure of the profound revolution in the general mode of thinking that has been effected by Russell, Einstein, Churchill, E. M. Forster[19] Somerset Maugham, and Popper, if not simply the shocking, tragic upheavals of our times, that this is no longer so: the most ardent defenders of democracy are no longer as optimistic as its past popular promoters were.[20]

---

18 Popper's negativist view of democracy is one of his greatest achievements, intellectual and practical alike; it is now the consensus. It is has met with too little scholarly attention; most experts still dismiss it as they have to if they wish to continue the traditional discussion, as they do. The traditional problem is that of sovereignty, so-called, namely, the question, what justifies governments and which government is justifiable. We should dismiss this problem, said Popper rightly. Incidentally, the very problem, or the presupposition to it, imposes choice between support of all governments and support of only the best, between possible tyranny and impossible utopia — in unintended disregard for democracy.

Jürgen Habermas agrees with Popper as he finds insoluble a problem that he ascribes to Hegel: how can a perfect system be improvable? He then adds [Habermas, (1999) 2003, 47, 209],

"the only thing that has made Hegel's problem more tractable is the fact that the proceduralist mechanisms of the constitutional state have turned the process of the realization of civil rights, through an institutionalized democratic practice of self-determination, into a long-term task. This is a task that, according to Hegel himself, should not even exist."

Carlos Nino says [Nino, 1996, 47],

"A central presupposition of the post-Enlightenment practice of moral discussion is that every authority or convention is subject to criticism, except perhaps the practice of criticizing. The role of criticism is associated with liberalism because this ... reflects the value of moral autonomy. In effect moral discussion is designed to overcome conflicts and achieve cooperation through consensus."

A tacit "presupposition of ... practice" is important, of course, but tacit assumptions often go with explicit statements to the contrary [Watkins, (1958), 1987]. Nino says he "goes beyond Popper" (38), If so, then it is in his justification of democracy. Popper rightly preferred efforts to improve over efforts to justify. Thus, a tyranny becomes more democratic by tolerating mass demonstrations although it still is a tyranny and a government by an élite becomes more democratic as it widens the circle of that élite. This observation, already made by Spinoza, got lost in the shuffle because of justificationism.

19 E. M. Forster says, "Democracy has another merit. It allows criticism ... This is why I believe in the press, despite all its lies and vulgarity" [Fadiman, 1939, 82], [Forster, 1965, 77].

20 There is scarcely a better expression of egalitarian optimism, than the famous one of Rousseau: "Man is born free but everywhere he is in chains." Russell rejected it [Monk, 1996, 318-19], as did Forster [Forster, 1965, 21], who translated the second half of a dictum into an observation: Man is born in chains. Popper said, we do not know: he advocated cautious optimism conducive to action but not to

If democratic rulers are not the best possible, who is? This is Plato's question. Plato's answer is simple: the wisest, the bravest, the most decisive, etc., etc. Unfortunately, responds Popper, this is obvious and no answer to the question: if not democratically, how do we choose the ruler? Robert Boyle, of the celebrated Boyle's law, said (*Occasional Reflections*, 1665), a theory of who has the right to political power is not helpful: whatever characteristic the theory will present as the one that justifies power, the powerful will acquire that characteristic by force. How sagacious. Popper agrees, of course [*OS*, i, 266].

Spinoza considered election the minimal right to participate in government; he argued that the rulers will never be the most suitable and that no known restriction of the vote is justifiable.[21] These negative arguments go well with the anti-Utopianism that both Boyle and Spinoza declared, yet the final plunge to the fully negative political philosophy remained for (Russell and) Popper. Bunge constantly rejects Popper's thoroughly negative attitude, stressing that we need the positive as well. This is a slip: negativism never denies the need for positive input; it is but the claim that the positive input is a gift, nevertheless to be suspicious of, to examine severely; the negative is more reliable.[22] Hopefully the process will improve the overall situation, but with no guarantee. No, says Bunge: we need not only negative arguments but also positive ones. That too is undeniable: we support our proposals by reference to our aims and by explaining how we hope to further them by our positive choices. But it is all tentative, hypothetical. No, says Bunge, we have empirical support for our hopes. Again, this is undeniable: often the law does not allow us to apply innovations in the open market until they gain empirical support. What then remains of negativism? Fallibilism. And possible pluralism. Bunge will not object, of course. This is why it is not clear to me where they disagree.

Fallibilism should liberate us from the constraints of tradition, from the demand to rely on tradition and from the demand to reject it. It also frees us from the need to choose between the best solution and no solu-

---

utopianism. For my part, it seems to me that granting the Lord the benefit of the doubt is sufficient inducement for action.

21 Spinoza, *Tractatus Politicus*, 11:2. In one passage (11:3) he qualifies his demand for universal suffrage and excludes those who are not accountable: foreigners, minors, and criminals. He then (11:4) also excludes women, although hesitantly: he adds that if we find one case history of a successful government by women, then they should qualify for universal suffrage. His small (and almost concluding) paragraph (11:2) is packed with ideas. It says, different systems of choice of rulers may lead to similar results; aiming at choosing the best rulers is not advisable; and the absence of rivalry among candidates for ruling leads them to lawlessness most. [Hence, elections are means less for selecting the best and more for preventing the cult of personality.]

22 Instrumentalists do not like to speak of refutations. Rather, they speak of the limits to applicability, which amounts to the same thing as they admit that domains of application are delineated by refutations. Inductivists should reconcile this with their doctrines or else relent. Their inductivism blocks this move as it requires of them to forget refuted theories. It is thus a reinforced dogmatism.

tion at all. The great methodological asset of the philosophy of Karl Popper is its avoidance of the polar options: it suggests most solutions as neither the best conceivable nor the worst [Sassower and Agassi, 1994]. And this looked to me sufficiently important that I wrote a book about it (*Towards a Rational Philosophical Anthropology*, 1977). I argued there that the classical and grandiose Greek polarization to nature and convention is responsible for our tendency to sway between extremes, such as optimism and pessimism. This violent oscillation is now useless: the polarization has been superseded by the Einsteinian revolution: Einstein did not claim that physical science captures the nature of things, nor did he admit the desperate view that scientific theories are merely arbitrary, accepted (by convention) as mere instruments for prediction. Rather, at best theories are approximately true. This is the Einstein-Popper approximationist philosophy of science; today most scientific philosophers advocate it, yet not consistently so; far from it.[23] It remains to say, institutions too are neither natural nor conventional but at times they stand in for the unknown natural arrangements; otherwise they are ways of doing things. In either case, they are means of coordination [Agassi and Jarvie, 1987, 147]. Popper nearly said all of this, but not quite: he published his approximationism only after his *The Open Society and Its Enemies*, and, not having admitted this as a huge shift, he did not come to recast his book in his new light.

It is thus worthwhile to scrutinize Popper's writings for consistency on these aspects and correct them when necessary. For example, as he rightly says, we do not know who should rule, and we are better off when we are able to replace a ruler than when we are stuck with one; nevertheless, we can and should discuss and decide who and how we should elect to powerful political positions among the available candidates and how we may try to prevent the elected rulers from using their power to specific ends that we prefer not to pursue. Although we do not know who the ideal ruler is and what the ruler's ideal task is, and although, as Popper has stressed, democratic control is more important, we may still guess and improve our guesses on the positive side too. This is vital, as the dismissal of a series of leaders who are unable to act when confronted with urgent tasks may easily lead to the loss of trust in democracy. This is the criticism that Judith Buber Agassi made of *The Open Society and Its Enemies*: the book offers a recipe for a minimal democracy, and as the first order of the day; this is excellent, but it should not be the only order

---

23 Russell took for granted the demand that past successful scientific theories should be special cases and approximations to their successors [Russell, 1996]. Popper's originality here is in his use of this idea, not in his statement of it. The same holds for Darwin's use of the idea of natural selection. As Shaw said, the idea is trite, since farmers regularly employ selective breeding. What he refused to appreciate was the force of Darwin's new use of it, that is, his use of it as a commonsense solution to problems that stymied his predecessors. The opposition to approximations is traditional — ever since Galileo and Newton. This shows the audacity and novelty of the Einstein-Popper view [Agassi, 1990].

of the day: there is a need for a more positive part of the theory of the leadership, to say what the leaders of a democracy have to do in order to perform their duties. In her doctoral dissertation[24] she suggested the demand for the involvement of as many individuals as feasible in the political process in order to insure that the government act in the public interest. (Popper would not object to this addition.) Meanwhile pluralist participatory democracy has made such demands; at the time, her dissertation advisers could hardly see the point (a case of time lag and of a loss of opportunity). Jarvie has generalized matters and said, though Popper's anti-utopianism is correct, since utopianism leads to fanaticism and to ruthlessness, we may nevertheless have images of utopia to compare and to examine critically, and the higher stages of the examination, then, may constitute careful, tentative attempts at its implementation. (Presumably, Popper would not object to this addition either.)[25]

The Open Society and Its Enemies, one should say in fairness, is only indirectly a book on democracy; it is primarily a contribution to the debate between the collectivists who prefer society closed and the individualists, who wish it to open up as much as possible; the democratic regime, then, is merely the political tool for the achievement of the desired end, for the purpose of keeping society increasingly open, making its members increasingly free. This is magnificent, of course; it goes better without the message, repeatedly presented in *The Open Society and Its Enemies*, that this should do: a democracy able to rid itself of wrong rulers should in addition demand of its rulers that they lead the people towards the desired openness,[26] very much in the way Spinoza had envi-

---

24 Judith Buber Agassi, *The Role of Local Government in the Working of Parliamentary Democracy: A Comparative Study of the British, Belgian and Dutch Systems*, Ph.D. dissertation, The University of London (The London School of Economics), 1960, unpublished.

25 Regrettably, Popper disliked Jarvie's advocacy of non-historicist liberal utopias. Josef Popper-Lynkeus, whom Popper mentions in his autobiography (§3) as an early influence and as the originator of the idea of the welfare state, may serve as an instance for this. I regret Popper never discussed all this. He advocated active, though minimal, state protection (especially Keynesian, not monetary but fiscal) state intervention in the market. He discussed it in his correspondence with Hayek early in the day and he recommended there that health insurance should be obligatory but private (in parallel with third-party motorcar insurance). Sadly, this idea suffers utter neglect. So is the idea that governments should not dispense student loans but guarantee them in ways that should be acceptable to all parties.

26 Popper admired Popper-Lynkeus's views about the welfare state. He deemed it advisable to leave them for another day. The two Poppers (incidentally, they were distant relatives) had very different starting points yet their views overlapped. The older one began with the demand from the state to cater for the citizen's elementary needs. He recommended avoiding radical changes. The nearest that anyone else came to this program is Abraham Maslow, and he fought more for the recognition of the needs of simple folk, especially spiritual needs, than for the question, who should cater for them. His discussion of spiritual needs had enormous value for the movement for the improvement of the quality of working life (QWL).

A significant suggestion of Popper-Lynkeus was to reform the conscription system: he suggested raising the age of conscription considerably, as means for raising public pressure on governments against trigger-happy belligerence. Advances in technology render this proposal increasingly reasonable.

saged: he considered legislation an instrument for the education of the citizenry for more enlightened way of life.

To put it in jargon, Popper's theory is the first to have offered a systematic negative philosophy that includes a theory of democratic leadership, assuming that both in knowledge and in politics the negative has priority over the positive. This is true, and in addition negative philosophy invites a positive theory of democratic legislation[27] and leadership — as a supplement to the negative one. It is easy to present one very much in line with his views on the matter, yet he preferred to ignore all this.

This omission would be no disaster were Popper to act in his capacity as an underground leader according to commonsense and intuition, as most people do most of the time, since the addition presented here is no deviation from his own guidelines. A consistent adherence to one's philosophy has its cost and its benefit: its benefit is that it enables one to use the best theory available (since commonsense often lags behind); its cost is its encouragement to use a theory that is possibly not as good as sheer commonsense (as commonsense is pliable and it permits *ad hoc* adjustment and flexibility even without the benefit of advanced theorizing). Russell systematically preferred commonsense to deep philosophy: he observed early in his career that holding a theory after bumping into some of its cruel implications imposes the choice between humaneness and consistency, and he opted for humaneness whenever possible [Russell, 1896]; [Russell, 1959]. Is it then more humane to follow a negative theory of leadership or to supplement it with a humane positive theory of leadership? This question is very difficult for all except the advocates of classical liberalism, of minimal government, as they reject all positive theory. In his *The Open Society and Its Enemies* Popper rejected classical liberalism [*OS*, ii, 125, 140, 330, 334], the theory of minimum government, and even declared it inconsistent [*OS*, ii, 124, 140, 179; see also 169, 327, 335][28]. Yet, perhaps under the influence of F. A. von Hayek, he endorses a theory increasingly favorable to minimum government. I ac-

---

27 Spinoza, *Political Treatise*, 11:2, accepted Maimonides' view that the educated have no need for laws. (Spinoza's *Ethics* is a non-sectarian version of *The Book of Science*, the first of the Fourteen Books of Maimonides' famous *Codex*. Both books take their readers' autonomy for granted, begin with some version of the ontological proof and then develop a naturalist metaphysics.) In their view of laws as rules of conduct, they display an oversight of their role as means of coordination [Agassi and Jarvie, 1987, 147].

It is easy to block democracy by the reasonable demand to limit it in emergency plus the fairly reasonable idea that emergency is lasting. This invites keeping on the agenda of every democracy the demand for some measure of normalcy and tranquility. The tolerance that Israeli citizens show towards continued emergency allows for the deterioration of the democratic aspects of Israel's regime.

28 [Radnitzky, 1982 and 1995] combined Popper's reformist politics with Hayek's idea of utterly free market — admirably not covering up the disagreement between Popper and Hayek, and declaring Popper's dissent from Hayek an embarrassing error even from his own viewpoint. Radnitzky recommended that we should do without democratic controls as they are weak. His arguments are sound but his recommendation is not; hence, they are useful means for the improvement of democratic controls.

companied him to the Austrian College in Alpbach in 1954, and we argued the matter in the discussion period after a very exciting lecture by Hayek. Popper responded to the critical assault that Feyerabend was unleashing with my help, defending Hayek as best he could. I knew then already that I would never agree with my mentor and master on political philosophy, even though by then I had not developed a theory to my own satisfaction. This came decades later, when I came to reject his identification of all nationalist movements and ideologies with the Romantic ones; moreover, I now consider the erroneous identification of all nationalism with Romantic nationalism the outcome of the classical liberal theory of minimal government: it leaves no room for the view of the citizenry as a nation, and its adherents insist that admitting the existence of a nation is support for the Romantic theory which demands from the individual conformity to the national ethos and destiny. Now the classical Enlightenment insistence on the claim that only individuals exist leads to the psychologistic theory of society which Popper rejected; it is also politically dangerous as it dignifies politics on some psychological grounds and thus it supports conformism as means of making individuals coalesce into a politically viable union. Popper stressed that a contractual consensus is better than psychological uniformity, both descriptively and prescriptively; yet he does not present this as a matter of the national consensus;[29] particularly since the contract is not a matter of ideal society but of the acceptance of the best there is, as Popper always stressed, it seems to me, for what it is worth, that the extant need for a sense of national identity in a modern nation-state is obviously basic.[30]

---

29 Popper-style political philosophy has to address the consensus, and to do so in manners different from those that the classical liberals employed. Tradition allows for only two options, both Rousseauvian and both unpalatable: a sum-total of individual aims, and a collective aim. Now sum-total of individual aims are fictitious, as aims cannot undergo summation without distortion. And averaging aims may, and often will, do injustice to all of them (as Ivan Krylov noted: "The Swan, the Pike, and the Crab"). As to the collective aim, it does not exist [OS, i, Chapter 7, note 23]. (The national aim is restricted to some shared aims such as peace and tranquility.) The consensus is an institution. Politicians treat it as such. Political theory should recognize this as a significant fact.
30 The appeal to the sense of identity is romantic and dangerous. Popper discusses this sense in his "Zur Philosophie des Heimatgedankens", 1927, that is the second item on his long publication list. He recognizes there the need and recommends diverting it to productive channels. Today we can go further and reject the Enlightenment indifference to it as well as the Romantic reactionary view of it as unique, not to mention the appeal to it to justify some inhuman raison d'état. Today sociology recognizes a multitude of identities that every individual possesses and of the multitude senses of identity [Banton, 1965]. This is problematic as it is certain to create inner conflicts: eventually one is bound to find oneself torn between different loyalties. Here Georg Simmel's philosophy is at its strongest at least on two counts. First, it is very astute in its insistence on conflict as a fact of life and in its opposition to the prevalent efforts to ignore it, not to say eliminate it, as these are utopian. Second, it is the crux of Simmel's intriguing theory of the webs of affiliation and his theory of the stranger (the individual who belongs to different groups, able to act as a mediator between them in times of crisis [Segre, 1974]). The theory of multiple identities assumes relatively stable social roles.

Banton is a former student of Popper and somewhat my senior. When we met in the 50's I failed to see its boldness and strength. I regret the delay of my appreciation of it.

The question of political leadership proper, of government, is but a part of the general question of leadership; the new views on pluralism and on participation help, of course, as they are best understood as the recommendation to take all the decisions that matter to a given collective as its politics, for its membership to decide and control democratically; the proposal to promote any group within the body politic (on the two conditions, that the national cohesion is not thereby seriously threatened and that individual participation in any political activity is at liberty, not an obligation) and to involve as many individuals as possible in any decision process (on condition that the democratic process is thereby not violated[31] nor severely weakened), makes democracy a much wider concept than it ever was, even wider than direct democracy (since in direct democracy the decision process is confined to assemblies and referenda). This seems reasonable, and it throws light on the interface between people's political opinions in general and their conduct in their social environment that is not political in the narrow sense of the word. Let me utilize this here, as I wish to discuss the philosopher's very ambivalent attitude — both philosophically and personally — towards his own place in his professional community (especially in the bygone Vienna)[32] that is a political matter.

The chief characteristic of Popper's attitude towards colleagues is disregard for those whom he thought poorly of and repeated complaint about the others for the absence in their writings of expressions of the recognition of his contribution that is his due. This seems divorced from the role of the intellectual leadership, it is a view of research in a political vacuum. For, he assumed that colleagues needed his contribution, and that they should in all honesty have used it with explicit gratitude, though he could (and did) go on doing research while ignoring their work as unimportant and their social status and activities as irrelevant — to philosophy proper or at least to his philosophical research. This is a social theory

---

31 A popular criticism of parliamentary democracy is that it is license to leaders to manipulate the consensus. The alternative to it is the communitarian democracy of Gustav Landauer and Martin Buber that Amitai Ezioni is popularizing these days. There is no knowing whether their system renders public opinion less open to manipulation or more. In any case, it is better to control manipulation than to seek a system that minimizes it. Indeed, Ezioni is rightly concerned with democratic controls (and more controls then democracy).

32 That all and only professionals are expert is a popular prejudice. Young Popper was an expert philosopher but no professional. (He was a substitute schoolteacher.) I think this hurt him much. He never spoke with me about his experiences as a teacher, and he always described intellectual encounters as if they were all informal. He made it clear that for him their import never depends on settings. He received official honors only in his sixties and later [Watkins, 1997a]. His complaints were never about their absence but about the absence (or inadequacy) of references to his output in the learned press. This is an odd view of that press, especially for one like Popper, who held most of it in contempt. Perhaps he would have corrected this remark of mine and say his complaint was about peers. He was bitter about the editor of *Mind*, G. E. Moore, as he had rejected his *The Poverty of Historicism*, preferring to it a longer work by a Wittgenstein adjutant that is now mercifully utterly forgotten. As Popper (like Russell) did not consider Moore very much of a scholar, he resented the rejection as such. I suppose he coveted popular exposure as many of us do and did not admit it, less alone examine it analytically.

of research that is unacceptable even according to his own theory of research as socially based.

Let me discuss, then, his ideas first and his conduct later, thus bringing this melancholy account to a close.

The defects of Popper's view of leadership are not easy to spot. The first criticism of *The Open Society and Its Enemies* that impressed me, you remember, was that of my wife Judith, who has expressed appreciation of the negative theory of leadership presented in that masterpiece plus dissatisfaction with his oversight of the need for a subsequent tentative positive theory of leadership. This gave me the incentive to work out the social and political philosophy expounded in *The Open Society and Its Enemies* in further detail. I thus found myself well prepared for the teaching of Hillel Kook, alias Peter Bergson, who was during the Shoah a young national political leader without a nation, who in despair attempted to arouse a mass movement to protest the Allies' indifference to the plight of the Jews of Europe, which he found both morally shameful and pragmatically harmful to the war effort.[33] He organized a new kind of Washington rally (of 400 non-Zionist rabbis [Wyman, 1985, 152]).[34] His teachings were regularly misunderstood in his country when he did have a nation — the Israeli nation, the citizens of which still refuse to view themselves as a nation, as they labor under the influence of the Romantic nationalist theory (of a nation defined by shared faith, heritage, language, land and above all destiny). I consider Judith and Hillel Kook my teachers second only to Karl Popper.

Young Popper's interests moved from education to the psychology of learning, and from this to the logic of learning in general — to the philosophy of science. He was then driven to political philosophy as a philosopher of science, and he suggested that his contributions to one field are significant for the other. His idea that science is explanatory conjectures and their attempted refutations has its parallel in his idea of democracy. This is how a new philosophy evolved that fits parliamentary democracy. And this is how the theories of historical inevitability and of utopianism play a role in his philosophy that is so important: they are irrefutable, and so their ill effects can easily develop beyond the point of reversibility of their ill effects, whereas piecemeal social reforms are easier to check as

---

33 The standard response of Western political leaders to the Shoah at the time was, the best way to help the Jews of Europe is to win the war fast. Bergson disagreed. He said, the war effort could benefit from a declaration that the rescue of these innocent victims was a war aim.

34 The first protest rallies were of suffragists. The first march on Washington was of unemployed workers. The freedom marches differed from both groups, as their participants and organizers were not personally involved. This way they resemble the march of the rabbis for the rescue of European Jews that Bergson organized during the War. The fact that the leaders of the first freedom marches were Jewish and Christian religious leaders links these two kinds of marches. Bergson's activities met with sabotage from all sorts of establishments, including the Jewish and the Zionist, so that the US Holocaust Memorial Museum recognized his activities only in August 2007. See [Agassi, 1999].

they are refutable: their ill effects may hopefully be limited, truncated, hopefully reversed. This is not a matter of intellectual nicety but a point that lies at the heart of democracy. Utopian leaders are not easy to dismiss, and they insist on their programs regardless of setbacks, as their ideas are irrefutable. The appointment of leaders that is reversible without bloodshed, then, is a part of the piecemeal reform system, in contrast with the appointment of a philosopher-king who knows the laws of history with certitude and uses this knowledge to construct utopian plans to construct an ideal society once and for all. Philosopher-kings may induce pain like physicians, as they know with full assurance that the pain induced is much smaller than the pain prevented. (This already links the Soviet leaders with Plato. Yet those who are so ready to condemn Popper for his comparison of Plato with some totalitarian thinkers forget that Stalinists and other totalitarians did argue this way. Even before World War II Russell [Russell, 1920, 29] and Crossman [Crossman, 1937] suggested very clearly that Plato would have approved of the systematic cruelties of the Soviet authorities.)

One need not endorse Popper's parallel between science and democracy in order to appreciate his contribution, nor take it to task. In line with his greatest contribution (the idea of appreciative dissent even in science) I see here both the importance and the limit of this parallel. It was, to take a concrete example, the achievement and the fiasco of the Israeli Kibbutz movement that it had attempted to create the New Jewish Man, the socialist individual of their dreams: the outcome was very impressive but far from perfect and too costly: Kibbutz youths have no attachment to money but also no sense of financial responsibility; they have a high sense of valor and a high rate of militarism and war casualty. Popper's claim that socialism is linked with the conception of perfect knowledge permitting the construction of heaven on earth is thus insightful but an error all the same: such conceptions, even when treated as gospel truths, are at times refutable and even refuted. Regrettably he wavered between asserting that the Marxist utopian dream is irrefutable and that it is refuted (and hence refutable, as Joan Robinson has noted).

What is the attitude of science to these conjectures and their refutations? The conjectures that science handles and tests are not random: there are criteria for the worth of hypotheses even before the decision that they deserve taking up to undergo tests: they have to be highly explanatory and/or solve some problems, for example, of theoretical or of practical import. It really does not matter here what these criteria are; suffice it to notice in the present context of the parallel between science and politics that there exist criteria for initial worth: we institute tests to examine not any theory but only the adequate ones. Analogously, we do not try any proposal for the reform of a defective institution. He did not propose or discuss in his classical books the idea of a criterion for attempting to implement social reform, except to propose that reformers should always attend to the worst and most pressing evils. This is no complaint: it is not fair to complain that authors have not executed this or that task unless they have promised their readers to do so. Nevertheless, we should note a lacuna here: Popper has suggested an analogy, and I am now following it up.

The lacuna in the theory of institutional reform as presented in *The Open Society and Its Enemies* is also the lacuna in the theory of the choice of leadership as presented in that classic work as well as in the theory of the body politic.

Perhaps the parallel can work the opposite way: as a social phenomenon, science is a proper object for scientific study. We can attempt to apply our social theory to its study. The social character of science needs no discovery, yet tradition systematically ignored it due to a number of significant factors, each of which is very important to our present-day culture. First among these factors is the accent that during the scientific revolution advocates of science (Galileo, Bacon, Descartes, Boyle) placed on the autonomy of the individual; it was the most significant contribution of the Enlightenment movement to our culture. The slogan of the Enlightenment became, "I think, therefore I am", not in the sense that Descartes intended, namely of proving the existence of the self (especially since Descartes' near-contemporary Spinoza showed35 that in his philosophy this was merely a preliminary step, a warming-up, a step that was strictly speaking neither necessary nor sufficient), but in the sense that thinking is what makes one's existence signify: in the sense, that is, that the ancient Socratic edict, the unexamined life is not worth living, came to mean: the value of life is in contemplation.

The Age of Reason viewed the individual rather than tradition as a source of knowledge. This led to the demand to consider all social and political theories as variants of psychology. This is known as psychologism. In the theory of knowledge, psychologism is the view of science as the activities and beliefs of the scientifically active individual; in political theory, psychologism then led to the view of politics as personal relations between citizens and rulers. In economics it led to the view of economic activity as individual production and trade.

These days psychologism is on its way out: most active students of social affairs (including psychology) admit that psychologism is a (noble) failure. This failure indicates that even science, under whose banner the most rational of activities are conducted, is a social phenomenon and thus in need of some coordination and so in need of a leadership for one end or

---

35 It is hard to assess the influence of this assertion of Spinoza that "*cogito ergo sum*" of Descartes is inessential to his own philosophy. Tradition said little about this.

another, at least as moderators. The inability of science to function without a leadership is the most obvious argument against psychologism. But the *locus classicus* for leadership always was, and still is, (national) politics proper.

The most significant political idea of *The Open Society and Its Enemies*, that the most urgent business of politics is the democratic control of leaders, holds primarily for national politics. He presented it otherwise, with no specification of the rulers, of the body politic, or of the set of the governed who should control their governors.

Today Popper's idea does not sound as revolutionary as when it first appeared. Soviet Russia looked to us less barbaric when it began to allow its former leaders to die in their beds. This was not due to the popularity of *The Open Society and Its Enemies*, but that book was the first to characterize democratization as moves in that direction. There is no need to learn this from *The Open Society and Its Enemies*; rather, a theory that does not include this proviso is deficient; and, remarkably, hardly any political philosophy or grand-scale political theory includes it. Why do most political thinkers cling to this defect? Because they cling to the classical dichotomy between extreme individualism (psychologism) and collectivism (sociologism). The theory in question clearly is neither a reduction of society to psychology (as it invites institutions for the safeguard of individual liberty) nor a reduction of the individual to the collective (as it values individual liberty). I have called it "institutional individualism". Most traditional theories ignore even the parliament as a body that deals with ideas — thrashing them out and examines them. Individualists viewed it with suspicion as it is not scientific; collectivists saw it as an arena for exercising power and wanted power less inhibited.

The Enlightenment Movement, says Preston King (yes, the same person mentioned way back in this account), had a horror of power. Yet, by the very ideas of the Enlightenment, the assessment of power as an instrument should be that it is neither good nor bad in itself but open to good uses or evil.

There is a subtle twist here. Viewed as the control of individuals against their will, power is inherently illiberal and so evil; viewed as an unavoidable social phenomenon, it is totally different. King has passed the divide: for him power is a fact of life to be dealt with as rationally and humanely as possible.[36] This means that it should be under control, but it

---

[36] Consider the idea that (political) power is morally neutral. In one reading it is obviously false, since its use limits rights; in another reading it is obviously true, as it is inevitable. Either way, the view of it as a mere tool is irrelevant. The urgent question is, when is its use necessary? Pacifists may judge it wrong even when necessary. This is a stern attitude that most of us judge irresponsible. Alternatively, they may say, it is never necessary, and the total abolition of its use is feasible. This is an obvious error, since the use of (political) power is a part of control (over people), and this control is the consequence of the unavoidable division of labor, as it includes the concentration of power, such as government

also means that it should be under good guidance. Popper was right to put the matter of democratic control first, especially in a book whose aim is the political aim of combating tyranny. Yet he was in error when he played down the second point. This led him to fall back unintentionally on a sort of quasi-Hayekian, quasi-liberal-conservative attitude, despite his sharp deviation from it (in the very same book). The result was that he also missed the corollary to his view that all social institutions have political aspects and so are in need of leadership. The classical liberal theory has masked this fact by two steps: it deemed science the model of rational conduct, and it deemed science as devoid of any leadership due to its well-publicized total impartiality. This suffices to make classical liberalism utopian — erroneously, since scientific impartiality is never total. There may be scientific schools of thought, of course, and these are notoriously governed by leadership (opinion leadership is the term coined by Elihu Katz and Paul Lazarsfeld [Katz and Lazarsfeld, 1955, 3]). Science, however, as the transcendental domain of true opinion, needs no leader: the truth is there for all to perceive, or else the reasonable will suspend judgment. Only to the extent that the classical view of science is true this is true too.

So much for the classical liberal view of science; the next move is from science to economics, where the manageable industrial or commercial institution is implicitly taken as the model and the single or chief owner (or at worst the chair of the board of directors) as the leader who is appointed by no one and answerable to no one. Even recognizing the need to change morality[37] from time to time, not to mention scientific societies

---

(Max Weber). Yet this is no conclusive argument: it should lead to the search for conditions under which the use of force is avoidable. These may not exist, as the wish to lead is as ubiquitous as the love of power. Not so: this love is not evil. The love of control is more problematic, although it often hides behind the love of power: the evil is action that the love for control for its own sake dictates, as well as its use as mere bravado. It is no accident that the better rulers are those who are indifferent to power, as were a few leaders up to and including Pierre Elliot Trudeau. Successful reductions of political power were due to some ingenious suggestions, and this should be encouraged. Preaching against the love of power and control is worse than hopeless; even education against it cannot be effective. So we need democratic control over the powerful and we have to educate for the insistence on freedom and for the readiness to volunteer for political activity. This is not easy, as the educational system, even if it rests on the value of freedom and responsibility, systematically violates the autonomy of its charges. That is to say, the exercise of power by the educational system is the root of needless applications of power. Einstein said, it is easy to reform education: all that is needed is to deprive teachers of their power. This is an exaggeration due to his pacifism or anarchism. The right slogan is, the power of the school authority must be subject to democratic controls (Janusz Korczak).

37 Russell said, we need a new morality. It is unthinkable, he said, that industrialists feel blameless over careless laying off hundreds of workers. Now this is a matter for the legislature to take responsibility for and perhaps force employers to be responsible for their employees. The decision to institute the welfare state was responsible; but was it right? We do not know, yet Russell's criticism stands even if the arguments of Radnitzky against the welfare state are just. Radnitzky is right: Popper's assertion that there is a better third way is metaphysical. But the same holds for the view of Radnitzky that the free market system is always the best. Its advocates have not answered the smashing criticism of Keynes (in the long run we are all dead) or of Popper (observed suffering takes precedence over the rosy future that some theories promise).

or trade unions or political parties[38], already breaks away from this antiquated model, yet the worst aspect of it is the dogma that, in science and in the market alike, there is no need for political parties. To maintain the parallel between science and politics it is preferable to admit that parties run each of them. They do [Agassi, 1981, 164-91].

What is leadership, what is its role, and how can it be made more democratic?

The absence of a discussion of these questions is inducement to cling to Romanticism: though the Romantic notion of nations and of leaders is not serious, as long as there is no known alternative, it remains very influential. For, generally, if there are no alternatives to an option, no matter how poor that option is, in time of decision it wins by default.

One role of the leadership — of a responsible leadership, that is — is to notice the absence of a contingency plan or the poverty of existing contingency plans and to cater for the development of one. And this requires discussion, and discussion requires political parties and a parliament with a legitimate opposition and a national consensus and caucuses and lobbies and local chapters and more. But I am going much too fast: I have not introduced the pivotal concept of responsibility.[39] Astonishingly, the first philosopher ever to have introduced responsibility into political philosophy was Karl R. Popper. If you disbelieve this claim, then I salute you: you have noticed how incredible my claim is. Perhaps you can even try to look up the philosophical literature and try to refute me. If you succeed, I shall be grateful. (Responsibility was properly introduced by Georg Simmel and Max Weber earlier, but it was merely a part of descriptive political sociology. This, too is quite incredible.)

The reason that responsibility was kept out of the discussion is that there was no room for it either in the Enlightenment philosophy or in Romanticism: the one declared leadership unnecessary — in science as in politics — and the other declared the leader the spirit of their nation incarnate. And this should count as the greatest criticism of both. To repeat,

---

38 Notoriously, Washington and Robespierre opposed the institution of political parties. So did Marx and his followers [OS, ii, 162-3], advocating the dictatorship of the proletariat to prevent a counterrevolution so that when that risk will vanish the dictatorship would wither away together with class society and all political power. Marx never worried about the possibility of a Stalin.

39 I do not know how responsible President Eisenhower was as he yielded to public pressure and ordered the removal of contingency plans to resist a possible Soviet conquest and to survive a possible nuclear strike. This plan remained, presumably with his tacit consent. For better or worse, we still do not know. It led to valuable things like the communication highways. Nevertheless, the high cost of the Cold War and the fact that the West won due to the superior standard of living of its citizens render questionable the very need for the Cold War, and thus for its contingency plans too — unless it was necessary for the prevention of a Soviet attack, and then the contingency plans were an essential part of it. Was there a risk of a Soviet attack? Here doubt requires alertness, and of the kind that puts at risk the very way of life that needs and deserves defense. This problem is perennial and the current global terror only sharpens it. But then, at least today this is a familiar problem regularly publicly aired.

already Spinoza viewed elections as a means for forcing rulers to be law-abiding.

Let me expand on this point. The Enlightenment thinkers saw responsibility as derivative of autonomy and concluded that inherently or initially responsibility is a personal or individual affair. Consequently, political leaders were viewed as delegates or representatives, as ones to whom citizens delegated their own God-given responsibilities as a matter of technical convenience: the delegate or representative represents a constituency. Of course, there is something naïve and noble about this idea that hardly needs discussion. Yet the constant presence of both responsible and irresponsible leadership in all sorts of societies and in all walks of life invites serious, responsible thinking.

Popper introduced into political philosophy the concept of responsibility as distinct from that of autonomy, and for the reason that it is a political rather than a moral concept. Regrettably, however, he spoke only generally of responsible citizens, not specifically of responsible leaders. He was missing something, of course, but he already achieved great results. He noted that responsibility is something different from either autonomy (intellectual, moral or legal) or obligation (moral, legal or political): autonomy is a sine qua non, and moral and legal compliance is the heart of democracy — as everybody will admit without hesitation. Indeed, today political philosophers and political scientists insist that civilized conduct and the rule of law are necessary but not sufficient conditions for democracy. Not so responsibility. Traditionally most political responsibility was hereditary, and it hardly made a difference whether the hereditary ruler was autonomous or not.[40] Though modern society must breed responsible citizens in order to function reasonably well, their responsibility is largely a matter of choice as far as they are personally concerned. According to Marxist tradition, and to the classical socialist movement in general, the demand from self-conscious individuals for responsible citizenship is the demand to be politically active for the right cause (and go to the barricades when time comes). The doctrine of the open society sharply differs from this. It promotes only minimal responsibility as obligatory (morally) to oneself and (socially) to the law; every other item of (private or public) responsibility it promotes as a matter of free choice. *A fortiori*, no one has to lead (as both the Hebrew and the Greek classics emphasize: *Book of Judges* 9:7ff.; opening of the *Iliad*). Also, one need

---

40 This is not the whole story, of course: notoriously, there is no shortage of cronies, minions and courtiers, ready to control rulers too weak to exercise their responsibilities. (Already the supremely mighty Pharaohs were often under the full control of priests.) The fact remains: every functioning system has its decision-making institutions, but only individuals can decide (*ex officio*) [*OS*, i, Chapter 7, note 23]. This fact remains outside both versions of reductionism, individualism and collectivism. And so, responsibility precedes autonomy – both logically and historically: usually, decision-makers follow tradition; they are rarely autonomous.

not be a responsible parent as one need not be a parent at all. (Hence, authorities in an open society should replace irresponsible parents; political philosophers regrettably overlook this — perhaps even irresponsibly.)[41]

How well can science function without leadership? I propose that Popper's moral commitment to hard work and to sincerity vitiates his marvelous theory of science and prevents him from developing a more realistic view of it. Anyone who wishes to develop a view of science that does not idealize it too much should develop a theory that allows for the possibility that science may develop a dangerously corrupt leadership. (Feyerabend is right when he observes that whenever science is powerful, any idealized view of it, including even Popper's, is downright dangerous. He should have added that Popper had never intended to idealize science — though he tried not to debunk it either — and that Popper's idealization is by far the weakest of the views that are found among the philosophers who appreciate science. Feyerabend vitiated his best observations, I suggest, by his debunking of science and of Popper, contrary to his own better judgment [Agassi, 2003, Chapter 5.2].

Having spoken only of the responsibility of citizens, Popper was at liberty to view the leaders of his profession not as leaders (who may lead their community astray) but as ordinary citizens: the complaints he launched against them were not in their capacities as leaders. When he did speak of leadership, for example, when he observed that Schlick or Neurath was the leader of the "Vienna Circle", or that Carnap was its intellectual leader (of some unspecified crowd), and so on, he clearly dissociated himself from the situation he was describing. (He said he would have been happy to accept an invitation to join the "Vienna Circle", but he said so only after he presented it as a private seminar; private; I can imagine what trouble he would have caused Schlick were he to have come to the "Vienna Circle" as if to a seminar; after all, this he did even to Wittgenstein the Terrible by treating the awesome Moral Sciences Club as a private seminar.) Clearly, his ideal of the Commonwealth of Learning is a seminar; perhaps an endless symposium in which neither flowing wine nor fleeting flute-girls frustrate the fun of witty and exciting discourse; or perhaps an entirely amorphous market-place of ideas akin to the market-place of classical liberal economics, with some odd-looking figure or another standing near a fruit stall pestering the passers-by with haunting questions: much as he has criticized the classical image of the market, Popper remained partial to it. The task of the leaders — modera-

---

41 Ignoring official responsibilities irresponsibly is common. Political philosophers may do so by skating around issues of responsibility. Consider John Austin's *How to do Things with Words* [Austin, 1962], a once famous Wittgenstein-style-ordinary-usage text. Its paradigm is "I now pronounce you man and wife." Of course, "Please, shut the door" is simpler and much more ordinary way of doing things with words. But perhaps he was discussing *ex officio* acts. If so, then the tacit thesis of his book is that some acts are decisions and some of these are *ex officio*. This requires responsible discussion.

tors, masters of ceremonies, or traffic controllers, as the case may be — would then be that of securing the smooth growth of the economy — of tactile and ethereal commodities alike.

Popper never advocated the naïve naturalistic image of science; he always said that science is a social institution. His pioneering study of science as a social institution proper, in his classic *Logik der Forschung* of 1935, was motivated and made possible by his refusal to idealize scientific theory, by his ascription to the theories of science the status of putative truths that often turn out to be falsehoods; his realism led him to a realist view of science as a phenomenon, and he then declared it dependent (not on personal factors, psychological or otherwise, but) on institutions fostering criticism. This initiated the study of the interface between science and other institutions. Today it is Robert K. Merton who is credited with this initiation, and he is considered the father of the sociology of science. Whatever Merton's motivation was, he still saw science as ideal: he assumed that its products are ultimate truths, that its institutions perfect, and that all participants in it endorse the critical attitude faultlessly. His realism came in the form of a historical study of science as an institution resting on some institutionalized metaphysical foundations, namely, puritanism. Was this metaphysics perfect? Max Weber, who first presented the history of economics by reference to the same instituted metaphysics, evidently supported the modern capitalist world and its puritan metaphysical foundation, but not the religious part of the puritan metaphysical foundations. What is Merton's view on this I do not know, just because he mixed idealizing with reporting. My question here, however, is, how far did Popper distance his theory from idealization? His rejection of the image of science as perfection allowed him to reject with ease the image of its institutions as perfect. In a sense, he did just that: the institutions of science are merely those that encourage criticism. This is too perfect as far as it goes and it does not go far enough. Its encouragement of criticism is not sufficient and there are additional roles to consider. Popper ignored these other roles, and with them other aspects of science, as he was satisfied with the possibility of criticism, and he always warned against utopianism. This is sufficient as a program for developing his approach further in the same direction.

What then are the institutions that encourage the critical approach that is so vital to science and how efficient are they? How do these institutions operate within the academy — in science as well as in fields of study not quite within the domain of science (such as the history of ideas)? These questions Popper brought to the surface, but he never studied them. He never asked, who are the guardians of the critical attitude and to whom can we complain that the job of guarding it is not as well attended to as it should? He took it for granted that the safeguard and maintenance of the critical spirit of science is not within science but within democracy at large. Did he deem the normal safeguards of democracy suffice for the maintenance of science or not? If he did, then the very possibility of a violation of the critical spirit of science without violating the law makes the safeguard inadequate. (Remember that the law does not forbid lying except under oath or as a part of an attempt to defraud — as Feyerabend repeatedly argued and illustrated by personal examples.) If Popper did not say that the normal safeguards suffice for the maintenance of science, then we may ask, and Bartley did ask, what would? When pressed hard about the tyranny of science or within it, Popper went so far as to declare that his theory of science is not a description but a mere proposal. A mere proposal concerning the way we should look at science. And with no description made, the field is open to those who wish to study the facts as they are.

This is it, then. Approaching democratic society on the Popper model of science, I found in the image of society nothing to parallel the important idea that science employs some criterion of choice of theories to scrutinize. Approaching science on the model of democratic society I found there nothing on the question, how does science discharge its political responsibility? What is missing here is a viable theory of scientific technology as distinct from science and from other social institutions: unlike science, technology involves the responsibility of the applications of theories; unlike other social institutions, scientific technology handles the best ideas available. (Politics will be scientific in the sense of scientific technology, then, when its institutions will favor putting on the agenda the best political ideas extant.)

How then are scientific leaders to be controlled? I doubt that Popper could view this question as secondary. Nor did he maintain that the model of economic free market is the very best there is, as he stressed that it needs government intervention to correct it. The less government intervention in the market the better, even if some intervention is unavoidable, he said. (It is unavoidable even on logical grounds, he added, as the extreme version of classical liberal economics is inconsistent [*OS*, ii, 124, 179].) Presumably, government intervention in science is likewise inevitable, though it need not be so huge as it lamentably was during the Cold War. And as other kinds of intervention are also needed, more criteria for intervention are needed too, and criteria for choice of leaders of diverse sorts. All inescapable intervention requires control. And international intervention is inescapable, since advanced societies can no longer ignore the rest of the world: even before the end of World War II, Popper observed, the question of exporting democracy was inescapable. John F. Kennedy succeeded in whipping up public enthusiasm for politics precisely because he offered leadership in uniting the earth this way.[42] (That the promise was not sufficiently warranted is a different matter altogether. This is not an assessment of Kennedy's leadership or of the success of his diverse programs but evidence for the popularity of the claim that there is a strong desire for a humane leadership to unify the world.)

And so, the scientific model of democratic society and the democratic model of science are both wanting, except possibly when we take seriously the classical liberal idea of the free market as a global model — as envisaged by Hayek and his followers — and these days this idea is quite untenable [Soros, 2002].

The parallel here tried out, thus, simply does not work; trying it out reveals lacunae everywhere, lacunae that want attention.

I am badly stuck. I do not know how to proceed. I have simply reached the end of my road, and it is a blind alley. Popper always declared,[43] you surely remember by now, that there is no problem due to a failure of communication, in the sense that any sincere and serious and honest effort of any two parties to communicate has to be crowned with success. He declares this with all the sincerity and seriousness it invites and while protesting honesty: he agreed to have his honesty questioned if he failed here. So I do not know whether this sincere and honest declaration is a descriptive statement or a proposal. (The proposal will read, perhaps, never give up hope to communicate; never relax efforts in any attempt to communicate! Try and try again, and try ever harder!)

View this as another way out: declare this more as a proposal than as a description. It will not do. If the declaration is a proposal, then, surely, it develops into a description of those who endorse it and attempt to live by it sincerely and honestly. And then, as the statement becomes descriptive, can it then also be tested? Can we test the seriousness and sincerity and honesty of the philosopher himself without thereby outrageously causing him a great, unjust personal insult? I do not know, except that, generally, discussion should proceed without giving offence, quite apart from the particular fact that I wish dearly to avoid insulting the person

---

42 The problem of exporting democracy is old. Before World War I imperialism was the received solution to it. Popper approached it through international politics, in his discussion of curbing international crime [*OS*, i, 107, 113, 151, 260, 288-91, ii, 8, 258, 270-2, 278]. He hoped to influence the writing of the charter of the United Nations Organization. Had its coalition heeded his advice and put Iraqi oil wells under international control after the first Gulf War, it would have prevented much suffering.

43 Considered descriptive, Popper's assertion conflicts with that of Kant: "many a book would have been much clearer if it had not made such an effort to be clear" (*Critique*, Axix). Of course, there is an answer to this: we need the help of readers who will play guinea pigs. Also, he said, when we do not know if we understand a text right, we may ask, what argument will make the advocate of the idea that it contains withdraw it: for, the content of a theory is what it forbids. Alternatively, we can offer alternative hypotheses as to the intended or possible meaning of a text. This is common in literary criticism. (The once popular methods of Wittgensteinian ordinary language analysis and of Gadamerian hermeneutics are far from commonsense: they boast ability to decide the proper meaning of texts, unlike the theory of literary criticism that invites critics to examine alternatives with no assurance.)

who helped me in ever so many ways and who was serious and sincere to a fault.

(Here is another important function of leaders. Unnecessary embarrassments, such as the one I am just now stuck in, a leader can and should eliminate, partly by following proper procedures and partly by the personal interventions that procedure allows people in authority to employ. When Adolf Grünbaum called Popper a myth-maker I approached Hempel, the only person in American philosophy of science more authoritative than Grünbaum, and asked him to discipline him some; remembering the way Feigl had asked Popper to discipline myself for much less, I assumed that this was acceptable in his circle. I was in error: Hempel refused, hiding behind the pretense that it was a matter of a personal discord. I found this bizarre and slighting.)[44]

We are all prejudiced, confused, befuddled; hence we fail to communicate despite good will. This exactly was the philosophy current in the heyday of Wittgenstein, whose popularity was at its zenith just after his demise, when I was the philosopher's apprentice, and it is precisely contrary to that alternative that Popper then issued his declaration and challenge to those led by their noses to endorse Wittgenstein's fable that poor communication is the great obstacle, and that it is a duty to overcome it through all round clarification by attention to good grammar.

Popper offered valid, deadly criticism and a better alternative. The criticism is that the wish to get rid of all prejudices and confusions[45] prior to reasonable communication will postpone all communication indefinitely and thus preclude the experience of reasonable communication and thus cause the loss of taste for it; not practicing the art of communication until one is good and ready puts one out of practice and in ignorance of communication of any sort. Popper's alternative is to begin with the good faith that requires the supposition that communication is unproblematic; when a discussion hits a communication barrier, then sincere and serious and honest efforts should help to get it back on the road. (Strangely, both Oxford philosophy doyen Gilbert Ryle, and Quine's most famous disciple Donald Davidson, made the first half of this proposal — to begin in good faith. Neither made any proviso for the case of communication failure, presumably leaving it to common good sense. This is not good enough.)

On second thoughts, Popper would concede to the Bacon-Wittgenstein view that there is a barrier to communication, but insist that it is not pervasive, and that it lies elsewhere: there is one pernicious prejudice to avoid carefully: the idea that since we are right we should win

---

44 See Grünbaum's response to me [Agassi and Grünbaum, 1992].
45 Russell, we may remember, declared sheer humbug the idea of total freedom from all bias. He also reported that he had deemed himself unprejudiced about the Anglo-Spanish War until he read a Spanish history book. He then recommended asking foreigners to write such books.

the debate at all cost. Other prejudices turn out then to be easier to attack. Indeed, he did assume that our view of ourselves as certainly right is the only serious obstacle to fruitful dialogue, to the elimination of error by losing in debates gladly or at least graciously.

The demand to be right tempts to apply some foul winning strategy: to confuse, to shift ground, and to accuse critics of partiality to cover up the lack of an impartial answer to criticism. There are other options. One is that of holding only views that are confirmable but not refutable. Metaphysical theories, according to Popper's *Logik der Forschung* and *The Open Society and Its Enemies*, are such; worse still, metaphysicians are prone to present the confirmations of their theories as scientific validation, even though they are irrefutable: they can never undergo sincere and honest tests. Hence, Popper concluded in 1935 (and never recanted sufficiently clearly as far as I know), metaphysics is pseudo-scientific; in 1945 he did admit explicitly that his earlier identification of openness to criticism with scientific character is too narrow. And in later years he did engage in frankly metaphysical discourse. But he did not take back his early use of "metaphysical" and "pseudo-scientific" as synonymic, and he repeatedly fell back on this usage. He still deemed metaphysics confirmable but not refutable, and rightly so. Is it then no longer pseudo-scientific just because it is also open to criticism? It seems so. Is Freudian metaphysics then open to criticism? It seems so. It surely still is unscientific. Did Popper no longer deem it pseudo-scientific? No answer. Pity.

Popper should have explicitly rejected his early endorsement of the positivist equation of metaphysics with pseudo-science; it is both incorrect and unnecessarily hostile. Some of the metaphysicians who advocate the most obscurantist and most brutal ideas around, as well as some of those who advocate the loveliest, most humane ideas, share the opposite characteristic: they show respect for science at a safe distance and do not display any scientific pretensions. The problem is general: how else can one present a metaphysical idea while avoiding pretense that what is offered is science? The answer Popper has offered in his two early masterpieces is, one should make a proposal, not a statement, as one cannot claim that a proposal is true. But one can claim that a proposal is useful, and that amounts to the claim that an untestable statement is true. Yes, we have met this already in the beginning of this very discussion; yes, we are going in circles; I did warn you, though: I told you I am stuck, and I am stuck far worse than I have shown you thus far. If you stay with me things will get worse, I promise.

Please permit me to weary you again with my distaste for Popper's expression of distaste for metaphysics; I like less his demand that we express views only if we have some possible refutations of them in mind. This idea is moralistic: much of what I say is not refutable, at least not to my knowledge, and I feel no obligation to suppress it, nor do I feel an obligation to work hard alone on what I wish to say until I find possible counter-arguments to it: it is easier and much more fun to discuss one's half-baked ideas in good company and have others come up with clever completions of them and/or find possible counterexamples to them. Barring the availability of immediate company, there are means of communication, such as letters and the press and the communication highways, as means of a search for possible company. Now all this may be read as the suggestion, fashionable today in some philosophical circles, that people say whatever comes to their heads and not worry about its value. Of course, such conduct is decidedly permissible and also decidedly very far from being recommendable, as it bespeaks boredom, at times even irresponsibility. Not caring about the value, especially the truth-value, of one's utterances, lowers the level of one's conversation. Yet worrying much about what one says is also unadvisable, as it stifles; heuristic is the proposal that at times it is advisable to pursue ideas and postpone for a while the worry about their value, including worry about their truth-value. But only for a while.

I confess this is a part of what I hope is distinctive of my own output, though it is in no way original with me, as it depicts Robert Boyle.

I know: fun and games is no moral category and science is no sport, they say. It took me a long time to realize that students are scandalized that the technical literature often calls "the rules of the game" rules of logic, of science, of intellectual activity of any sort: they find morally offensive the very suggestion that these lofty activities are deemed games. It is easy to mollify these students by calling this a mere metaphor, an analogy: students love analogies (they are the stuff magic is conjured from, says Sir James Frazer). Yet to serious gamblers things look differently, need one say. Bernard Shaw has noticed (*Androcles*, Preface) that being scandalized at the mixing of the mundane and the lofty is the wish to ignore the lofty when peddling the mundane. (Mysticism, on the contrary, sanctifies the mundane.) Yet things go further than that: the idea that fun has nothing to do with ethics is puritanical and objectionable. Popper found this idea in Schiller's celebrated critique of Kant (for its suggestion that the mix of the right and the enjoyable is not quite right) and he said it is a rare case of valuable, non-trivial ethical argument.[46]

---

[46] Popper repeatedly insisted that ethics is trivial in the sense that it leaves little room for serious controversy. His admission of one marginal example (Schiller's criticism of Kant, whose view of all enjoyable moral conduct as non-moral erroneously prevents treating friends morally) need not be serious criticism of Popper's view. Yet the following passage of Popper is [Popper, 1957, 159]:
"Even the emotionally satisfying appeal for a common purpose, however excellent, is an appeal to abandon all rival moral opinions and the cross-criticism and arguments to which they give rise. It is an appeal to abandon rational thought."
This assertion is correct and far from trivial: it is contested seriously. In any case, we should never deny the right to try to contest seriously any idea, be it trite or not, regarding an imperative or not.

Strange. I hope you remember that in a way I think that the rift between us is due to his refusal to discuss with me this very matter.

(Here is another role for responsible leadership: it should set public agendas and platforms for their criticism that may lead to their modification. It should also encourage and orchestrate the expression of some half-baked ideas and offer platforms for their critical discussion among the interested. This is particularly important nowadays, as beginners are often vulnerable to intimidation of people who pretend to be offering more than half-baked ideas and who display the tough-and-no-nonsense attitude. Today, when leading scientists conceive research projects and force normal scientists to them, the leaders of science should attempt to secure for all the right to follow their own bent and to express and discuss half-baked ideas that may interest them. But I digress too far.)

More should be said about the philosopher's observation that good will and good faith and hard work suffice for the overcoming of obstacles to communication. This observation of his abides by the feature which he pejoratively confers on metaphysical doctrines: it is confirmable-but-not-refutable: the observation is not refuted when communication fails, since possibly with more good faith and more hard work and more ingenuity communication might succeed, and it is confirmed whenever communication is restored. The characterization of metaphysics by its adherence to this feature is in itself generally sound. Yet the condemnation of metaphysics for it is not. Already Galileo said, a doctrine able to win but not to lose is as unfair as the game of heads-I-win-tails-you-lose. Popper repeatedly used it as an indicator of a form of cheating.[47] He relaxed this standard when, later in life, he endorsed the (Cartesian-Kantian) idea (which he explicitly rejected in 1935) that metaphysics might play a systematic role as heuristic (as the half-baked idea that may invite diverse completions and/or possible criticism). Yet he was right to alert us to a source of confusion here: metaphysical half-baked ideas are easily misread to mean their completions or near-completions, and then their confirmations look like proper empirical support. This explains the tremendous usefulness of

---

[47] When Newton's theories triumphed, the received opinion was, he followed the sure method and so he was assured of success. In his *System of the World* Laplace sees the success of Newton as crucial evidence for Bacon and against Descartes. He was shocked to learn that Newton's optics could give way to a better theory as it did (in 1818). This revolution was a kind of an earthquake [Agassi, 1971, 29 ff.]; [Agassi, 1975, 214-16, 230]; [Agassi, 1981, 390-1, 393, 413-14] as it raised the suspicion that Newton's theory of gravity might face the same fate. Nineteenth century thinkers argued that there is no such risk. Sir John Herschel claimed that the inductive foundations of Newtonian mechanics were secure [Agassi, 1981, 388-420]. This invited a differentiation between the stable inductive foundations of his mechanics and the feeble ones of his optics. This was the achievement of William Whewell. He said, testing of the former was a kind of cheating: it did not face the possibility of refutation. (He could not say this as clearly and succinctly as Popper could, but let this ride.) Popper rejected Whewell's claim that past success of a theory assures its future success. Rather, he said, corroboration shows that the tested theory is valuable. True, but the converse does not hold: Newton's optics was important, although it never underwent a severe test. Whewell's and Popper's views are too stern.

his theory of corroboration, of empirical backing, as failed refutations. (On this he expressed an improved version of the view of Whewell, of course, except that Whewell still dismissed refuted theories.)

In 1945, you remember, Popper suggested to present metaphysics as mere proposals. This is irrelevant to the present discussion: if we propose that good will should act as a panacea against obstacles to communication, then we may expect that it will facilitate all communication barring the dishonest, the insincere and the unserious. And then, whenever there is a possibly scientific communication, science will be inevitable. Hence we need no institutions for the safeguard of criticism — except the democratic right of speech as secured, say, by the First Amendment to the Constitution of the United States of America. The sole possible exception is the case of reinforced dogmatism, where one may feel trapped by one's inability to criticize a doctrine that seems to be clearly false. And we may preclude this by deciding not to argue against the reinforced dogmas or propose a contrary one (Shaw, Back to Methuselah, Preface). But then who will insure that our decision is observed?

Reinforced dogmatism is any opinion such that any piece of evidence whatsoever confirms it. At the very least, it is any opinion that is easy to confirm but hardly open to criticism. There is a concrete example here: vulgar Marxists, observes Popper, take any criticism of their views as evidence for the viciousness of the intellectual hirelings of the capitalist system. Agassi praises science, writes Feyerabend, because the capitalists pay him to do so. They also pay Feyerabend, and even more handsomely; this, however, shows that even the capitalists cannot ignore the good doctrines that Feyerabend spouts. True or false, this makes Marxism only confirmable, not refutable. The same, continues Popper, goes for the Freudians who say, your opposing Freud's theories is proof that you need psychoanalysis in accord with them.

Various thinkers have declared reinforced dogmatism wicked; it is easier and more effective to see that it is boring (Shaw). As there is no need for a prohibition against boredom, there is no need to institute rules for the purpose of preventing dogmatism. Popper might be sympathetic to this: he says that he loves criticism, not that he is obliged to welcome it by the rules of honesty and sincerity and seriousness. Why then does he not ascribe to science this characteristic? Why demand that scientists exhibit honesty and sincerity and seriousness and resolution rather than that they act out of the love of truth and that criticism is a powerful instrument for the search for the truth? This is very comfortable, as it explains the easily available observation that scientists who hate criticism and sincerely declare their hatred for it nonetheless succumb to its allure: it is so because of their love of truth. This will explain how scientists could use the method of criticism by empirical evidence for centuries when they officially clung (and most of them still cling) to the method of proof by empirical evidence. But this is a mere philosophical nicety. The question (Schiller's problem as he posed it to Kant) remains: what is preferable, love or obligation?

What comes first, what is more basic, more vital, love or obligation? This question was indeed central to Kant's moral theory. He said, love serves better, provided we are all serious, educated adults; as long as we are not, obligation should serve us better. Love will replace obligation, he promised, after the task of the education of humanity will be accomplished [van der Pitte, 1971, 25]. And then he proved his theory of obligation, forgetting that in the world of proof there is no room for haggling and for going for the second best, certainly not for hoping that the best will one day be available. For us Kant's negotiation is useless anyway: the morality that it concerns is not public but private. His system did not have room for public morality.

Public morality, not morality, we learn from *The Open Society and Its Enemies*, is what makes for both democracy and science. Popper says correctly and emphatically that both in science and in politics, whether one's motive is the quest for the truth or the passion for fame, one will play by the book because one lives in orderly society. So the positive effects of sincerity, honesty, and, above all, seriousness, are not merely effects of personal dispositions but also of public attitudes: back to David Hume, my favorite: frivolous motives may serve the public just as well as noble ones, if not even better.

There is a response to this: democracy is, indeed, the set of rules that save us from the examination of motives. Yet, nevertheless, democracy is defective, as everything human is, and so it needs some guardians who are motivated by the love of freedom and of truth; they must be serious and sincere and honest.

Fine. These, then, are, in a very specific sense, the leaders, the underground leaders (in the American sense of the word) to be sure, yet nevertheless the true leaders of democracy. (The Cabbalist tradition calls them the thirty-six righteous; they are anonymous and they keep the world going round.)

This discussion displays the difference between private and public criticism — in politics and in science alike. It also tells us that the criticized may personally wish to dismiss some criticism yet consider it all the same for fear of having to face it in public. (This, says the philosopher, is his response to Schlick's accusation that he, the philosopher, presents science as masochistic.)[48] Thus, aptly, both private and public criticisms

---

48 See note in Chapter Seven. In his seminar Popper used to elaborate on this. Of course, you wish to be right, he said, but you know that you are fallible, and so you may prefer to be the first to know of your failings. This (unpublished) remark concedes too much to Schlick. A much better reply to Schlick ap-

play some towering role. (This tallies with institutional individualism or the autonomy of sociology: the private-individual blocks reduction to collective sociology, and the public-institutional blocks reduction to individual psychology.) But then, motives are again out: in private or in public alike, it does not matter any more why anyone dishes out or accepts criticism; and so, seriousness is not all that important for serious situations, at least not in all of them.

I have offered the philosopher criticism — some in private some in public. In private, not in public, he responded; at least while I am still alive and eager to listen to any criticism; at least my loyal students and close friends should come to me with their criticism. If they go public instead, then they show that they do not take my seriousness seriously and/or that they are after fame. (Thus again lust for fame is the wicked motive which, says Bacon, interferes with proper research.)

Fine. Query: was this (Baconian) response of the philosopher open to criticism? I do not know. I do not think any of his loyal students and close friends ever managed to discuss this with him in private. This may be my *cri du coeur*, but not a complaint: there can be no ground for complaint here. And certainly there is no right to demand more of one who has given us so much, especially not for one who personally owes him so much as a loyal student and a close friend — regardless of whether once and always or only once and never more. But I am badly stuck all the same: the philosopher was at pains to make amends, but, as I told you, he sometimes forgot:[49] we are all human, after all. And he was at pains to have things straightened out with his professional peers and friends, and yet they misunderstood him — as did others. For decades on end. What is to be done about the situation? Is it ill will? Is this explanation not too psychological? Is it not simpler to accept refutation of his moralistic theory of the source of all misunderstandings?

No; there is another way. (There always is, if you still have the patience, before the debate degenerates into trivia.) His professional peers misunderstood him, even while the general scientific public increasingly admired his thoughts and at least took them seriously. (Here, indeed, is another role of bold leadership: the recognition of valuable items despite taboo and professional jealousy and any other irrelevancies. Hence overspecialization is the enemy of responsible leadership. This is so despite

---

pears in his published works. It is, indeed, his main thesis: we learn from making conjectures and refuting them. If we must attend to personal motives, then, let me add Einstein's empirical observation: learning is great fun and often personally useful. Also, to refute a received doctrine is to hit the jackpot.

49 Popper's short memory and short temper often caused confusion and raised bursts of anger. Also, these fed his suspicion. He said he remembered clearly that I had called him a greedy old man. So my denial is proof that I was lying and calling him a liar. He usually had a terrific ability to think up alternative explanations for odd situations. This ability vanished when he explained some situations in accord with his favorite default assumption that a cosmic conspiracy against him was afoot.

the fact that irresponsible leadership repeatedly hides behind the demand to prove oneself responsible by staying within the confines of one's recognized specialized expertise.) Did the philosopher's present attainment of the admiration of the scientific community at large vindicate his constant complaint about being wronged by his professional peers? This question is immaterial, as even the following is: if he was wronged, then how should we rectify matters? The right question is this: what control mechanisms should we institute in order to prevent the recurrence of such wrongs?

Leadership is the ability to bring to the agenda of any public — national or professional or club — the issues that are important for them and find or encourage the search for some reasonable solutions for them. This is no recipe; it is the opposite of that, since it is the clam that leaders should be open minded and encouraging all forms of consultation.

Thus, in addition to predilection, the personal qualities required of leadership constitute ability and proficiency. I do not wish to discuss here the pedestrian skills of proficiency of moderating a meeting and of instituting proper search committees and such, even though it is my observation that democracy fails in many places for want of people proficient in these pedestrian skills. And, contrary to the theory of the scientific leadership that Michael Polanyi and Thomas S. Kuhn have propagated, and that are now in vogue, leaders do not have to be opinion leaders; not even within science: suffice it if they encourage thinking and offer platforms for any interesting opinion.[50] This is the most valuable lesson Niels Bohr learned from Albert Einstein: leaders need not endorse the opinions they encourage: they only have to be able to sift the wheat from the chaff, and even this not perfectly well but better than most. The leaders, in addition to their skills, may be gifted and desirous of leading positions and able to attain them. The leaders then are either people in positions of power or individuals able to discharge simple duties responsibly, or preferably both, at least to a reasonable degree. And the list of their duties should stay open, especially since the urgent problems of the day prescribe bold and rapid changes. Once this is clear, the road is open to a theory of democratic leaders that is neither as over-optimistic as the Enlightenment view of them as mere delegates, nor as dangerous as the Romantic depiction of them as individuals destined to great deeds.

This should lead to the theory of scientific leadership. The scientific community is a complex social structure involving so much prestige and economic and political power, that it is naïve to ignore its politics. It is important to view the role that intellectuals in general play on the international political scene, in the national political scene, and in the internal

---

50 Einstein praised H. A. Lorentz (in his obituary on him) for his ability to guide fruitful discussions, ignoring his staunch unwavering allegiance to Newtonian physics.

politics of diverse intellectual endeavors — whether in the international intellectual community or in the national one or even on the college campus or in the local chapter of the learned society to which they happen to belong.

Things are out of hand. The agenda of the intellectual institutions and the agenda for their role in national and international politics, all these were neglected. This is notorious: the leadership has created a new academic specialization to meet this neglect: science policy studies, a specialization to which I belong, somehow, and even as an honorable member. The members of this specialization have not yet noticed that they can have no influence except on bureaucrats; and bureaucrats will never interfere in matters of academic injustice: they simply are not equipped for that. The leadership of the community of philosophy and of the social sciences, especially in the philosophy and the sociology of science, have all maltreated Popper; that should be rectified both for the specific case, by bestowing on his memory some official recognition, and in general, by creating some new safeguards. But first the charge I lay must be examined. Can my charge be checked now, not only by some future historian? How? Some intellectual leaders and some political leaders have used Popper for their own advancement and that led to fame that likewise could do with examination. Popper himself came up with a proposal to institute a sort of Hippocratic Oath for all intellectuals and to campaign against the flow of low-quality intellectual publications [Popper, 1969]; [Popper, 1994, 122-4]. This is a proposal for self-censorship. Hence, it is impracticable. Hopefully it can stay off the agenda without debate; otherwise, let there be a debate of it, by all means. The agenda must be instituted by some authoritative philosopher who cares enough to organize a new sort of institution, some sort of a grievance bureau or an ombudsman, if we can have one that will not be flooded by trite complaints. Reform of great institutions is possible, but it requires ability and dedication that neither the philosopher nor any of his entourage has exhibited. Do we need new blood?

No. I did not mean the question seriously. Perhaps I was carried away; perhaps I tried to portray an attitude which differs from the one behind the philosopher's just complaints, one that a proper leadership may institute after it is recognized, and its duties spelled out, and its modus operandi is made explicit and put to critical discussion. With no sincerity and severity and seriousness; at least not to excess.[51]

There was nothing like that in the philosopher's circle. He was a leader, of course, willy-nilly, and he did want to function as a leader, as

---

51 This is no forecast. It is the denial of Popper's assessment that his view goes so much against human nature that efforts at its implementation must be a burden ("the burden of culture" and "the cross of civilization"). If it is a burden, then we may seek ways to reduce it.

an opinion leader and as a personal example. This was insufficient: it also was disastrous. With exceptions like Wisdom, Jarvie, and Miller, we all, yes, particularly I, were concerned with challenging his priorities. I repeatedly told the philosopher that he need not make acknowledgements to me, especially when he was busy writing ones. And it was because I was both convinced I had earned more than I was given and because I found the acknowledgements he gave me disturbingly inaccurate. At times they were ungenerous, but they were generally quite generous and yet disturbing all the same. Even the one he gave me in the prefaces to his *Postscript* was much too generous and still embarrassing. I now am very critical of the help I gave him at the time (it was done in good will, of course, but out of a major communication blunder). I remembered every item I felt he owed me, especially those he had promised to acknowledge and did not. It was silly of me, I know. Was my wish for more recognition or for less? I could not answer. Much later I understood: he was following the Baconian, not the Popperian theory of acknowledgement[52]: the philosopher was too eager to receive the acknowledgements that he deserved to attempt a reform of the system of acknowledgements, though, of course, this reform should and will be his triumph.[53]

Watkins admitted to me, in a moment when he was trying to convince me to suppress some of my criticism of the thoughts of the master, that he, too, had been moved by ambition to offer some critique of the master. The idea is Bacon's that if one's ambition is too high or too low than one is blind to the proper value of one's contribution [Bacon, 1620, I, Aphs. 63, 67, 87-8], [Bacon, 1625, Essay 36, On Ambition]. (It is originally biblical [Deuteronomy, 15:20].) Lakatos learned the trick very fast and soon everybody in sight had stolen ideas from him. Feyerabend, at least, never accused anyone of plagiarism, and he freely engaged in it, not viewing it a crime, as he rejected the rules against plagiarism: a remark of Mozart (though I think made in jest, as a substitute for some complaints) seemed to him sufficient evidence. But this is not the end of that.[54]

---

52 Bacon suggested honoring innovators — discoverers and inventors – by erecting public statues for them, for example. Faraday said, new problems are laudable innovations too. Popper said the same of criticism. Indeed, according to his view, the best criticisms are new discoveries and the best theories yield only to these.

53 Popper's complaints of intellectual theft invite some rule of recognition of ownership. At times he referred explicitly to some standards of recognition but only vaguely, and he never tried to specify any. Rather, he spoke of historical truths. Alas, responded Munz, these are not always self-evident.

54 To my surprise, Feyerabend vehemently withdrew his scanty acknowledgments to Popper [Feyerabend, 1987, 312]:

"doesn't Agassi remember how often he begged me, on his knees, to give up my *reservatio mentalis*, fully to commit myself to Popper's 'philosophy' and, especially, to spread lots of Popper-footnotes all over my essays? I did the latter — well I am a nice guy and quite willing to help those who seem to live only when they see their name in print — but not the first."

My memory of Feyerabend's "*reservatio mentalis*" is vivid. He expressed reservations very clearly whenever we met. We always had a terrific time arguing. He once came to Boston to give a lecture early in a sabbatical of his, and he stayed in a hotel there for the rest of the semester, visiting my semi-

Lakatos had been schooled in the art of political leadership, and in a culture where, under Stalin's corrupting influence, the question of responsibility had no place. And so he was at first incredulous as to the openness and vulnerability of the system — both the English system of higher education and the English philosophical community, not to mention the philosopher's entourage. And then he took over, first the philosopher's school and then he went for the whole of the profession; the last time I met him he was planning some large-scale schemes with the ambitious president of Boston University, where he and I were alternating colleagues. (I told you he followed me everywhere.) And he used as his first lever the fact that the philosopher did not recognize power struggle among intellectuals, and he manipulated the philosopher's natural weakness for suspicion — until the two of them fell out with each other. At first I tried to dissuade him and gave him advice. He misused it regularly and very creatively. I stopped.

Lakatos was a great thinker, but this is not the place to talk about his trailblazing work on mathematical methods. He also loved intellectual activities, and he could forget his politics while thoroughly enjoying a debate on an interesting question. But mostly he was a born leader — old-style; he was talented and energetic and equipped with an excellent memory for boring but useful details (especially ones useful for the unscrupulous); he was burning day-and-night with the ambition to control. He used threats and promises freely and he spread vicious rumors with moral indifference to the damage they might cause. He received a grant that enabled him to bring together more leading logicians and philosophers of science than anyone else had ever managed. (Those who were absent, Alonzo Church and myself, he boasted, were the disinvited.) He changed the status of the Popper school and of the master himself within the profession considerably and for quite some time — despite his early demise. I did not care for what he had done: after decades in which the master refused to compromise with the establishment regardless of cost, Lakatos made the master kowtow to the same establishment all the same. This is how he showed that he was the boss. Not quite, though: he still harbored resentment, and few leaders can stay established and constantly blame everybody in sight; Lakatos could not furnish true leadership, as he himself was in the last resort an irresponsible, spoiled brat.[55]

---

nar (on biology) and my home, arguing all the time. It was terrific.
55 The damage that Lakatos caused to the philosophical public should not count too much against him: it is better to blame the general educational level for the success of demagoguery. Unfortunately, Lakatos also caused much personal damage, and this is more serious. His mitigating circumstance then is his obsessive character. In Communist Hungary he obsessively betrayed friends and teachers [Long, 1998, 267-71]. He was quite obsessive even in his efforts to monopolize conversations. When he first heard me lecture on the philosophy of the social sciences, he left with a severe headache. He soon could feign proficiency there too: he had total recall and much experience, and he was very intelligent. All this should not lead to a misconception: he cared deeply for scholarship [Long, 1998, 302-4]. Lowering his

Nor was the success of Lakatos purely due to his personal ambition and ability to lead: as an intellectual leader he could not succeed without great intellectual abilities and without comprehension of the politics of the intellectual situation in which he found himself. He had no doubt that after the dust settles Popper's contributions must come under serious consideration, as they are the very best, and the considerations must be critical and bring some improvement, as they are imperfect. And on this I agree with him. The starting point should be the social and political philosophy he has offered us and the adjustment it invites for which Lakatos had no understanding, as he was free of all feel for democracy.

Popper's social and political philosophy requires adjustment on at least two factors. First is the fact that politics is nationally run, whereas he did not recognize nations (out of his valid if not very philosophical hostility to Romanticism, for sure, but this is throwing out the baby with the dirty bath water)[56.] Second is the pluralism that is the outcome of the horrors of World War II, and the idea of pluralist participatory democracy, on which he has said nothing, and which his philosophy encourages *grosso modo* but denies in details that invite fine tuning.

Remember: all leadership is voluntary. Intellectual leaders are not in the least to be the political leaders of their peer-groups (*pace* Polanyi and Kuhn). Hence, the philosopher had no such obligation either, and his intense concentration on his researches and his distaste for power and for committee work made this impracticable anyway. He was also unable to do what was needed in order to acquire and hold a power base. Yet he did have one, the British Society for the Philosophy of Science which he co-founded its organ, *The British Journal for the Philosophy of Science*. Lakatos took it over and then he surrendered it to the inductivist camp with incredible indifference in exchange for some political favor or another. I do not know if I would have ever come to his seminar and become his disciple and collaborator for years but for the fact that I bumped into *The British Journal for the Philosophy of Science* and found there the magnificent "The Nature of Philosophical Problems and Their Root in Science" — the outcome of the philosopher's lecture to the Cambridge Moral

---

intellectual standard was his great sacrifice to his obsessive need to control. He repeatedly admitted to friends, me included, that he was impatient to return to scholarship proper, as he intended to do soon after consolidating power. But for his early unexpected demise, he might have done that.

56 The slogan of the French Revolution, Liberty, Fraternity, Equality, was always problematic, of course. That liberty and equality easily come into conflict is obvious (we all too often curb liberty in the name of equality with sad results) and we have no valid general prescription for handling this conflict. It is important to take the sting out of the conflict by reasonable compromise and by seeing to it that people will have many options at each turn. The demand for fraternity is more serious an obstacle. Popper repeated the classical liberal maxim: imposing good will on others is wrong [*OS*, ii, 235-7, 240, 244]. So did the once-famous comedian and civil-rights activist Dick Gregory when he criticized Martin Luther King: you need not love your neighbors, he said; respecting them will do. But the wish to belong that the demand for fraternity expresses is very strong. Popper erroneously ignored it [Eidlin, 1999]; [Hall, 1999]. The wish to belong to a fraternity needs looking after in as democratic a manner as possible.

Sciences Club, and his presidential address to the British Society for the Philosophy of Science, you remember — and decided, even before finishing reading that essay, that he was a teacher for me. This is the wrong way to go about things.[57] Were the leadership[58] in philosophy half as efficient as the leadership in mathematics or in physics, then the philosopher's works would be anthologized at once and otherwise be made available to students. On this Thomas S. Kuhn is right: the better intellectual disciplines do have better leadership in this very respect. He is in error about leadership: [59] he said, all of intellectual disciplines is the intellectual leadership of the discipline and that they do not tolerate dissent. Here he espouses the myths of the establishment. And he is cheap when he says that unanimity secures quality. And, since his work is popular and pertains to intellectual politics, it must undergo effective criticism. For, the matter of intellectual leadership is urgent agenda. The agenda depends on the rules of the game; it also determines them, of course. This sounds like the trouble with the hen and the egg, and again the hen and the egg need not be problematic, as they can evolve from each other. And the study of scientific method is just the theory and the etiquette of the commonwealth of learning.[60]

---

57 The discussion here concerns pluralism in politics. Popper did not discuss it as he was offering a magnificent minimum program for democracy. He ignored even the place of political parties in democracy, and rightly so, since his book is powerfully as general and programmatic as possible. And in defense of this he observed that the expressions of democracy outside political parties might be vital too, such as the ability to impeach a politician. Regrettably, Popper did not discuss the positive role of political disagreements. This important task remained for Watkins to perform later on [Watkins, 1957-8]. Popper's discussion was also needlessly constrained by two interesting but erroneous ideas; first, that institutions are hypotheses (rather than social facts) and second, that in science agreement prevails. This depends on the preference for the most testable hypothesis. Whereas in science the choice of a hypothesis is for testing it, in technology choice is for implementation. In technology consensus is on permissions, not obligations, to implement innovations. Moreover political systems are less open to criticism than science. This is obvious, since science is a relatively new institution.
  Nevertheless, Popper did open the road to intellectual pluralism when he suggested in his *The Open Society and Its Enemies* that we may approach history in diverse ways, all legitimate. This important item is under dispute among historians. Later on Popper spoke about clashes of cultures [Popper, 1994, 117-25], thus bringing his political philosophy nearer to the spirit of his own critical realism.
58 Lakatos wished to study scientific etiquette, but he postponed acting on it. He planned to amass sufficient power in order to be able to do much good (by his own lights). To that end he postponed doing good and allowed himself to do evil (by his own lights): he even advocated false ideas (by his own lights). "Scientific method, as softened by Lakatos, is but an ornament which makes us forget that a position of 'anything goes' has in fact been adopted", said Feyerabend [Lakatos and Musgrave, 1970, 93]. This is a half-truth. Personally, Lakatos upheld traditional scholarly standards and detested irrationalism. He disliked, for example, Feyerabend's recommendation to take magic seriously. Yet publicly he was very vague. Robbins read him very differently from Feyerabend [Robbins, 1979], and it is impossible to determine who was right: the vagueness was cleverly inbuilt. But on some points Lakatos simply had to admit irrationalist moves, even though, he convinced himself, only temporarily.
59 In addition, Kuhn was a poor leader: he declared he was in agreement with everybody and he did not produce an agenda: he had none to offer.
60 A word about Feyerabend. He was not power hungry; he craved publicity. By he acted not much differently than Lakatos. Publicly he advocated anarchism and privately admitted he was a conservative. The label "anarchist' is one that Noam Chomsky and Howard Zinn, the leaders of the students' mass movement in the late 60's, adopted to denote a non-Stalinist (pseudo-) Marxism. There was an

P. S. One passing incident stands very sharply in my mind. For years it puzzled me that it puzzled me. We were walking in the neighborhood of his home, the philosopher and I, in Buckinghamshire, where he lived until Lady Popper died, on an errand that I do not remember, engaged in a conversation that I do not remember either, and he expressed his indifference to the national question. If Buckinghamshire would secede from the United Kingdom, he said, it would mean nothing to him. (He was not alluding to *Passport to Pimlico*.)[61] Though a passing remark is no evidence that he held the opinion it expresses for more than a fleeting moment, this kind of remark shows what little sense of national identity he ever had and what little empathy for it. Granting that many atrocities were committed in the name of nationalism will not justify this indifference: after all, the philosopher has stressed (in his classic *The Open Society and Its Enemies*) that the many atrocities committed in the name of religion did not make him hostile to religion or insensitive to the strength of the religious sentiment. Atrocities, as Feyerabend later stressed, were committed in the name of reason too.

Suppose Buckinghamshire were to secede from the United Kingdom. Popper said, he found it of no import. But as the British government would not tolerate it, he could not be so indifferent had he chosen to be

---

effort to recruit him and me to the movement, organized in a lunch at Hilary Putnam's home. Feyerabend was silent as Judith and I argued about the tactics of the movement, since we shared their views about the Vietnam War. Later he was in Berlin and gave his students the full right to speak and act in his name without checking. As to me, my friend Abner Shimony tried to recruit me again in a party in which all the leaders of the movement were present. I said I could not support their lies. Incredibly, he stopped everyone and asked them all for a comment. Alice Wartofsky stopped him dead yelling, we are trying to save women and children from napalm bombs. She had a point, and not one that can be discussed in the middle of a party.

In response to Feyerabend's slogan, anything goes, I said, not anything, not Auschwitz. Against this he spoke of unspecified intellectuals as criminal megalomaniacs, adding [Feyerabend, 1987, 313], "there exists no difference whatsoever between the henchmen of Auschwitz and these". He said he could not take me seriously, as I praise them: "How can I take a person seriously who bemoans distant crimes but praises the criminals in his own neighborhoods?" He meant me.

Feyerabend's initial response [Feyerabend, 1978, 125-38]; [Agassi, 1988, 405-16] to my review of his major book [Agassi, 1988, 33-404] was quite civil, and my reaction to it was gentle [Agassi, 1988, 416-17]. He concluded this with the postscript to his [Feyerabend, 1978, 138-40]. His slightly later text [Feyerabend, 1987, 3131-14], quoted above, is a little less friendly. But having asked for it I do not complain. Still, I should respond to it now: I should admit that my report on his autobiography might be in error: having heard it from him is no guarantee that it is true: it may be a memory failure, his or mine, for example. Also, he relied there on Edward Rosen's criticism of my *Towards an Historiography of Science*, adding (312), "he rarely knows what he is talking about even when he is trying to tell the truth". I do not complain about this slight, either, especially since elsewhere Feyerabend repeatedly referred to my *Historiography* respectfully [Feyerabend, 1981, 75n; 1999, 152n, 158n]. As to Rosen's criticisms, they have no bearing on the present discourse. I elaborate on them elsewhere, in my forthcoming *Science and Its History* (2008).

61 The 1949 movie *Passport to Pimlico* of Henry Cornelius with Stanley Holloway has a small borough in Greater London secede in response to Whitehall bureaucracy.

politically active.[62] National unity and the right to self-determination thus clash, and we should not overlook this fact.[63]

Of the philosopher's stand late in life on the national question in general and on the Jewish problem in particular I learn from Herlinde Koelbl's superb *Jüdische Portraits*. It has portraits and interesting interviews of both Popper and Gombrich. They express similar sentiments, and Gombrich speaks of his Jewish origins as a sort of cultural heritage and of his religion as universalist. Popper expresses an optimist view of current politics peculiar to himself, and hostility to all nationalism. A semantic difference between the two ought to be cleared here: it is hard to compare attitudes to nationality unless the term is clear, especially when one compares views of closely interacting individuals. Thus, when the President of the United States of America delivers his traditional address on the State of the Nation, he does not say things that in principle Popper could take exception to. In what sense, then, did he deny the existence of nations?

Popper did not address this question. He could tend to see a country not as a nation proper but as a collection of individuals, in accord with classical liberal philosophy. To the extent that he did this he violated his own opposition to the reduction of social science to psychology. He could object to this criticism by observing that he did not reduce institutions to individuals. This response is valid if and only if under the banner of institutions one includes not only markets and law-courts and schools and parties and parliaments but also economic, legal, cultural, educational and political sets. The economic set is called the (national) economy, the educational set is called the (national) culture, the traditional set is called the society and the political set is called the nation. And then the state will be seen as the instrument of the nation and the nation will be recognized no less than the state. The nations that Popper says do not exist are indeed fictitious. But, in whatever sense we refer to nations regularly, in that sense nations do exist all the same. Alternatively, all institutions are fake to some extent yet also real to some extent. Popper would not deny this, of course. Yet all too often he ignored it, and this was an error.

---

62 Popper followed Jefferson's doctrine. Lincoln disagreed: although the Confederates started hostilities, it was Lincoln who waged the civil war, and as a response to secession. Now the right of secession of any group is to declare itself a nation, the right for national self-determination. Popper admitted the right of secession and rejected the right for national self-determination. This is no inconsistency: he spoke against this right [*OS*, i, 288, ii, 55, 63, 244, 306, 312, 361] as the right of a group to declare itself a nation in the Romantic sense, not in the ordinary, liberal sense. Hence, he was against nationalism in the Romantic sense, not otherwise. This is sadly missing from his writings.

63 The right of secession invites more study. It is no accident that in our times it is a major cause for war, especially when the seceding party is a national minority with a separate religious affiliation, more so when that national minority stands to gain much wealth from secession (Chechnya). This raises the question, what group has this right? This question draws the discussion into the Romantic considerations that Popper rightly opposed. We need a rule that should invite the parties in conflict to negotiate.

The national sentiment exists; people like Popper condemn it rather than handle it like leaders. The leadership has to decide if it is worth opposing, or if they can use it legitimately as a lever to bring about more liberty and more openness.

Popper saw nationalism as but an instance of collectivism. Now collectivism comprises of two assertions, one erroneous as it belittles the individual and one right: collectives do exist, as distinct from individuals. What makes collectivism a false and evil doctrine (as I have learned from Karl Popper) is its anti-individualism. By the same token, nations as distinct from citizens do exist, yet collectivist nationalism is likewise a false and evil doctrine (as I have learned from Hillel Kook). What is at stake here is a different matter, the matter of individual affiliation to a collective that enlightened political leadership should take account of and regulate. This is a new item on the agenda of students of political affairs. The following is the starting point for the discussion of it. Political leaders in the more liberal societies are usually tremendously superior to their peers in the lesser fortunate parts of the world, yet even there they are still allowed to practice scarcely legal methods. This leads to the defense of *Realpolitik* (the doctrine that might is right, so magnificently criticized in *The Open Society and Its Enemies*), and thus to general demoralization. This is an urgent matter to place high on all political agendas.

And approaching the matter of national leadership in such a manner, we can see clearly that we urgently need two more items to place on the agenda, the matter of the intellectual leadership in national politics and the matter of the leadership of the intellectual segment of the national population. That the intellectuals have played a great role as the national opinion leadership and that this role has been damaged by a century of irresponsible conduct of leading intellectuals is clear, and requires correction. That the intellectual world has both opinion leaders and political leaders was masked by the ideology that is hostile to all leadership and by the identification of these two kinds of leadership. The requirement of opening intellectual world to the general public and the democratization of its institutions comes next.

CHAPTER NINE: EPILOGUE: THE FUTURE OF PHILOSOPHY

> We still have a silly habit of talking and thinking as if the intellect were a mechanical process and not a passion.
> (Bernard Shaw)[1]
> The discovery that the human animal is so constituted that it responds emotionally to the practice of intellectual honesty is just as great as the discovery of the human race about itself, that it responds emotionally to music.
> (P. W. Bridgman)[2]
> There exists a passion for comprehension, just as there exists a passion for music.
> (Einstein)[3]

My apprenticeship began in January 1953 and terminated a few times. Officially, in my role as his student and assistant, it terminated in June 1956, when I passed my doctoral exam. Contractually, in my role as his close pupil, it terminated in May 1957, when I told the philosopher I was his pupil no longer and when he would not discuss ethics with me. Personally, in my role as his closest associate for the time, it terminated when the philosopher accused me of improper conduct in 1962. Socially, in my role as his former student, it terminated in 1964,[4] when the philosopher declared unwillingness to communicate with me. Yet, in a sense, my apprenticeship, like any, is never ending, and my apprenticeship to Popper ended fairly close to his demise. The strong interaction, chiefly intellectual but personal and emotional as well, could not terminate all at once, especially since, as I have told you, to the end we did not quite let go of each other. The interaction had to loosen. The philosopher would not discuss with me my criticism of his views and he would not criticize mine. He would not engage with me in a written public debate either. From time to time, I discerned in his writings responses to my criticism, or rather of what I surmised that he had surmised was my criticism. But I never know,[5] as he never said so explicitly, and this prevented him from presenting my criticism as fairly as he could, as was his

---

1 Bernard Shaw, Preface to *Parents and Children* (*Misalliance*), The Pursuit of Learning.
2 [Bridgman, 1955, 361].
3 [Einstein, 1954, 342].
4 During my stay at the University of Illinois (1963-5), the editor of a prestigious periodical sent me an invitation to submit a paper and withdrew it almost at once with an apology: Professor Popper had first suggested it and then relented. This delighted me as a quaint initiation rite: I was now on my own: I could no longer suspect that my praise for Popper is self-serving.
5 Whenever Popper criticized an opinion of an unspecified author, it was very hard if not impossible to surmise who that author was. During my days of close cooperation with him, I learned that these omissions were intentional. This was never to my liking. When I was the target of the criticism, this I found this particularly distasteful.

habit.[6] Responding to his comments thus became increasingly difficult and pointless. He was fighting the old battles that he had won long ago with no time for new challenges such as I would gladly have put to him. So I had to make do without his stimulus and aid and support.

I could not have achieved my present position without his stimulation and aid and support: I evolved slowly from a political revolutionary to a philosophical one — or at the very least to one who uses philosophy as a weapon. What I have learned seems to me to have culminated in the following concluding message.

The leadership of almost every society and sub-society, be it spiritual, artistic, social, political, or military, is on the conservative side, usually on the presupposition that the conservative is safer than the adventurous.[7] Almost all societies suffer the strain of constant conflict and repeated clashes between the realities of cautious leadership and the ideals of valor, of courage and loyalty. Such conflicts may be expensive, yet little is done to minimize them, perhaps because they are politically extremely useful: when caution will not do, the truly brave take over — in mutiny if need be — and this would be impossible were there no permanent problem of balancing between safety and valor.

This picture or analysis of traditional society, be it true or false, certainly does not obtain for us now: too much is at stake, change is too swift, and the longevity of established leaders makes stagnation endless. The major political problem is now no

---

6 Popper's enviable clarity is largely due to his habit of strengthening and defending as forcefully as he could the ideas that he discussed, and as solutions to significant problems — regardless of whether he finally rejected or endorsed them. He was rightly very proud of this and he said so repeatedly. This incidentally makes it easy to sift his usually serious discussions from his rare less serious ones: the latter he never expounded, let alone defended. When he dismissed Hegel's works without debate, he did so openly and clearly. He rightly dismissed romantic nationalism without debate. When he dismissed the nationalism of Tomáš Masaryk in his Prague lecture, 1994, however, he was not serious about it. (This is no complaint.)

7 The conservative preference for playing it safe may be a psychological matter, but as such it is seldom relevant to pressing public issues, simply because public affairs seldom rest on individual psychological characteristics: a leader is seldom so very astute as to be able to carry out decisions on public affairs without depending heavily on others. The cause of the conservative preference for playing it safe is in the very role of political leaders as traditionally perceived. This role is the maintenance of social stability. This suffices to make conservatism in politics the default option. This renders very important and basic Popper's idea that democracy rests less on social stability and more on the reliability of social and political controls. To see this, it may be useful to contrast the traditional view of the role of politicians with the traditional view of military leaders. Their role as responsible for national defense forces them to consider as the default option for the military aspects of their jobs the most daring attitudes they can think of. (Their attitude to other aspects of their own jobs may still be conservative. For obvious reasons it is all too often arch-conservative, as they prefer armament to disarmament: distrust is in the nature of their role.) Conservatism is inadvisable under threats of collapse of a nation; politicians who ignore this invite rebellion or grass-root revival movements. All this is very sketchy: decision makers may brush aside default options, and these are easily given to great varieties of applications, and in principle applications may depend on personal styles and on *ad hoc* assessments that are often colored by diverse unforeseeable factors. This is where famous national leaders impress their personalities on their nations in moments of stress — if and when they emerge out of the distress victorious: in national politics learning from failure is rare and then hardly ever the result of analyses of what went wrong.

longer merely national, as it is concerns human survival. This is too urgent to allow leaders to stay conservative for long [Agassi, 1985]. Established philosophy can hardly contribute to the effort to overcome the present global crises that threaten our very survival: it has its own mess to clean up. And it is impossible for me to offer a general view of the problems, even without being able to offer a solution [Agassi, 1985, concluding chapter]. So here is my analysis and proposal intended to help philosophers get on their feet fast and join the action, in the hope that they will collect enough occasion and enough courage to do what is necessary. For, philosophy is in a crisis due to the loss of nerve, to use Gilbert Murray's expression [Murray, 1925]; [Agassi, 1977, 332, note 19]. You are cordially invited to criticize what follows, and to present your alternative to it to the public.

Since the dawn of philosophy, most philosophers dismissed traditional thinking as myth-ridden.[8] They have offered some sporadic alternatives to this or that traditional notion and in this they did very well indeed. They also offered some systematic alternatives to the traditional way of thinking. There are three such alternatives and each of them comprises a few items: *logic, scholarly etiquette*, and *research methods* — and *education* (concerning them) too, of course. These three alternative methods are the *inductive*, the *deductive*, and the *dialectic*;[9] and each pertains to the three (or four) items mentioned. (The Greek word "method" derives from "*hodos*", meaning way.) This invites comments on nine (or ten) points. The *inductive method* is hopeless: there is no way out of the heaps of data that it encourages collecting, and all the data-retrieval systems of the world cannot help here, since induction reinforces the given intellectual framework, yet we desperately need a new one. *Deductive logic* is just wonderful as far as it goes, and it is useful both in mathematics and in every situation in which the application of mathematics at all makes sense. But not otherwise. *Dialectic* remains the only useful practical logic. This is the message of the leading masterpieces in contemporary philosophy of

---

8 The word "myth", etymologically meaning story, was used pejoratively with few exceptions, including Plato's presentation of myths — traditional ones and ones of his own making. Commentators always viewed these as mere presentational devices. Anthropologists had trouble with myths. At times the positivist proposal was popular to see myths either as parts of hopelessly superstitious cultures or as parts of ritual system, as symbolic in nature. Anthropologists who favored myths, like Claude Lévi-Strauss, were rightly suspected of irrationalism. Popper, who first used the word in this traditional way, broke new grounds as he later changed it with his idea that science evolved out of myths.
9 Aristotle rightly divided logic from dialectics. They are isomorphic but serve different functions: logic transmits truth from premises to conclusions and dialectics retransmits falsity from conclusions to premises. The isomorphism is the theorem in the theory of inference that the transmission of truth is equivalent to the re-transmission of falsity. Modern logic unifies deduction and dialectics either on the strength of this trivial theorem (first stated explicitly by Popper) or out of disregard for disagreement coupled with a high regard for form. This latter attitude is erroneous, since it encourages engagement with useless forms. It is (like scholasticism and like induction) a version of cargo cult.

science, Popper's *The Logic of Scientific Discovery*, *The Open Society and Its Enemies*, and *Conjectures and Refutations*. This presents philosophy with a built-in crisis: we do not know and have not tried to explore the immediate corollaries to this old-new idea. We have scarcely developed its application to *education*; what considerations are thereby suggested by decision theory as to what responsible leadership is to do, what is too rash to count as permissible? We have not studied the corollaries to this as to what reforms of the institutions of leadership are called for. We have not attempted to apply this to educational practice.

Attitudes to dialectic find varied expressions in the philosophical and social literature as well as in fiction. They are unbelievably obscure and ambivalent: criticism often exudes a sense of irritation. The reason is simple: critics are all too often disrespectful of their targets. The reason for this is also simple: except for Popper's philosophy,[10] all current ones offer recipes for the avoidance of error. Inductivist philosophers argue about induction on the supposition that their activity is productive, yet their inductivism is the view that dialogue is anything but. Sir Francis Bacon said, criticism and respect do not mix, as criticism is the expression of contempt (Preface to *The Great Instauration*). This runs contrary to Plato's expressed view (*Gorgias, Laws*). We should discuss the question, what makes a dialogue fruitful. (This is the variant of the problem of rationality that appeals to me on many counts.)

A popular eighteenth-century saying, under Bacon's influence, no doubt, says, dispute emits more heat than light. If so, then we should learn to improve it[11] as was done with lighting proper, which improved from the fireside to the cold fluorescent light. In the preface to his *magnum opus*, *The Elements*, Lavoisier invites his reader to criticize his book in the conviction that attempts at criticism will fail and thus bring about assent. Experience partly corroborated this, but only partly so. My late friend John M. Roberts has found out empirically that there is a constant conflict in our educational attitude to dialogue, so that intellectuals who are under the strain of this conflict tend to escape from it to some intellectual activity in which they see no conflict, such as engagement in classical music [Roberts and Ridgeway, 1969]. This is true even of Russell, who confessed that writing fiction was a tremendous relief for

---

10 This may be unfair to Mario Bunge, perhaps also to some other philosophers. If so, then I apologize to them. My shortcoming here is due to my division of methodologies to deductive, inductive and dialectical. If Bunge allows for this division, then if he deems his philosophy dialectical, he is included as agreeing with Popper here. I cannot say.

11 In the eighteenth century, invite contesting parties were repeatedly invited to air their differences in public once or twice and then desist. This rule is good for debates for which the public loses interest fast. This is not always the case, though, especially not for those perennial debates that positivists condemn just because they are perennial. Wittgenstein's assertion that undecidable questions are no questions at all rested on no theory of questions and bred much contempt [Joad, 1950].

him from the strain he felt writing non-fiction [Russell, 1969, 35]. I hope this shows you how much the field of inquiry about dialectics is still neglected.

I should now discuss the fact that *etiquette, research methods* and *education* are in the same mode of discourse, but I have said enough about that. Let me also mention that the matter is still open and I have discussed elsewhere the question how much education is given to the dialectic method, especially in the academy, even in the study of mathematics. Let me repeat that it is here that I feel most indebted to Imre Lakatos who practiced the dialectic method of teaching when we first met [Agassi, 1976]. With this I turn to the final item on my agenda: leadership.

It is clear why the leadership of philosophy could not concede to Popper: as both Lakatos and Bartley repeatedly stressed, any concession to him, however small, is the admission of the need of a wide-ranging philosophical revision. (It is no accident that those who feel obliged to concede something to Popper often mention his terminological innovations[12], which is no concession at all, or they concede to him something insignificant.[13] And they prefer it to be in fields other than those they care about. This is a recurrent phenomenon, as I have illustrated in my *Towards an Historiography of Science* [Agassi, 1962, 1967, Chapter 3 and notes]. Moreover, contemporary philosophical leadership is in a particularly unhappy position these days, being torn between two higher establishments, the progressive and reactionary, for and against reason, for and against science. These philosophical leaders share the inability to assess the situation critically, since they grant their respective views scientific status, and traditionally ideas that are scientific are non-negotiable.[14] Here, again, Popper is in a privileged position: scientific establishments recognize the significance of his work before the philosophical establishment did.[15] This was an

---

12 The new technical terms of Popper that commentators love to ascribe to him are "basic statements" and "corroboration". (He invented them in order to leave behind the inductive flavor that their more familiar cognates have. It matters very little: etymologically the terms hardly differ. As the use of term "corroboration" for positive evidence is usually legal, it is somewhat superior to its cognate "confirmation" whose flavor is naturalistic. The US Supreme Court has recognized the superiority of Popper's theory for practice [Edmond and Mercer, 2002].

13 The common trite idea that commentators love to ascribe to Popper is that science progresses; it was reasonably innovative when Bacon announced it over four centuries ago. Another is that observations are theory-laden – unless they ascribe it to Norwood Russell Hanson. Galileo expressed it clearly. He gave two memorable examples: as you stroll down the streets of Florence on a moonlit night you might see the moon follow you like a cat jumping from rooftop to rooftop, and, you might as well say that the tickle is in the feather.

14 The disposition to criticize ideas of peers often appears as the challenge to the scientific status of these ideas. The purpose of this is to open debate about them. Unfortunately, all too often admitting that an idea is unscientific does not open it to debate but encourages their dismissal. It is much better therefore to ignore scientific status altogether and open respectful critical debates. It is still better to introduce the idea that scientific status is openness to criticism rather than the traditional opposite.

15 Today the scientific establishment has less of an official philosophy then it used to have in its golden past. When inductivism came under heavy fire and they could not let go of scientific certitude, they refused to view the situation as in serious jeopardy. Search is under way for a new, satisfactory version of inductivism and there is no deadline for it.

unspeakable embarrassment to the philosophic establishment (except that establishments never collapse under the weight of embarrassment).[16]

So now is the time to have a grass-root revolution in philosophy. Whether you are a philosopher or not, if you wish you may contribute. You need not agree with anything here advocated. Indeed, post-modernism, the current fashion of expressing agreement with every party to every dispute is the worst philosophy ever; it is much better to be consistent and disagree with any opinion that clashes with the accepted one or else change one's view. If you merely ask students of philosophy or of social science who come your way what they think of the matters here aired (see the abstract of this book placed in its opening), and if you follow the response you receive in a way that looks intelligent to you, then, it is very likely, we are contributing to this revolution in philosophy. This should be very exciting.

November 1991
November 2007

---

16 Things changed when certitude gave way to verisimilitude. This is not to say that science needs an official philosophy: for a century now it has none. Science administrators and law-courts do need something to go by. It is hard to say that they follow a coherent philosophy. It is therefore hard for them not to admit that Popper's negative philosophy is most suitable for them, as it is obviously minimalist. Hence, the situation requires very little, namely, the lifting of the philosophical establishment's unofficial ban on intelligent discussion of his philosophy. Someone should speak up.

CHAPTER TEN: POSTSCRIPT

> ... nothing is more difficult than to describe a person — and the more unusual and outstanding that person, the more difficult it becomes ...
> (Sir Ernst Gombrich)[1]
> We need personal stories — whether biographical or autobiographical. Personal education, personal religious commitments, personal relations between social life and personal experience (dreams included) are what we want to know.
> (Arnaldo Momigliano)[2]

A: As Time Goes By

Sir Ernst Gombrich died in November 2001. He was the last of the first generation of the family of critical rationalists Popper-style. He responded to this book censoriously. His last letter to me, however, written shortly before he died, was explicitly friendly and conciliatory, possibly because that correspondence referred to art and not to this book. Gellner died years earlier, in November 1995. He had read this book with interest. His only comment on it was very critical and concerned a minor point: my censure of Sir Isaiah Berlin's great insult to Popper. He thought I did not bring out forcefully enough the unusual and intolerable outrage that he considered it to be. Perhaps he identified strongly with Popper, sharing his feeling of rejection; the philosophical establishment maltreated him no less: his treatment of their ideas was as fair and straightforward as he could make it, and sincere enough and clear enough and argued well enough to deserve the honest treatment of which they deprived him. Despite my sympathy, my discussion of Berlin remains unaltered: for the sake of keeping moral demands minimal, we should censure only the worst transgressions. Admittedly, Berlin's conduct was far from proper, but not the worst.[3] Watkins died in July 1999. He had expressed disapproval of this book: he said it bored him. As he had enjoyed the notes to some of my older pieces, he might have disliked this second edition less. (He also said, this book is unfair to Lakatos. It is easy to show that his favorable view of Lakatos is an expression of his just appreciation of Lakatos' intellectual abilities and achievements and of his sense of loyalty, not

---

[1] Ernst Gombrich, letter to me in the Appendix to the Prologue above.
[2] [Momigliano, 1994, ix]
[3] Carnap's ascription to Popper of a demarcation of language is more outrageous than Berlin's ascription of Popper's ideas to Wittgenstein. So is Hempel's unexplained, flat rejection of my effort to correct Carnap's distortion that he repeated in total disregard for my observation on the facts of the matter.

of his critical acumen and capacity for impartial assessment of character. On this there is a wealth of material, including the recently published correspondence of Lakatos with Feyerabend [Motterlini, 1999], [Agassi, 2002b], and his archives as well as those of Popper and of Feyerabend, not to mention the detective work in "Lakatos in Hungary" by Jancis Long [Long, 1998].) Watkins kindly shared with me many hours and had had many lengthy discussions — with Gellner in a conference on Popper in Warsaw, and a week after Gellner died in a conference that he had organized in Prague, and also, just before Watkins died, as we participated in a Popper conference in St. Petersburg. We discussed many things on these occasions, including world affairs, philosophy, his own work, and his profound disappointment with Popper, his beloved teacher, whom he finally viewed as a liar. This is a dreadful misjudgment; it is atypical: his judgments were usually realistic, generous, and loyal; he expressed his disapproval of Popper's conduct with visibly great pain.

I wonder if either Gombrich or Watkins ever revised his views of this book. I did. My revision is not of my story but of my assessment of Popper's assessment of his own work, on his conduct as a leading philosopher and the disputes that engaged him, and on his refusal to argue with his former students.

B: Popper's Makeup

I do not know what a character is and whether it is at all alterable. But it seems to me obvious that we do characterize people in our environment, and it seems to me clear that I should say something about Popper's character as I see it.

Despite all the changes that Popper underwent, personally and intellectually, it seems the broad outline of his character was determined very early. (He accepted without question his father's admiration for Schopenhauer and for Darwin.)[4] Of course, intellectual progress is by definition unpredictable. But one can fix one's scope of interest and style surprisingly early. This Popper did. He told me that he had refuted his father's essentialism when he was five years old. Five. This can hardly be true, but it may be somewhat indicative. (Somewhat: there is evidence to the contrary, namely, his published report of a discussion with his father [Popper, 1997, 14]. It had happened when he was 13, naïve, and quite in

---

[4] The view of Schopenhauer or Lamarck seems to be in conflict with that of Darwin, since the one includes the assumption that the will to live decides on survival and the other includes the assumption that what seems adaptive may be causal. Not so [Popper, 1972, 267]: the survival that the one spoke of is of individual creatures and the other of species. This allows for Darwin's famous effort to wed the two ideas (Preface to the fifth edition of his Origin). Popper tried to do so with his theory of genetic dualism, as he called it: the evolution of the skill for using a new organ precedes the evolution of that organ [Popper, 1972, 279-80]. This idea came with no problem to solve; also, it is obviously false [Agassi, 1988, 285-6, 311]. It is not even clear that it is helpful in any significant way. It may thus bespeak his loyalty to his father who admired both Schopenhauer and Darwin.

error, he admitted.) He repeatedly reported of his meeting with Alfred Adler when he was 17, at which occasion he dismissed a passing remark of Adler's that was inductive: he announced this dismissal as his refutation of inductivism, no less. This is false: his earliest publications, when he was 23, are inductivist. He developed his revolutionary ideas in 1933, while writing his *Die beiden Grundprobleme der Erkenntnistheorie* (published in 1979 and in 1994; English version, *The Two Fundamental Problems of the Theory of Knowledge*, 2008) as John Wettersten argues in detail [Wettersten, 1992]. Popper received a contract for it on the understanding that he would abbreviate it. He rewrote the book into a version that incorporates this great insight and then abbreviated the new version as much as he could. His uncle finished the task. The outcome appeared in November 1934 as his *Logik der Forschung*, his first vintage and his *magnum opus*. My view goes further than Wettersten's. As I see it, for many years Popper developed forcefully his great philosophy by simply learning about the revolutionary character of his first, truly great central idea. He later acknowledged that until he met Bartley (two decades later) he had not seen himself as an important revolutionary thinker (only that his ideas were much better than those of his Viennese friends). In his first book he followed Russell's revolutionary way of presenting Hume's problem. He went further. How can theoretical knowledge (= knowledge stated in universal statements), he asked, follow from information (= stated in particular statements)?[5] And he answered this question by observing that the *modus tollens* is such an inference. (A simpler and more traditional example is from the Aristotelian square of opposition: a particular statement entails the negation of a universal one: "some swans are non-white" and "not all swans are white" are logically equivalent.) Popper made this even easier, as he used the same traditional rule to present all universal statements as negations of existential ones ("there is no non-white swan") so that its refutation is more obvious (the observation report "a non-white swan is observable here and now" entails "there is a non-white swan"). He recommended wording all scientific hypotheses as negative existential statements ("there is no perpetual motion machine"): he said, laws are prohibitions. This suggests that the knowledge of the empirical scope and limitations of a theory ("all swans are white" does not

---

5 Inductivist readers of Popper are blind to this influence of Russell: they take observation reports to be "direct" (about observations), not factual (describing the outcomes of observations) like reports on the space-time coordinates of a planet. Popper allowed for both readings, admitting that personally he preferred the factual one. Around 1953 he rejected the "direct" one. Russell allowed for it, but merely as a hypothesis, and even as a possibly fictitious one, that he found necessary to maintain his commonsense view of the world that he took as more significant than his sensationalism. Now the reason that so many philosophers prefer "direct" evidence is that the realist one is uncertain. But then the "direct" one is no more certain. They use the Duhem-Quine thesis (a refutation may be of the tested theory or the initial conditions or some auxiliary hypothesis) to show that realism is uncertain, in oversight of the fact that both Duhem and Quine said, there is no "direct" evidence.

hold in Australia) is theoretically informative. This information, then, is empirical theoretical knowledge. This idea is revolutionary and very powerful, as it breaks many central traditional moulds, especially that of the theory of rationality. Popper noted this break in his *The Open Society and Its Enemies* of a decade later. He then realized that this is a new realist philosophy: tradition is too narrow in its demand of realism to declare scientific theories true: suffice it that realism should take theories literally — as true-or-false.[6] He thus shed the positivist stance of his first book. (This positivism was due to his respect for Mach. He endorsed then Mach's proposal to stay aloof from all metaphysical controversy. Incidentally, he respected Mach's position even after he relinquished it. He always derided Wittgenstein's "logical" positivism, the thesis that metaphysics is but verbal muddle. It is regrettable that he used the word "positivism" to mean "logical positivism". It clouds the great influence that Mach had on him early in his intellectual career [*Autobiography*, §3]. He even claimed at times that in the last resort "logical" positivism and positivism are identical [*OS*, Ch. 11, n51]. Yet he could not possibly identify the two kinds of positivism, that of Kant whom he always earnestly admired and that of Wittgenstein whom he always earnestly derided.) In his early days he was under the spell of Mach, but he never yielded an inch to Wittgenstein. Never. This difference is significant enough.

Popper's great insight has predecessors, of course. He scarcely mentioned them. First, the idea that there are no theory-free observations (Galileo, Kant, Whewell, Duhem).[7] This is problematic for traditional empiricism, as it renders the reliance of its theories on observation circular (Bacon). It is no problem for the philosophy that does not rely on observations, possibly because it does not aspire to prove anything. This holds both for instrumentalism and for *theologia negative*. Although nothing that human language can describe can characterize the divine, said Maimonides, it behooves us to assert of Him all that we can, and then try to refute it. Joining *theologia negativa* and Spinoza's *Deus sive Natura*, is *philosophia negativa*. This is a framework for a whole metaphysics, for a whole worldview. One may approach it very closely, however, without endorsing it, at least not as metaphysics. This is what Popper did. He was ambivalent about metaphysics, and it took him a long time to jettison hostility to it, and longer to publish his view of the objective character of abstract entities — problems, ideas, social institutions, works of art and whatnot (his World 3). For years he was reluctant to

---

6 This is much due to the influence of Frege's theory of meaning as truth-value.
7 Regarding observation reports, Popper mentions [*LScD*, Section 26] Reininger and Neurath about their revocability, not Whewell and not Duhem; his proof that they are all theory-laden by reference to his theory that all nouns are dispositional goes further than any of his predecessors, and this may explain his lack of any reference to them.

publish it, especially since he had nothing more to say than that these are objective. He had nothing more to say of this objectivity, except that the objectivity of ideas is manifest in our ability to mould them like pieces of clay. He finally published it all the same. He did not say what problem it solves, however, not even by what train of thought he came to conceive it.[8]

Traditional ontology allows at most for matter and for minds; it has no place for many things, especially social institutions, as it denied that these are substances. Now Russell (and Popper) rightly ousted this ontology: they gave up the theory of substance as a bad job. The traditional identification of metaphysics with the theory of substance lent force to the rejection of all metaphysics. But then it is unclear what Popper allowed existence of and what not, and on what grounds. Admittedly, he could assume that all things that populate our theories about the world are real. This is more-or-less the ontology of Mario Bunge [Agassi, 1990b]. Not so Popper. He postulated three worlds, with world 3 that houses abstract things that are creations of human mind.[9] But then, are plants mental or physical? Before we can try to answer this question, we should ask, what problems does his theory of World 3 come to solve?[10] I asked him

---

[8] Popper declared that his dualist solution to the mind-body problem depended on the existence of World 3. I do not know why, unless he was alluding to the point made in the next note.

[9] Spinoza's theory of attributes was erroneous: he identified the world of ideas as an attribute in parallel to the material world. This way he erroneously conflated ideas with minds. Undoing this conflation yields three attributes instead of his two: material objects, minds, and ideas. These three attributes are Popper's three worlds. Spinoza's parallelism is a theory that correlates between the two attributed. It assumes coordination between attributes without interactions between them. This proscription on interaction was the outcome of the view of mind and matter as substances in the light of the definition of substance that renders it utterly independent. But as attributes are not substances, Spinoza did not need parallelism. The same holds for Leibniz, yet his parallelism goes further and allows no interactions between any two individuals. The reason Spinoza advocated parallelism is that he was a materialist. The reason Leibniz advocated parallelism has to do with logic, not with metaphysics. This won him the admiration of Spinozist Russell.

[10] An excellent example of a problem-oriented text is Popper's *Logik der Forschung* that centers on the problems of demarcation and of induction. A rather poor one is his presentation of his improvement of the theory of language of Karl Bühler. No problem and no criticism, not even in the background. In an effort to remedy this, we may perhaps begin with the common claim that human language differs from languages of other animals as it is conceptual, and note that concepts are problematic. Perhaps Bühler tried to do away with concepts by using instead the functions of language for this very differentiation: only human language is descriptive. Did Bühler insist that bees' language is signaling rather than descriptive? If so, why? No answer. As the division of the function of language to signaling and description is not exhaustive, he noticed that a lower function is sheer expression. But then, is bird song only expressive or also communicative? Freud said, it is communicative and Lorenz agreed. Is the very first baby sound only expressive or also communicative? Charlotte Bühler said it is communicative. When then is language merely expressive? (Remember, they all postulated that functions are hierarchical, that the higher function includes the lower one.) Popper declared the critical function higher than the descriptive. This is an intriguing observation: some people are unable to criticize. But this still lacks a problem, which is puzzling, as language has many other functions, inviting diverse classifications. Roman Jakobson, for a conspicuous example, also distinguished between the functions of a message from the viewpoint of sender of a message and from that of its recipient, thus introducing falsehoods and thus ranking between Bühler and Popper. Is it important? For this we need to know the problem-situation that serves as a context to this discussion.

this, to no avail.

In Popper's entourage some had observed that I am crazy,[11] literally certifiable. This observation appears first in a letter from Popper to Watkins that is now in his archive. For all I know it still is Lakatos who should be credited with this observation. It was the imprimatur from the Master, however, that counted — evidently not only in an odd letter. The letter was a response to my notes on Popper's response to Bartley's claim that Popper had surreptitiously altered his views, reported in Chapter Five above. I did not publish these notes. Popper did not mail the letter to Watkins, I surmise. This was the fate of many of his letters. Thus, he did not mail the aggressive letter to me that Edmonds and Eidinow cite [Edmonds and Eidinow, 2000, 143]. The reason for his not having sent these aggressive letters seems to me to be shame: they are unkind, and he must have been at least ill at ease about having harbored any unkind sentiment. In all of my efforts to comprehend the person that this great philosopher was, meanness was the last quality that one can ever ascribe to him. He was never mean.[12] Surprisingly, some of his unsent letters — not only to me — are rather unkind. But then they are unsent. And they expressed genuine pain. What pained him most about me is that I endorsed Bartley's observation that his translation of his *magnum opus* was not as honest as it could be, although, to repeat, unlike Bartley I deem this defect quite marginal. The book includes many admissions of error, obviously made reluctantly. And they sound odd when seen in the background of his late realism, whereas they would have been natural against their positivist background that he tried so hard and so needlessly to make little of. Let me offer the tiniest example for this that I could find. Popper rightly read Heisenberg's views as more of a program than a definite theory, and as a positivist program to boot. He tried to improve upon it and failed. He openly admitted the failure. Yet the title of the section in his *magnum opus* (§76) that should read "An attempt to eliminate metaphysics by inverting Heisenberg's Program" reads instead, "An attempt to eliminate metaphysical elements by inverting Heisenberg's Program". This criticism of mine is minuscule. But it is valid. And it does not speak well of its object. It is heartbreaking that so minute a point was the main reason for his inability to patch up our separation — much as he would have loved to. He mentioned twice my claim that his translation of his *magnum opus* was inaccurate, both furiously. I addressed his fury before I could

---

11 The reference to my being crazy is in Chapter Four.
12 This is not to deny that on occasion Popper was inconsiderate and even unkind. But this was always due to insensitivity rooted in powerful obsession. Once, early in the day, Lakatos said to Popper nastily that *The Logic of Scientific Discovery* is inferior to *Logik der Forschung*, and Popper responded at once: "you are a nervous wreck." This was painfully true and unkind as I later I said to Popper. He was surprised and asked me to tell him my view of him unsparingly. I complied, observing that he was consistently unkind to Lady Popper. He never forgave me that.

answer him. My effort was successful only once, and then, not furious he dropped the whole matter. Pity.

C: Popper on Communication

The appendix to this chapter discusses Popper and the Wittgenstein crowd, including the "Vienna Circle". I explain there why he would have done better had he ignored them. It is his views on the philosophical questions that he discussed that matter, not his arguments against these people who claimed that there are no philosophical questions to discuss. My criticisms of Popper's views on these questions do not prevent me from declaring myself his disciple, as they seem to me marginal. My criticism of his conduct as a philosopher is more severe and perhaps unfair: he wasted his time arguing with his scholastic Viennese peers, but possibly he could not help it. Still, inasmuch as his conduct rested on a view, that view can tolerate criticism: he constantly expressed high appreciation of criticism while regrettably overlooking his own disapproval of the continuation of critical debates after they deteriorate into scholasticism.[13] In his seminars he repeatedly spoke of deterioration, admonishing his audience to avoid it by keeping in mind the problems that engage them and showing readiness to replace them with better ones. Scholastic debates, he said, are not serious: they treat minute problems that have lost their value just because the significant views that they pertain to are no longer agenda. He wisely proposed to shun them. He did that with one exception, which is a good record. This exception was in part due to his mistaken oversight of the demarcation of scholasticism within science: his theory concerned not scientific problems but degrees of falsifiability of scientific theories: he said it is best to test (= try to refute; try to criticize) the one most open to test. He also said, it is best to treat problems that one cares for. These two proposals clash often. His theory of degrees of falsifiability is a solution to the problem that he barely stated: how does science reach unanimity? My answer to this problem is, unanimity is the default option: when there is one good answer to a tough question, it is easiest to take it for granted and it is better to try to test it. Usually[14] it is also easier to discuss it critically then to devise an alternative to it. This seems to me to close the discussion about unanimity in science, but here let me ignore it altogether and replace it with an empirical observation: unanimity in science is partial. It appears at most within one discipline: no one expects researchers across disciplines to agree. Discussion of this

---

13 Lakatos always hated scholasticism. As he distorted my work on research programs to claim that every research must follow a full-fledged program, he faced a new, rather silly problem of demarcation: which one is useful? As he could not possibly solve it, he followed Popper here and divided them to progressive and degenerative. This, said Feyerabend rightly, is not an answer but a renaming.
14 The exception is one of the most brilliant observations of Lakatos: when efforts to refute a theory fail, it may be useful to seek an alternative to it or an explanation of it — whichever comes easier, I suppose.

matter is almost nonexistent. What then is a discipline? Popper said, at worst it is an administrative convenience and at best it is a tradition of studies anchored in a cluster of problems. (He did not ask, what keeps some items in one cluster. The answer is obvious: it is a metaphysical assumption, whether it belongs to meta-physics, to meta-biology, to meta-psychology, or to any other field.) The unanimity, then, at most exists within one such cluster. The appearance of these clusters is a historical given, so that it is subject to historical examination, hardly to empirical test. It is thus no surprise that in one given discipline diverse general views, metaphysical and methodological, may appear as dissent about these issues grows with no swift settlement in sight; disputes about them can linger for generations on end, possibly leading to a split of the discipline.[15] Popper did not examine this process. The same goes for his useless dispute with his scholastic peers from the old "Vienna Circle": it concerns basic issues[16] that engaged him for half-a-century and more. He said he killed their doctrine, namely the view of Wittgenstein that all statements of metaphysics are ungrammatical in principle. This doctrine deserved to be dismissed off hand — at least ever since the demise of the search for the ideal language. For, as Popper observed, even were there a language that complies with Wittgenstein's proscription of metaphysical assertions, it would be wisely dismissed – unless it is the ideal language. The last word of Rudolf Carnap was, he still hoped that such a language was possible. That is to say, he hoped for a refutation of Popper's proof that such a language is impossible. Are there people who still share this dream? Does this matter? How do intellectual coroners pronounce doctrines dead? Who are these coroners? Who appoints them? As long as Wittgenstein or an heir of his is a coroner, the corpse will remain pronounced alive and if need be kicking as well.[17] After all, conventions[18]

---

15 The inability to resolve a dispute within a discipline breeds incentive to split it that is incentive to cheat. A conspicuous example is child psychology, where Freud's and Piaget's competing views appear as different topics taught in different sub-departments.
16 The "logical" positivists declared Popper's demarcation of science meaningless, since is neither analytic nor empirical. This holds for their view [Wittgenstein, 1922, §6.54], not for his; yet here they forgave only themselves, not him.
17 Neurath endorsed the behaviorist proposal that science should ignore the inner world. He assimilated it to "logical" positivism by limiting the language of science to that of physics, whatever this is. To that end he claimed that physics ignores the inside of atoms. (He labeled this profundity "physicalism" and "the unity of science" while endorsing flippantly Duhem's instrumentalism that declines even the unity of physics, as it is the proposal to consider competing physical theories separate languages.) In the International Congress for the Unity of Science (Copenhagen, 1935) Popper said, Tarski had wrecked the unified language. Wittgenstein rejected Tarski's view; Neurath responded to Popper shabbily [Agassi, 1987]. All this is ancient history: behaviorism and ideal language died together; physicalism followed.
18 Popper wisely spoke of the endorsement of refutations as a convention of the scientific community. He wavered between viewing this idea as descriptive or as prescriptive. This holds rather generally, as in the case of good grammar: we recognize the conduct of our betters, whom it behooves us to emulate. This Popper did not say, perhaps because it smacks of élitism. It need not be elitist, since we decide who our betters are and why. This is a bootstrap process, for sure, and so not given to justificationists.

regulating the pronouncement of the death of ideas should stimulate some radical reforms in our intellectual environment.

Popper's assessment of his own output was ambivalent. He repeatedly confessed to me that at times he thought it trivial and at times innovative in the extreme. This is not surprising, as it is very hard for great thinkers to admit their greatness, especially in the light of the popular morality that preaches distorting the value of one's contributions by efforts at humility. Philosophy, any contribution to it, is particularly hard to assess. Commonsense plays a constantly confusing role here: for a philosophy to abide by it is advantageous since this is the default option, and it is an obstacle to those who refuse to submit to it. Aristotle fell back on commonsense whenever he met with trouble, showing little concern for consistency. The great Russell admitted that commonsense presented a serious dilemma for him, and on a few counts. Obviously, however, although it is distinctly advantageous for a solution to any problem to accord with commonsense, when it breaks new grounds it violates commonsense — for a while: if it succeeds, then commonsense wisely follows suit. Even if when the commonwealth of learning rejects a new doctrine in the name of commonsense, and even if that rejection is sheer commonsense, the new doctrine may still alter commonsense, and then people will wonder what was new in it and why did it meet with opposition.[19] And finally, when a solution accords with (the extant or the new) commonsense view on its material, it is easy to avoid granting that solution the praise that it deserves, and to do so by dismissing it as mere commonsense. Even such forceful thinkers as Shaw could make this kind of mistake. He dismissed Darwin's theory of natural selection as commonsense, ignoring its role as a solution to a problem that had led his predecessors astray. Alfred Adler knew better: as critics dismissed his contribution as mere commonsense, he was not defensive. Rather, he invited them to share his views.

The obstacles to Popper's effort at a proper self-assessment were two. One was his repeated identification of appreciation of his ideas with assent to them. This identification is Baconian, and Popper regularly spoke against it, yet he did not take sufficient care to avoid it. The other obstacle was his repeated singling out the "logical" positivists to express dissent from them without explaining why they deserve this gesture of appreciation.[20] So let me conclude this discussion with a general observation about them.

The "Vienna Circle" excelled in two connected things, expressing

---

19 This is wisdom after the event. My discussion of how to avoid it [Agassi, 1963] won me compliments (even by Merton), although it is sheer commonsense.
20 In his 1959 Preface to his *Logic of Scientific Discovery* Popper complimented the Wittgenstein crowd, calling them fellow rationalists. He dismissed my misgivings, saying no one around is better.

contempt for what they rejected as metaphysics,[21] and their engagement in clarification. There is no need to discuss the hostility here, especially since Popper repeatedly dismissed it as trite. (In the American Philosophical Association Eastern Division meeting of 1983, Hempel admitted, the hostility of the "Vienna Circle" to all metaphysics was an exaggeration.) Their engagement with clarity is different: no true lover of learning can be indifferent to its value. The "Vienna Circle", however, did not clarify; on the contrary, they launched confusion that still prevails, as it rests on their endorsement of Wittgenstein's theory of clarity: his (unreasonable) view of his ideas as "unassailable" led him (reasonably) to deem clarity and obscurity as black and white, in the denial of the possibility of any gray in between. (He struggled with all his life against the commonsense arguments to the contrary.) He and they never freed themselves of this extremism, as they were busy modifying it in small steps and surreptitiously. Wittgenstein fans in England, by the way, declared themselves adherents to (an unspecified version of) commonsense, and, as Russell observed, they thereby rejected the best of science and clung to their pet prejudices. Popper performed a miracle of clarification, but they were all too busy maltreating him to be able to see this, especially since his clarification was not Wittgenstein-style verbal, but an outcome of a study in the traditional style of the inner logic of science.

Here is a characteristic statement from a popular newsletter concerning the highly corroborated hypothesis of utter symmetry between matter and antimatter:

"more sensitive tests might show some slight asymmetry, which would help to refine our understanding of the fundamental structure of nature, and may also answer the puzzle of why there is an imbalance in the amount of antimatter and matter in our universe."

Here is my translation of it to Popper's frame:

"The corroborated symmetry hypothesis renders puzzling if not impossible the observed paucity of antimatter. Better tests might hopefully refute it and help develop a better alternative to it."

The authors of the newsletter in question will refuse this kind of translation: the admission that a corroborated hypothesis may be false looks like a slight on science. Realizing that this is unnecessary defensiveness, we can easily see the advantage of the translated version. Here is an example for this advantage (*Science*, 23 March 2007):

"Suppose you are a particle physicist. ... When, after 2 decades of

---

21 Since the claim of the "logical" positivists was that metaphysics is ungrammatical but not that the ungrammatical is metaphysical, they should have said what renders an ungrammatical sentence metaphysical. All they could say was, whatever tradition deems metaphysical is ungrammatical. Why tradition is as we know it, was in principle beyond their comprehension! Since they had no grammar, they had to accept metaphysical assertions and seek reasons to discredit them. This had to be *ad hoc*.

preparation, you get ready to switch on your rig, you would fear nothing more than the possibility that you were wrong ... , right? Not exactly."

"Many particle physicists say their greatest fear is that their grand new machine ... will spot the [expected] Higgs boson and nothing else. If so, particle physics could grind to a halt, they say." This is in the spirit of Popper's methodology. It appeals to me greatly, as it illustrates Popper's contribution to the workings of science, unlike the view of traditional methodologists who never hoped to be able to help scientific research workers in their work. Of course, Popper was not the first to have contributed positively. Plato's contributions to science are famous since Whewell discussed them in the early nineteenth century; Einstein acknowledged his debt to Hume; and many biologists and psychologists acknowledged their debt to Kant. My contrast here is between Popper and Wittgenstein.[22]

My disapproval of Popper's persistent criticism of the "Vienna Circle" holds also against this chapter. It continues his battles with the heirs to his peers. This chapter is the last of my efforts this way. I have no intention to insist and persist to the last as he did. To the extent that they, or their heirs, have relinquished the "logical" aspect of their positivism, he won the battle. Alternatively, his weapon (criticism) is useless against their irrationalism, since it is impervious to criticism, being too blatant and brazen for that. They could nevertheless have valid criticism against him, of course; they had none. They offered two major arguments against him. First, his ideas were very similar to theirs and he exaggerated the differences between them out of his tendency for self-aggrandizement. This is a shamefaced admission of defeat. Popper said that this rested on their refusal to take seriously his view that metaphysical statements are meaningful [*Autobiography*, §xx]. This is sufficient reason to stop arguing with them: there is no point in criticizing those who refuse to listen to criticism. This is problematic, of course: both parties to a jammed debate may blame each other, and they often do. Nevertheless, to continue is useless. Something else may help, but there is no guarantee.

Their second argument against him was that a difference between them existed, after all, but of interests, not of opinions: they differed in the questions that they were trying to answer. As time goes by, an in-

---

22 Wittgenstein called metaphysicians slum-landlords and appointed himself police officer [Edmonds and Eidinow, 2001, xx]. The logic of the situations of slum-landlords is better understood than that of metaphysicians, but then Wittgenstein had no interest in sociology. Quine called him "the prophet". Gellner saw him as the "night watchman ... of philosophy" [Gellner, 1959, Chapter I:3]. Rorty called himself a his follower but he said he refused to join the police force, preferring to view philosophy as conversation, and on the authority of Michael Oakeshott. Neither Wittgenstein nor Oakeshott is to blame for this flippancy.

creasing number of commentators[23] view Carnap as the representative of the "Vienna Circle" in their dispute with Popper. And he finally placed the dispute about the existence of philosophical problems with the one on induction. This is progress, since it has nothing to do with Wittgenstein and his heritage.[24] Popper proved Carnap's view on induction inconsistent. Carnap responded that they had different concerns. Such responses do not resolve inconsistencies. Still, let us consider the difference that Carnap said he had with him and Popper. Carnap wanted a guide for lifeand Popper wanted to characterize science. Did Carnap accept popper's demarcation? He did not say.[25] As a guide, Carnap offered one rule [Carnap, 1950]; sooner or later, he promised, business people who obey the rules of the calculus of probability might expect profit.[26] Now as a socialist Carnap cared little about the level of profit [Parrini, 2002] and less about raising it. So what is the advice he was giving and to whom? This question is not for me to answer. Rather, let me say, whatever is the right reading of Carnap's text, on this he was right: Popper did not care about it: he cared about science as a spiritual adventure. So if Carnap's claim that he and Popper were talking at cross-purposes is right, then Popper had no business quarreling with Carnap or with his peers.

D: Popper and his crowd

It flattered me no end that Popper wished me to be his heir and successor. To this end, he showed me great good will and much fondness and he offered me tremendous help, including the procurement of a job offer in the London School of Economics. I decided to go away all the same[27] — even

---

23 Commentators (incudlign Popper) view Neurath as the most original member of the "Circle". This is a different matter.
24 Wittgenstein declared the principle of induction not given to wording. He worded it thus: a scientific theory is not descriptive ("*Gesetze ... handeln ... nicht von dem was die Netz bescreibt*"; where "*Netz*" is his name for a coordinate-system) [Wittgenstein, 1922, 6.35]. When he challenged Popper to present a philosophical problem, Popper named the problem of induction. Wittgenstein dismissed it. Carnap endorsed the received view of induction as probability. To up-date it Wittgenstein's way, he presented his work as verbal clarification. Popper had proven that the rules governing the probability of hypotheses (their empirical support) are different; Carnap preferred to ignore this.
25 In private Bar-Hillel expressed agreement with Popper on this (Chapter Five above). He should have said so openly but then he realized, I suppose, that the consequences of this would be too far reaching.
26 When gamblers exhaust their funds, they drop out of the game. This refutes Carnap's rule. Gambling away all of one's assets, said Dostoevsky (*The Gambler*; *The Brothers Karamazov*), is irresponsible on any odds. Bar-Hillel told me that this kind of argument troubled Carnap in the last hours of his researches. This, he added, made him change his view of Carnap: he felt cheated.
27 I cannot adequately express my gratitude to Judith for her having allowed me refuse the LSE job offer around Easter, 1960, while I had no alternative job for the following academic session. A job opening in Hong Kong appeared due to the tragic death of a philosophy lecturer there. Popper mobilized the School's Director to help me and this tipped the balance in my favor. Of the tragic story of the lecturer's death I learned there soon after my arrival in Hong Kong. He was appreciated and could live there happily without further ado. Nevertheless, he felt obliged to submit a doctoral dissertation to Oxford in the style then current there. It examined the use of moral words in *Macbeth*. It seemed to me no worse than the usual philosophy publications issued from there. He failed, I have no idea why. He then joined a cult that happened to oppose modern medication. Sadly, he suffered then from boil infection, refused antibiotics, and paid with his life to prove the sincerity of his new faith.

before I had an alternative job offer. I could neither let him impose his advice on me nor quarrel with him about it. I thought that our relations might have improved had he discussed ethics with me. And then I might have accepted the job offer. I no longer think so. His refusal to discuss ethics with me is the expression, not the cause, of our inability to adjust to each other. I saw his conduct as obsessive, but this holds for both of us: I should have been more indulgent but was not capable of it. At the time, indulgence to obsessions seemed wrong to me; this is an exaggeration, as the ability of David Miller to get along with him has taught me.

Popper was unusually open-minded and quick to learn. It took a lot to realize that this was so only within certain fixed limits. No; this is not what I mean, as we are all open-minded within limits. So I should retry: Popper exhibited an unusual combination of two unusual and conflicting traits. One was the unusual scope of his open-mindedness: it was very wide indeed. The other was the immense stubbornness that he exhibited when he passed the limit of his open-mindedness. Consequently, I seldom found an opportunity to see how stubborn he could be. This is strange, as the instances I did notice were all too clear. Some examples stand out. One was his gluttony for hard work; another was his craving for compliments; still another was his persistent expression of hostility to Gellner, whom he obviously loved. His hostility to contemporary art, especially contemporary music, also stood out. He did argue about this matter, and so he was not closed-minded about it, at least not obviously so. Yet, even then he had his eccentricities. An example is his attitude to Arnold Schoenberg: he was seldom ready to express any appreciation of him, and when he did, he declared his work, especially his *Pierrot Lunaire*, theatre rather than music. I found it funny when I read this very excuse in *The New Yorker*, whose music critic for decades, Winthrop Sargeant, was famous for his utter loyalty to strict tonality and uncompromising hostility to cacophony. Not so when I heard it from the great philosopher. Of all the composers past Wagner, he was forgiving only to Anton Bruckner. This is common: we all make exceptions. But he also staunchly refused to discuss this. And refusal to argue was not Popper's distinctive trait. Perhaps this has to do with his youthful memories: he belonged to a set that obviously adored Bruckner. (Their whistle was the theme of the scherzo from his seventh symphony, ironically, the one dedicated to Wagner.) The most pernicious of his ironclad principles was that usually his disciples criticized him in public out of spite. I have already discussed these items. So let me comment on his stubbornness. I now consider it less idiosyncratic and more in observance of academic custom.

Consider Edmonds' and Eidinow's best selling *Wittgenstein's Poker*, [Edmonds and Eidinow, 2001]. It discusses in detail the famous clash that took place between Wittgenstein and Popper. Its point is right[28]: Wittgenstein said, to ask a philosophical question is sick; and his view is now *passé*. This should be obvious, yet it is repeatedly denied — implicitly and explicitly. The current leaders of the analytic school who discuss Wittgenstein, Kripke and Hintikka, mislead their public [Agassi, 1995 and 2000]. Hopefully, *Wittgenstein's Poker* marks the end of an era. But this is a different matter [Agassi, 2005]. I am discussing now idiosyncrasies. The incident would be normal, had it taken place within the German tradition on some neutral ground. It took place on English soil, however, and on Wittgenstein's turf. Even a typical German professor would have behaved better than he.

Traditional academic etiquette in Germany and Austria notoriously differs from that in Britain and North America. Let me mention David Hilbert, the great German mathematician of the early twentieth century. He alternated unpredictably between great friendliness and frank hostility, brave fairness and blatant bias, an admirable open-mind and a lamentable closed-mind. He loved to shock and to interfere with his students' private affairs. He spread his interests to a superbly wide range, always fruitfully, even if at times unreasonably. He utterly refused to acknowledge debt or express gratitude to anyone he disliked. He was hostile to critics (including even Gödel). And in his studies he displayed attitudes that had been formed very early in his career [Reid, 1970].

In all these characteristics, Hilbert's personal mannerisms remind me of Popper's. Yet there was a striking difference between them. Hilbert was an established leader, surrounded by a crowd of admirers, all illustrious in their own right; he behaved in accord with the accepted rules of conduct for great professors as understood in his milieu. Popper was the odd man out both in his English milieu and in any philosophical company. He was egalitarian to a fault, and so his mannerisms did not reflect the required sense of self-importance as those of Hilbert did. So I took him to be peculiar but still consistent. I noted his peculiarities as minor oddities. After all, no one is fully and systematically and consistently true to type. But certain things stood out. His hostility to Gellner may be due to disappointment, since their relations began in great intellectual affinity and in the excitement that kindred souls experience, especially when they suffer isolation as they share great and unpopular ideas. [Jarvie and Pralong, 1999, editors' preface]. His hostility to John Wisdom was different. He disliked him from the start. He had him hired, he told me, because he was

---

28 This refers to the philosophical issue, not to etiquette. Russell and Tim Smiley's comments deserve mention here. Russell wrote to Popper the day after the event that the rudeness of the Cambridge philosophy crowd present had surprised him, not that of Wittgenstein, which he was used to. Smiley is the only commentator to have bravely noticed [Smiley, 1998 and 2004] that Wittgenstein's leaving the session in a huff was violation of basic etiquette as he was then in the chair.

a fellow refugee: he had taught in Egypt and political upheavals forced him to leave. He was a gentle soul, a terrific moderator, and a very original philosopher. He was humble to a fault. It took me years to see this although he generously granted me a fairly close friendship. I was familiar with his output early in the day, but I failed to appreciate it, partly perhaps because Popper was too unfair to him for me to notice to the full and immunize myself against this misfortune. He showed that the criticism that Berkeley had launched against the calculus was highly productive. He did so long before he met Popper or Lakatos, long before he invited Popper to write on Berkeley's instrumentalism — in the paper that was part of Popper's self-liberation from positivism and so also of the development of his later, more explicitly realist views.[29] Already in Egypt Wisdom wrote that modern logic undermined essentialism and thereby much of traditional philosophy [Wisdom, 1947]. He saw this as the forte of analytic philosophy and the reason for its popularity, though already then he judged it excessive.[30] Popper ignored all of Wisdom's merits. Perhaps he could not forgive him his respect for Freud. (In his *Postscript*, Popper spoke differently, comparing Freud to Homer, no less.[31]) He disliked Wisdom and he actively undermined his career, especially by refusing to recommend him for promotion and instead recommending Watkins, who was Wisdom's junior. (As it happens it was all to the good.

---

29 Popper's "Berkeley as a predecessor to Mach" (1953; now in [Popper, 1963]) breaks new ground. He had expressed his admirably balanced view if conventionalism in 1935, and conventionalism and instrumentalism are twins. Nevertheless, his opposition to instrumentalism came only in 1953, when he struggled to consolidate his dissent from Mach as a response to Wisdom's invitation. Is it not easy to see the difference between conventionalism and instrumentalism. They must differ, however, since partial instrumentalism is ubiquitous but it takes sophistication to conjure partial conventionalism. (I said this in my contribution to *The Philosophy of Karl Popper*, 1974. And though his response was curt, he kindly complimented me for this point.) See next note for a case of partial conventionalism.
30 This is an astute observation. It explains why the analytic philosophers whose view of science was either essentialist-inductivist or conventionalist-instrumentalist were stuck. There remained the option of partial conventionalism (as mentioned in the previous note).The philosophy of Richard von Mises [Mises, 1951] is an example for it. He suggested that the status of having passed tests gains scientific status (as true) quite irrevocably. Hence, he had to conclude, theories that lose their status as truths by nature become truths by convention. This view permit the rescues of refuted theories from oblivion, and so it is obviously better than inductivism. It is also better than instrumentalism that says, some theories deserve rescue from refutation, but it offers no criterion for detecting this asset. Von Mises did: it is having passed empirical tests. Moreover, instrumentalism is the comprehensive view of all scientific theories as mere instruments and thus as mere *façon de parler*. Mises said, this holds only for refuted theories. He limited science to all and only empirically successful ones. His idea is very sophisticated. Not surprisingly, it never won serious attention. As it happens, it is easy to refute empirically, as many once empirically successful theories are gone. This objection holds more forcefully against traditional instrumentalism: neither version can explain well the choice of theories to rescue. The dominant view on this today is Einstein's: science salvages those refuted theories that are good approximations to their successors. This obviously goes well with Popper's magnificent suggestion that we should not forget all old errors but remember the important ones as errors while specifying their importance that renders them memorable.
31 Popper did not reconsider his attitude to Wisdom after his view of Freud mellowed [Popper, 1983, §18] as he should have done, and possibly he might have done had he lived longer. (The literature hardly notices the change in Popper's attitude to Freud.)

Soon afterwards, Wisdom went to North America, where he met with much more appreciation than in Britain, as peers there viewed him as the leading student of psychoanalysis that he was and showed no hostility to his biographical study of Berkeley.) Much later, I asked Popper why he refused to recommend Wisdom for promotion. He said he could not swing it; I said, this is irrelevant, as he could have tried to recommend him anyway, since the absent recommendation bespoke absent appreciation; he said he could not bring himself to recommend Wisdom: it felt wrong. This kind of reason he usually (rightly) dismissed as Romantic and unbecoming.

I made only a cursory search in his archive, and I will not trouble you with it. Let me mention two items.[32] In File 154.27 # 33 there, item dated January 19.71, Popper repudiates as myths two assertions. First, there is the assertion that the ideas expressed in his latest works belong to Agassi, Bartley, Feyerabend, and Lakatos. Second, there is the assertion that that he never changed his mind.[33] He then mentions his having learned from Lakatos about mathematics as the sole exception to his observation that he learned nothing from his students.[34]

A second, more depressing item is from File 266.24, with no date.[35] It responds to Lakatos' report to him that I ascribe sensationalism to him, which is outrageous, of course, and to his having showed him some passages of mine that are just ridiculous.[36] Popper then reports that Lakatos had said to him that my main claim was that everybody before me had been a sensationalist[37]. He then told me on Lakatos' advice to

---

32 Another item appears in a note at the end of Chapter Four above.
33 No one in Popper's entourage ascribed to him these assertions. Everyone there noted how often he said, he loved criticism, yet the received view he was ambivalent about it and he distorted some past events in a manner that covers up some of his changes of opinion.
34 To repeat, the trouble with Popper's mode of acknowledgement is not limited to his students, from whom he learned anyway less than from his august predecessors. This becomes interesting upon examination of the idea of influence. I discuss this in my "Karl Popper: A retrospect", in [Agassi, 1988]): criticism of older ideas that open the way to new ones display influences of originators of the criticized ideas. Popper made this point in the abstract, never in relation to his own intellectual ancestry. His report of his criticism of Alfred Adler's inductivism makes no mention of the option that if true, then it was Adler who triggered his anti-inductivist stance. It is questionable, though, as Popper's early publications are inductivist. See [Wettersten, 1992].
35 Popper did not send this letter. Since it invites me to withdraw my "Sensationalism", it is obviously from before 1966.
36 Years later, Nathaniel Laor visited Popper in Kenley and tried to persuade him to make peace with me. He refused but expressed readiness to receive a letter from me. They held very long conversations, and Popper referred then to the text in question. It is from my "Sensationalism", now in [Agassi, 1975, 110]. It comes after my assertion about the agreement that one has to admit any scientifically acceptable observation-report as true. Here is the relevant text in full.

"Popper has stressed that this acceptance must be tentative; *but even he* agrees that accepting a report is, for the time that it is accepted, considering it to be true."

Popper's evidence was the three words that I italicize here. I am respectfully adamant. I suggest that he never read this text, not even the whole sentence (see next note) but rather trusted Lakatos' deceptive report on it. See next note.
37 Popper's error here defies my imagination: I never claimed that he advocated sensationalism, and I

withdraw the paper,[38] threatening to attack me in public if I refuse. All this reflects the very tense atmosphere that Popper lived in day in day out. I deeply regret having contributed to it and hope that some things I did ameliorate to some extent for the pain I caused him, no matter how unintentionally.

E: Coda

I browsed through Popper's correspondence one sunny morning in the loneliness of the huge, luxurious, nearly empty reading room of the Hoover Institute archives. He was dead then for years, as were of most of the individuals with whom he ever corresponded. The bitterness that most of his letters emanate hit me hard. It should not have surprised me so much. I should have taken better cognizance of the fourteenth international congress of philosophy in Vienna, in summer 1968, on which occasion we met for the first time after years of estrangement at his behest. He saw me in the audience of his lecture and he expected me — so he told me years later — to stand up in the middle of the lecture and leave demonstratively in disgust. When I approached him afterwards to congratulate him on his lecture, he showed amazement. One particular picture of him that stands out in my memory should have told me more than it then did. The organizers of that congress managed to get good, cheap tickets to a performance of Mozart's *The Magic Flute*. It was propitious. The production was excellent: high-spirited, folksy, and intelligent. The music and the acting were fine. We had excellent tickets, in the boxes nearest to the stage. As luck would have it, the Poppers sat in the box opposite us, on the other side of the large theater. They did not see us. He was absorbed, presumably in sad memories: it was his first return to his hometown[39],

---

repeatedly praise him as the first empiricist philosopher fully free of it. Popper's comment here on my view that his solution to Fries' trilemma is redundant is also erroneous. He declares this a poor restatement of an idea of his. It is his, of course: observation reports invite explanations. Except that he said we should explain them as true, and I say (following him on this too), not necessarily so: we may explain them as true or as approximately true or as errors of this or that sort, or even as plain lies. (Scientific fraud is rare in the annals of science, but it is naturally not totally absent there. The famous English novelist C. P. Snow has at least three novels in which it plays a central role.) My effort was to show that Popper's philosophy is even more remote from sensationalism than it might otherwise appear. I suppose he was able to take it as a slight only because he never read it.

38 "Sensationalism" was in proofs for four years during which Lakatos did his best to force me to withdraw it. (I still see all of his efforts in this direction as a great compliment. He once rejected a paper by Jarvie and then used its content in a paper of his own.) He explained to me that publishing "Sensationalism" was bad for my career; and he forced people to put pressure on me to withdraw it. He also said it was immoral, suggesting that I paid a compliment to Ryle there as a means for getting it accepted. I was so glad that the compliment to Ryle, the then editor of *Mind*, was an insertion made after he accepted it for publication, since otherwise I might have doubted myself. After four years that the paper was in proofs, I wrote to Ryle, promising not to withdraw it come what may. He apologized and published it at once. This, incidentally, is not a record for an editor's delay. My "Who Needs Aristotle?" waited for publication a quarter of a century after the editor accepted it for publication.

39 Popper refused to return to Vienna and he would not entertain the possibility of receiving an appointment there. Lady Popper told me she wanted this very much. He went to Alpbach, Tyrol, as often as he could, since it was a summer school organized by Viennese intellectuals who were bona fide anti-Nazis

and it was his favorite opera. He sat motionless, ashen faced, painful to look at. Lady Popper sat motionless too, looking utterly at a loss.

I do not know if bitterness ever left him; I hope it did. I have as evidence for this his posthumously published study of Xenophanes,[40] with whom he strongly identified. He described Xenophanes as not only his predecessor but also the discoverer of most of his — Popper's — great ideas; which is incredible for one who all his life so jealously guarded his priority. The essay includes a poetic fragment of Xenophanes, one that Popper retranslated just before he died. Xenophanes describes himself there as a weary old man, a guest comfortably inclined by the fireside, sipping sweet wine and nibbling nuts, and narrating about his long life of travails in reply to queries, evidently from a friendly host who gave him shelter for the night. Popper's essay ends with a comparison of his hero's and his own backgrounds. He refers to the clash of cultures that took place in Asia Minor in the youth of his hero, to the war of aggression that he witnessed later and that caused him to emigrate to an exile community of "highly learned and educated refugees". He once asked me, I vividly remember, very painfully, why don't you settle down in England and have Tirzah (my daughter) grow up here like the children of other refugees? Other refugees. I did not answer. The right response, I am no refugee, seemed to me too harsh. Perhaps he suffered less pain as he grew older; I cannot find out. I think he died in peace.

One last point. Popper was very critical of the academic tradition, and he rightly disregarded much of it that he found no less than contemptible. That an academic position was so important for him was merely a matter of expediency: were his family not financially hit by the war as it was, he would have no doubt preferred to live the life of leisure as a *Privatgelehrter* with no job at all. He had no idea of a university as a power base. He did tell me that he would have preferred a position in Oxford or Cambridge or Princeton, but this was of marginal importance for him. Of course, the university position offered him more than money, but he pretended that this was of hardly any importance. He did benefit at the very least from having graduate students, but in his own view[41] he treated them more as younger colleagues than as students, and more as private students than as members of the University of London. You remember I hope that

---

with no hope of securing a university position, and it was for many years the nearest that he came to Vienna. There he met Feyerabend. In the late fifties he went to Vienna for an emergency operation on a detached retina. In 1968 he returned to Vienna for a conference. Some time after his retirement, he returned to Vienna for good, and the mayor received him with great honors. I was not then in any contact with him. I heard from others about his understandable failure to reintegrate in Vienna and his having left again his beloved hometown for England — where he stayed to the end.

40 See final Appendix below.
41 I illustrated his view of his relations with his students in my story above of his response to my fictitious dialogue between him and a student.

he accepted me as a student while advising me not to enroll in the London School of Economics. But the thing that he received from the University and the School that was of the greatest value for him personally was his seminar: it was a great alternative to the academic system, an island of alternative practice, all of its shortcomings notwithstanding. It was for him a remote replica of his image of Plato's Academy at its best.

It is thus very hard to view him as a victim of the German academic tradition. Yet the same holds for Freud, for example, regarding both the academic system that was unkind to him and the Victorian ethos[42] that he helped destroy. We would expect Popper to exhibit in his conduct the willingness to be a target of respectful criticism. He was not. Nor is it so easy. I have recently reread my review of his *Objective Knowledge*,[43] and its harshness surprised me. It is more hardnosed than I remembered. For,

---

42 Speaking of Freud as a victim of the Victorian ethos that he was fighting, I have in mind his avoidance of all talk about sex except for one place (*The Ego and the Id*, 1923), where he said, copulation involves a little death (of the semen, as it turns out).

43 In 1974 I visited Popper's home for the last time. I was preparing a critical review his *Objective Knowledge* of 1972. He argued for hours in attempts to dissuade me without reference to my (unstated) criticism. (My hints, I supposed, sufficed for him to surmise it.) He evidently was ready to discuss my criticism only upon my withdrawal of my plan to publish it. In 1978, when we last met, (in Alpbach, Tyrol), he expressed fury about my having published that review, which, in an unsent letter to me he declared scandalous because it was a personal attack. This declaration and the appearance of a revised edition of the book (1979) offer me opportunity to present an update of my review. The rest of this note is my review of the new edition.

Style. I regret the tone of my initial review of the aggressive style of the book. It looked to me reasonable, especially as it contains high praise of his ideas. But I was in error: at a superficial glance the praise seems marginal in a nasty review. Perhaps a second glance confirms this impression, suggesting that I harbored some resentment.

Content. The best in the revised edition of Popper's book is the added final appendix. It contains replies. The first reply is an excellent appeal to inductivists, offering them two pieces of advice that they can hardly refuse without severe loss. First, whatever principle of induction is under consideration, it is a conjecture and so be advised them to try to render it testable and test it. In my view it is testable and amply refuted. Second, some may see his theory, or his advice to test, as a satisfactory theory of induction; so be it: this idea, though not his, is the best theory of induction extant. In my view this theory, too, is too conservative, and abiding by it may cause stagnation. As far as inductivism is concerned, these lines are perhaps Popper's deepest as well as kindest, although not the most brilliant: much more brilliant ones appear in the latest new appendices to his *Logik der Forschung*. Russell had considered Popper's view of induction defeatist: "Should we, perhaps, adopt the somewhat despairing theory of Professor Popper that supposes scientific laws can be disproved, but never proven or even rendered probable?" [Nicod, 1970, 164]; [Russell, 1997]; [Wettersten, 2006]. Arguments against induction, then, would have not moved him to withdraw this rhetorical question. The appendix to the revised edition of *Objective Knowledge* just might. (This is not to endorse Russell's rhetorical question. The message of Jorge Luis Borges runs deeper and more in accord with Russell's own sentiment: the best starting point is utter despair, as after that, all else is sheer bonus.)

In the opening of the revised edition of Objective Knowledge Popper thanks all those who had offered him criticism of the book as he had requested in the opening page of its first edition. And he ignores Feyerabend and me while answering some criticism that he deemed poor. That of Feyerabend is serious. Popper's idea that a scientific theory modifies what it explains conflicts with his assertion that explanation is the deduction of what it should explain. So is Popper's approval of Galileo's expression of admiration for Copernicus for his disregard for certain refuting evidence: it should have made Popper admit that we may reject refuting evidence (even before it is refuted), observed Feyerabend. Why did Popper ignore this criticism? In the new appendix he still discusses properly reported observation statements as if they were all true. He knew much better.

Except for that splendid new appendix, the changes in the second edition are all minimal.

in it I declare that he made quite a few errors of judgment that he should have avoided. And this is less respectful than to criticize him for his magnificent errors, such as his theory of explanatory power as equal to testability, or of all testable deductions as scientific explanations. Even the positivist aspect of his *Logik der Forschung* I find terrific, no matter how alien to my taste positivism is and how erroneous I consider it. He demanded of me that I say of all of his published ideas that they are both admirable and true. This is somewhat excessive. When I said so he agreed that this demand is excessive and added an adamant denial that he had ever made it. So we disagreed about that. Except that imputing to any person excessive demands despite express denial is not very pleasant, especially as he saw in this a charge of dishonesty. (He always saw a moral issue when he was personally involved, in contrast to everything he ever said that pertains to such matters when these did not involve him personally or when he could put his personal concerns aside.) And when he viewed me as casting doubt on his integrity, he repeatedly said, all that remained for him to do then is to cease being my friend. He once said so as we were sitting in his garden during a visit of mine to England, when I came to see him at his invitation. So I shook hands with him, expressed my unqualified love, and made for the gate. He stopped me. I was touched, but found it nonetheless a jot too Kafkaesque.[44]

A: I dreamt I was a butterfly, said Chuang Tzu.
B: I dreamt I was human, said I.
A: I dreamt the world was my oyster, said Falstaff.
B: We do the best we can, anyway, whatever it is.
A: I wish I were more human, say we all.
B: I wish I were a butterfly, said Borges.

---

44 The most forceful way to portray the sad aspect of the interaction reported in this account may be to report this. Popper told me repeatedly that he envied me my having won the friendship and appreciation of my teacher. He could not understand the detachment with which I took this great gift. Obviously, it is something he lacked badly, could never acquire, and never ceased missing very much to the very end: it is the next best thing to a mother's love. Freud said, those who have it differ for life from the rest of us. Perhaps both Freud and Popper were right, who can tell. For my part, I see these two great unhappy souls as suffering from an excessive need for approval, in great contrast to Einstein, and projecting it to everyone else, in great contrast to Russell.

APPENDICES

My defense of Popper's ideas against the criticism of his Viennese peers and of many classicists is hardly apologetic or sycophantic, as my systematic criticism of these ideas may testify. Their criticism has no saving grace and his incessant clobbering of them was erroneous: they did not deserve his attention. Happily, towards the end of his life he changed.

The story of Popper's position in the commonwealth of learning is too intricate to sum up here; some observations on it should suffice here. His popularity grew less among his philosophical peers than with the educated public. This is unusual, as professional intellectuals command popularity. (The guardians of academic disciplines are much more jealous in keeping heresies out of the trade market than out of the professional literature.) They ignored the support that Russell and of Einstein lent him. Professional recognition of him finally emerged, first in the scientific-research community.

What makes it hard for professional philosophers to take Popper as seriously as they should is the popularity of Heidegger and of Wittgenstein Here I will ignore Heidegger. I will also ignore here responses to Popper among political scientists. I will discuss here mainly the Wittgensteinians and marginally the classicists. The heirs of "logical" positivism and the ensuing analytic movement, Adolf Grünbaum, Jaakko Hintikka, Hilary Putnam, Bas van Fraassen, and Michael Friedman, all offer the mixture as before – inductivism, instrumentalism, or a mix – with a touch of verbal embellishments somewhat reminiscent of the Wittgenstein heritage. The heirs of the commonsense philosophy of the ordinary language school, G. J. Warnock, John Searle, and Saul Kripke, defend commonsense. Popper paid less attention to them as he considered them not sufficiently serious. One appendix here explains why. Another appendix reports the discussion of Debra Nails of the Socratic Problem, as it seems to me the best response to Popper on the problem (although she barely mentions him there).

Appendix One: Russell on "Logical" Positivism

Russell was very critical of Wittgenstein and of the "logical" positivists. His overall assessment of that movement is his 1950 "Logical Positivism" [Russell, 1950a] (and it includes material from a 1945 publication by the same name plus a powerful argument, cited below, from his 1948 book). It is broad and balanced. It expresses sympathy with the "logical" positivists: modern logic has dispelled much philosophical nonsense. This success was a cause of the popularity of "logical" positivism, says Russell. He lauds their seriousness, but all the same he finds their central doctrine sham, no less. They declared that the clarification of metaphysical locutions would definitely prove that no correctly worded metaphysical assertion is possible. Yet they had no sufficient idea about meaning. Only recently, let me add, did heirs of the "Circle" admit this, which is no small matter, as it is an admission of bankruptcy.[1]

Russell always deemed commonsense-realism the most basic assumption of all philosophy. The "logical" positivists did not, and he saw this as their faulty epistemology and as their downfall. To be precise, it was their verification criterion of meaning, said Russell, as it conflicts with commonsense: some assertions are not verifiable yet obviously significant (373), such as purely existential statements not backed by observation: we all admit the meaningfulness of assertions about the possible existence of things that have vanished before the human race appeared. (Some may hope for help from subjunctive conditionals regarding the possibility of verification. But it makes poor sense to say that were I alive then, I could observe what has happened then (374): this move only raises new problems.)

A few refutations of the verification criterion follow. I will report one, as it disposes elegantly of the popular idea common to most views of empirical backing. It is this. A theory is empirically backed if and only if there are instances that agree with it and none that conflict with it. Consider then A, things that have no observed instances (mermaids) and B those that have (the benevolent). Add that an instance of B (the Good Samaritan) to A to form the new set A' (the set of all mermaids plus the Good Samaritan). "All A' are B" has empirical backing. So does then its corollary "All A are B". Thus, there is empirical backing to the claim that mermaids are benevolent.

Russell added a caution (380-1): "there is a danger of a technique

---

1 See, for example [Soames, 2003, 2, xiv-xv],
"at a certain point ... philosophers who were convinced that philosophical problems were simply linguistic problems came to recognize that they needed a systematic theory of meaning. However, it was unclear whether such a theory was possible, or, if it was, what it should look like. ... we have a historical development with a considerable irony."

which conceals problems instead of helping to solve them". Carnap's use of the distinction between using a word and naming it as a means of elucidation is an example for that. (This criticism, let me add, is not of a marginal point, as it attacks Carnap's method of dismissing metaphysics.)

In the penultimate paragraph Russell expresses regret that the "logical" positivists neglected the study of non-linguistic aspects of meaning and yet respect for their preference for rigorous piecemeal studies over grand theories. The final paragraph ascribes to them "uncompromising empiricism" and rejects it. End of Russell's paper.

Russell's critique never met with serious efforts to rebut it. Nevertheless, its very presence throws doubt on Popper's boast that he is the one to have slain the dragon – assuming that the dragon is dead, of course. Is it? We do not know. Considered seriously, the very appearance of the dragon of "logical" positivism on the philosophical scene should have been prevented by the mere appearance of Russell's 1922 Introduction to Wittgenstein's first book; considered a mere fashion, it is hard to assess, since we know too little about what makes fashions come and go. "Logical" positivism played a significant role in the climate of opinions of the time, when the public debate for and against religion played a great role in politics, for good and (mostly) for ill. And, as Joad has observed [Joad, 1950], theology was the chief target of "logical" positivism. In a discussion with me in Boston in 1983, Hempel still saw in the attack on religion by the "logical" positivists their chief asset (see above, page 301).

Appendix Two: Popper versus the "Vienna Circle"

Popper's incessant criticism of the doctrines of the "Vienna Circle" is a compliment to them; they did not deserve it; he would have done better to stop referring to them after he demolished their doctrines in his first two classical books. Admittedly, all of his critical comments are valuable, since practically all of his output scintillates. Still, except for the brilliant asides that his persistent one-sided dialogues with them include, these dialogues are of little value.

The major targets of Popper's criticism of the "Vienna Circle" were the contributions of Wittgenstein, of Schlick, and of Carnap – in strange disregard for Carnap's modifications of Wittgenstein's views [Carnap, 1937] in the light of Russell's criticism of it [Russell, 1922]. I will postpone this to the next appendix. Carnap should have said that he was offering a modification of Wittgenstein's ideas.[2] But then, an admission of the possibility of a philosophical dialogue would be self-defeating. Admittedly, Ayer admitted that the "Circle" had tolerated disagreement [Ayer, 1959, 6]. His aim in doing so, however, was merely to dismiss Poppers claim that philosophical disagreement is legitimate and that this comprises a major disagreement with the "Circle" [*LdF*, Preface].[3] There is truth in this counter-claim: the "Circle" did allow some open disagreement between its members (Carnap and Neurath, about the status of observation reports), even though it forbade it in the name of logic. The conduct of the "Circle" was inconsistent. And so, had Popper thrown into the pot his own share, perhaps they would not have minded. He would then have hardly made a difference. In my view this was Schlick's intent in his acceptance of Popper's book to his celebrated series.

Schlick was the authority to adjudicate on disputes between members of the "Circle". He was not up to the task. My view of him is of a tragic if pathetic figure, a broken soul of sorts. He was an established leading thinker when he bumped into Wittgenstein, whom he graciously recognized as his superior. He realized that Wittgenstein's demand to add to every sentence its proof left no room for dissent; so he tried to maintain unanimity in the "Circle" that he led; he could not [Hacohen, 2002, 261, 310]. Alberto Coffa explains his fascination with Wittgenstein [Coffa and Wessels, 1991, 421] as resting on his adoption of the idea [of Bacon] that since it is hard to admit having committed an error, it is better to avoid error at all cost. He soon realized that this demand to avoid error was erroneous [Hacohen, 2000, 210,

---

2 Wittgenstein called Carnap a plagiarist [Stern, 2007]: he evidently deemed his modifications minor.
3 This same distortion that Ayer exhibited in one sentence Nancy Cartwright presented in a full-length book [Cartwright, 1996]; [Agassi, 1998]. Stadler did it in a much greater length.

223] but he could not bring himself to recant publicly. As he learned about Popper, he found him much more congenial than Wittgenstein, but he did not have the energy to make one more major switch. He expected Popper to help him return silently to his older position, I surmise. Popper blocked any such move forcefully, quoting Schlick's most Wittgensteinian assertion in a most prominent part of his book. This seems to me the reason for the anger that he expressed after he read the published book (that was very different from the submitted draft[4]), and for the hostility to Popper that he generated afterwards and that still abounds. This surmise of mine makes some sense, although still not too agreeably so, of the rather silly claim that out of sheer self-aggrandizement Popper stressed differences instead of similarities. My surmise does not explain how this silly, disagreeable explanation won popularity, of course. Perhaps the other members of the "Circle" had no better excuse and they needed one badly. As Hempel told me, he disliked Popper's repeated denial of their gossip that he was one of them. (As Gödel was one of them, they reported this, but not that he emphatically dissented from their anti-metaphysics. His case, however, is quite different: he was too famous for gossip to tarnish his reputation.)

Popper also presented to the world a new image of the "Vienna Circle". It is a posthumous brief (8-9 pages) preface to *The Collected Works of Hans Hahn* (1995). It has a refreshingly new tone: friendly, relaxed, and quite appreciative. It is largely autobiographical and charmingly self-effacing: Hahn was a brilliant lecturer and a leading mathematician; Popper was an unofficial student. Hahn founded the "Circle" and its members were all interested in science. Schlick, too, receives here a great compliment: he had written a "really excellent book – I think the best book on the theory of science since Kant." Schlick's membership in the "Vienna Circle" was marginal at first, but he "became very active after he met Wittgenstein, whom he venerated as a sort of demigod."

This is the pre-Wittgenstein, realist "Vienna Circle" of course, and this, by my surmise, is what Schlick had hoped that Popper would describe in his first book. Now there must have been some logic to the process of Schlick's becoming a Wittgenstein fan, since Hans Reichenbach shared it. Initially, they shared Einstein's critical *a priorism* or critical realism. They found unsatisfactory his uncertainty-in-principle. They had no criterion to differentiate between valuable and valueless uncertain conjectures. Popper did. He sidestepped the question whether realism is a precondition for

---

4 The draft was *Die beiden Grundprobleme* that later on Popper tried hard to read as a precursor to *Logik der Forschung* [Wettersten, 1992].

science: he confessed he was a realist but added (in dissent from Einstein) that he allowed nothing to depend on it. Later on, when he changed his mind and advocated bold realism, he explained why he valued it. This reduced the small distance between Popper and the early Schlick.

Einstein had a high view of the early Schlick and he then expressed readiness to support him [Born and Einstein, 1971, 38]. When Schlick publicly defended Einstein's realism, he won a fan letter from him that praised his philosophical clarity [Stadler, 2003, 173]. And then he gave it all up under Wittgenstein's spell — because he wanted as strict a cure to self-delusion as possible, said Alberto Coffa. His allegiance to Wittgenstein soon made him consider statements of natural laws meaningless [Ayer, 1959, 82-5], thereby deserting realism [*LScD*, 13, 37, 40, 141, 145, 312].[5]

This is tragic, as Schlick had been a passionate realist. In one of his last lecture courses [Schlick, 1987] he discussed metaphysics (19-22) and science (42, 83). Realism is true, he said (47; 181, line 2 up), only in the scientific sense in which it is verified, not in the metaphysical sense. This is impossible to square with the views of Wittgenstein, and it is unclear what Schlick did about it. Nor does it fit his earlier chapter on metaphysics (19-23). And, of course, Popper cited him to have given up scientific realism altogether. Wittgenstein's influence on Schlick was tragic.

Before he met Wittgenstein he held [Schlick, 1979, 338-9] that science has inbuilt uncertainty. He notices [*ibid*, 282, 286] that trouble began with "the cleavage" between reality and appearance. He said, there are many kinds of real things but only one all-embracing reality. He had to explain how this claim helps overcome the trouble. He never did this; rather, he succumbed to Wittgenstein's sensationalism, following the (true) traditional view of verification without synthetic *a priori* knowledge as sensationalist.

---

5 I know of no response, direct or oblique, to Popper's just snub to Schlick for his view of scientific laws as meaningless, except perhaps for Frey's [Frey, 1982, 154-5]. He offers it while laboriously reinterpreting a sentence from Schlick's "Positivism and Realism". The sentence is this:
   "The verification is logically possible, whatever might be the case as regards its actual execution."
Frey says, "This doubtlessly is a slip" and he reads it thus:
   "verifiability (or falsifiability) of a claim is logically possible (or logically impossible)"
This, Frey explains,
   "can only mean that the verifying (or falsifying) actions are in accord with (or conflict with) the theoretical and practical knowledge that is accepted at the time."
Does this not present Schlick as a relativist? This he certainly was not. As if to answer this question, Frey adds here a censure of both Popper and the "Vienna Circle":
   "they all seem to forget or ignore the fact that problem-solving ... is not to be restricted to merely cognitive tasks."
This censure rests on the assumption that verification is essential for action. He indeed does suggest this idea, and he ascribes it to Schlick:
   "The quest for absolutely certain knowledge appears herewith as a necessity of life ..."
The quest for certainty is a necessity of life because action is, says Frey. True or false, this is not Schlick.

In his early phase Schlick supported the idea of synthetic *a priori* knowledge minus certitude, yet he felt he needed certitude because, we remember, it was all too easy to deceive oneself that one has contributed to the growth of science [Coffa and Wessels, 1991, 421]. Popper's view of science as a social rather than a personal enterprise made this point irrelevant.

Giving up certitude was insufficient; the addition of verisimilitude into the picture was necessary to render synthetic *a priori* knowledge without certitude progressive. Verisimilitude is realist: it is the hope for the new scientific theory to be truer than its predecessor. This hope is reasonable, but it is not a proof that the better explanations are truer, (nearer to the truth). This we cannot know for sure.

As Russell has observed (see above, p. 314), the "logical" positivists' theory of meaning is partial and their view of it as verifiability ignores realist common sense. As Popper has observed [*LScD*, Preface, 1959], they promised new-style analysis of meaning of statements, but offered old-style analysis of concepts. This had worried Schlick, whose response to it is [Ayer, 1959, 58], " the decisive epoch-making forward steps of science ... always ... signify a clarification of the meaning of the fundamental statements." He thus obscured matters defensively: forward steps in science concern theories, not meanings.[6] He tried to link the two, offering the conjecture that they improve together. Even were this conjecture true, why prefer meanings to theories? What is their advantage? It is that meanings are always right, whereas theories can be false: the traditional horror of error creates a preference for analysis over criticism.[7]

It is hard to say how much Wittgenstein is to blame for Schlick's muddle. Not every commentator sees his influence on the "Circle" as crucial. Paul Engelmann says, they misread Wittgenstein: they saw his book as dealing with science and he saw it as dealing with art [Engelmann, 1967]. A quotation from Wittgenstein that he brings throws doubt on this: he preferred Luther's translation of the Bible over Buber's because the latter was too clear; yet in his book Wittgenstein demanded full clarity; hence the book is not about art. Elizabeth Anscombe caused a similar confusion. She seem-

---

6 Confusing vagueness with confusion and admitting, as Francis Bacon did on a rare occasion (*Novum Organum*, ii, 19), that "truth emerges quicker from error than from confusion" should together tip the scale in the opposite direction. This seldom happens.

7 Traditionally analyses were conceptual. Frege and Russell, perhaps already Boole, were the first to analyze statements. Yet, as Popper has observed, Wittgenstein and his crowd reverted to tradition and analyzed concepts and spoke of the meanings of statements in a slogan that identified it with verification. The naïve grasp of propositions and of their truth or falsity was always much superior to that of concepts and their adequacy or inadequacy. Thus, for a conspicuous example, Aristotle's grasp of the concept of truth is naïve, realistic, and still viable, whereas his view of concepts and their adequacy is his methodological essentialism that no decent logician or mathematician will entertain.

ingly contested Popper's understanding of Wittgenstein's term "atomic sentences" [*OS*, Ch. 11, nn46, 51], yet she seems to be in agreement with him – except when she relies on Wittgenstein's claim that he had left their contents open to interpretations. This conflicts with his demand for absolute clarity.

Wittgenstein's message is his refusal to recognize any philosophical opinion other than the most commonsense ones (whatever these may be). The ("first generation") disciples of Wittgenstein considered this his cardinal idea and his cardinal asset. He was painfully aware of the limitation that this imposes: it leaves no room for development within philosophy. He therefore expressed the wish for all his disciples that they abandon philosophy. Alternatively, he expected them to repeat with acknowledgement what he had said, and he allowed them to innovate only by the use of verbal analysis in order to discredit some allegedly significant philosophical innovations. He enjoined them not to offer alternatives to the theories that they were debunking, as they should have shown them not false but senseless. It is no surprise that he viewed the contributions of some of his most celebrated disciples as plagiarism from him.

The alternative to the proposal that Wittgenstein had exerted the chief influence on the "Vienna Circle" is the proposal that Ernst Mach did. This is hard to judge, since both were metaphysically neutral monists: they both reduced the world of (scientific) experience to sensations. They shared an antimetaphysical bias, but Mach's bias was within traditional positivism and that of Wittgenstein was of his new "logical" brand. Mach kept aloof from participating in any metaphysical controversy; Wittgenstein denied that there was anything there to be aloof of: he claimed that grammar renders metaphysics impossible in principle. In the "Circle" only Neurath claimed they were following Mach and he rejected the views of Wittgenstein as metaphysics, no less. Nancy Cartwright and Friedrich Stadler take Neurath to be a representative of the "Circle". Paper is tolerant.

Modern scientific philosophy is traditionally inundated with hostility to metaphysics. The early hostility to metaphysics aimed at Aristotelian metaphysics. Perhaps Hempel was right and in the twentieth century the object of the hostility was theology, not all metaphysics; but this was never official. Sir Francis Bacon offered the most enduring reason for the hostility. He opposed the metaphysical method, which is speculative. He assumed that speculations are often erroneous, that error leads to prejudice, and that prejudice stands in the way to the proper performance of good research. But no one is free of all prejudice

Bacon hoped for a scientifically founded metaphysics. He said, it will be the pinnacle of science. Descartes followed Bacon's demand to eschew all prejudice and yet he offered a metaphysics of his own — on the supposition that he had demonstrated its truth. The same holds for Kant. He wrote a *Prolegomena to Any Metaphysics that in Future Will Claim Scientific Status*; he spoke vehemently against all unproven hypotheses. The metaphysical foundations of natural science he declared proven on *a priori* grounds. Modern logic is largely the creation of Frege and Russell who worked much in reaction to Kant's view of the very possibility of *synthetic a priori* knowledge. E. A. Burtt took it for granted in his path-breaking *The Metaphysical Foundations of Modern Physical Science* (1926) that the metaphysical foundations of science are unavoidably conjectural.

Mach's neutral monism is neutral regarding the choice between metaphysical doctrines that postulate the existence of substance; this is insufficient, as it is but a different metaphysics: phenomenalism. Russell's theory of definite descriptions came to dispose of Frege's Platonism. He wished modern logic to obliterate Platonism systematically. He admitted defeat (*An Inquiry into Meaning and Truth*, [Russell, 1940, final chapter]). Quine declared allegiance to Platonism. Efforts to do away with it continue.

The heirs of the "Vienna Circle" agree: it had no theory of meaning [Quine, 1977]. Hence, they had no reason for their dismissal of metaphysics as meaningless. They inherited two central ideas, of Russell and of Wittgenstein. Russell's theory of definite descriptions came to dispel Frege's Platonism.[8] Its success was partial (as it covered only nouns, not the whole of language). Wittgenstein dismissed theology [Wittgenstein 1922, §6.53; 1953, §§13, 15, 26] due to its inability to give meaning to the name of the divinity. Popper has refuted this [Popper, 1963, 276] by the use of Russell's theory of definite descriptions (by the use of divine attributes to substitute for the divine name). Quine had gone further as his version of logic has no names. Naming is problematic anyway, as Kripke has shown, and so Wittgenstein's reason for proscribing theology is questionable.

---

8 Russell's 1905 paper that announced his theory of definite description is also the first important paper in language analysis, as it was anti-metaphysical, i.e., anti-Platonist, and as it says, we analyze sentences regularly, for example the sentence "I thought that your yacht is bigger than it is" that is hard to take literally. Wittgenstein never criticized that theory, and his followers preferred to leave it alone, except for Strawson [Strawson, 1971] whose comment met with one devastating criticism [Linsky, 1967. 85-99]

Appendix Three: Carnap and Reichenbach in Retrospect

Carnap's was renowned because of his effort to accommodate Russell's criticism of Wittgenstein's first book that appears in his celebrated preface to it.

First, Carnap openly rejected the naturalist view of language that was a cornerstone of all logicism from Frege to Wittgenstein. This forced him and his followers to declare metaphysical statements impossible in any adequate language that anyone will ever construct. This invited proof. He could not possibly tackle such an awesome task. He later asserted [Carnap, 1956] that he still hoped that one day someone would construct a language that has no room for metaphysical assertions. That is a downbeat ending.

Second, he rejected Schlick's view of universal statements as not genuine [Carnap, 1937, 101, 186]. He did so without showing that they are verifiable. Decades later, he replaced proof with probability [Carnap, 1950]. He then discovered that no amount of information will render a universal statement probable by his measure of probability. So he fell back on Schlick's view.

Third, he attempted to present a proper system of logic: starting with a meta-language [Carnap, 1937, 186, 295-6] and defining identity [Carnap, 1937, 49-50, 139, 303], existence [Carnap, 1937, 140-1, 61], and infinity [Carnap, 1937, 46, 101, 321], the functions that Wittgenstein had banned as metaphysical and Russell proved necessary for mathematics. Carnap wished to block their smuggling metaphysics into logic. To that end he endorsed extreme formalism, the crazy view of mathematics as ink stains devoid of all meaning. This idea seemed to have the blessing of the great David Hilbert, the inventor of formalism. Not so: Hilbert developed formalism only as a technique, in order to answer certain needs within mathematics: it is easier to prove a theory consistent by *pretending* that its formulas are mere signs and its rules of inference are moves in a senseless game. But this is not mathematics; Carnap himself did not mean his suggestion when he made it [Reck, 2004]. Once we consider mathematical formulas as no more than ink spots (or chalk marks), we are bound to notice that we identify signs by their shapes, and that shapes come with Plato's ontology.[9] It is no surprise that

---

9 Hilbert both viewed formalism as a tool rather than a view of mathematics and he limited it to the naïve level of students of Euclidean geometry or the like, as this sufficed for his purpose [Agassi, 2006]. Hence, the objection applied here against Carnap does not begin to apply to Hilbert. Hilbert's procedure is this:
   syntax = semantics ⊖ meaning;
Carnap's, by contrast, is this:
   semantics = syntax ⊕ meaning.
That is to say, he (unbelievably) suggested that the construction of a mathematical system is the wording of its axioms and the subsequent building of a model to fit them.

Carnap wavered between formalism and logicism [Agassi, 1988, 86-8]. Amusingly, Hilary Putnam glosses over this trouble. He says [Putnam, 2000, 5], according to Carnap, "sentences are marks and noises." He does not mean that, of course, since he recognizes sentences that are neither — semaphore or Morse signals, not to mention those that are memorized in order to say but never said. Putnam means to refer to any signal, of course. For, he adds, "(I take this ... from Richard Rorty, but I think Carnap would have no objection.)" He knows that Carnap did have an objection: he wrote a whole book [Carnap, 1942] to explain his shift from pure syntax to semantics. Hence, like Carnap, Putnam meant something else. This is unobjectionable, except that it is against the religion of the "Vienna Circle".

No matter how often Carnap changed his mind, his assertions were all very definitive. Thus, in his above-mentioned *Syntax* he says (280), the "Vienna Circle" has "established in detail and investigated in all its consequences" the proscription of metaphysics, even though he also noted that Wittgenstein and Schlick were in error (102) about it.

What were Carnap's contributions? This is under public discussion. The latest example is [Bonk, 2003]. The most significant essay there is the first, by Ilkka Niiniluoto. He ascribes to Carnap two ideas. One is that theoretical terms possess partial meaning. Pierre Duhem has expounded this idea in detail. It is the refuted idea that all meanings [of descriptive terms] derive from the meanings of names of sensations. It says, a theory's meaning is that of the data that it covers, and so it grows with the growth of its fund of observations. (The standard reference to its criticism is Quine's classic "Two Dogmas".) The second idea is that of possible worlds. It is hard to say what it is good for, and Carnap's contribution to it was nil [Agassi, 1988, 88-9, 97]. There remains Carnap's study of dispositions that was once very influential [Moulinas, 1991]. The reason for this is never made clear. It is this: Popper argued that no observation report is free of conjecture, since all nouns are dispositional and so permit predictions that may be refutable. Thus, reporting about a glass of water, one allows the conclusion that it is breakable and this entails a prediction. Carnap tried to reduce dispositional terms to observational ones and failed. The other contributors to that volume struggle with the fact that Carnap's major message was anti-philosophical and is now *passé*. The same goes for Carnap's *Die Logischer Aufbau der Welt*, unless one views it as a part of the dogma that Quine has criticized. C. Wade Savage suggests that it is possible to put the *Aufbau* to test by constructing a machine whose sensations imitate ours and its processor follows the rule that the *Aufbau* describes. In this sense, let me concede, the *Aufbau* offers a refuted theory of perception. This illustrates again the fact that not all refuted theories are valuable; only those are that stood up to the criticism of their predecessors.

Popper was friendly with many Wittgenstein fans [*Autobiography*, §39]; [Stadler, 2004, 474-97]. He did not complain about Reichenbach's refusal to shake hands with him [Popper, 1990, 4]; [Stadler, 2004, 493]. His chapter on probability [*LScD*, Chapter 8] has a few unflattering comments on Reichenbach's output on probability. As to his output on quanta [Reichenbach, 1944], Popper evidently deemed it too poor to discuss. Feyerabend agreed [Feyerabend, 1967]. Wesley Salmon did not [Salmon, 1994]. Reichenbach had called unobserved, extrapolated events "inter-phenomena". He ascribed to them special status, presumably that of fiction. Salmon denied this as it clashes with what he calls "the principle of common cause" that he declares a "major contribution" of Reichenbach.

This so-called principle appears in the essay on it in the Internet *Stanford Encyclopaedia of Philosophy*, adorned with a score of learned references. Here is its abstract in full:

"This idea, that simultaneous correlated events must have prior common causes, was first made precise by Hans Reichenbach (Reichenbach 1956). It can be used to infer the existence of unobserved and unobservable events, and to infer causal relations from statistical relations. Unfortunately it does not appear to be universally valid, nor is there agreement as to the circumstances in which it is valid."

That "the idea ... was ... made precise" is puzzling. It seemed precise enough already in its biblical version: "Can two walk together, except they be agreed?" (*Amos*, 3:3) Yes, they can. So the Internet *Stanford Encyclopaedia of Philosophy* admits, the idea, "the principle of common cause", has many exceptions. "Unfortunately" — for Reichenbach fans, that is.

Einstein responded to Reichenbach with admirable deftness [Schilpp, 1949]. He was regularly much more careful than Popper in the matter of choice of opponents, but here he had to comment on a substandard work. He most elegantly limited his comments to the least objectionable part of it. He also criticizes Poincaré there rather than Reichenbach, and even him not all the way — out of respect, he explained.

Did Reichenbach say anything that is of any value? His most famous discussion concerned what he called inter-phenomena, namely, the events between couples of observed ones: he denied that they are real. He wrote on probability, on quanta, and on the arrow of time. These are forgotten. What is left is his teaching, and his popular work, some of which is fine, though unavoidably somewhat outdated [Röseberg, 1998, 204].

Appendix Four: The Picture "in Essentials"
Wittgenstein published in his lifetime his first and famous book and a paper [Wittgenstein, 1929] that is forsaken, although it surfaces repeatedly, as, for example, in the Wikipedia item on Wittgenstein:
"Wittgenstein published only a single paper, 'Remarks on Logical Form', which was submitted to be read for the Aristotelian Society and published in their proceedings. By the time of the conference, however, he repudiated the essay as worthless ..."
Consider the philosophy of Wittgenstein's first book the way Russell and Popper understood it. By their reading he was a phenomenalist who conceived of the presentation of science as a finite map of data within a system whose coordinates are space, time, and the set of the various sense perceptions [Russell, 1968, 199], [*OS*, ii, 293]. Theories, then, are nothing but simple groupings of the data [Wittgenstein, 1922, 4.11, 6.32, 6.343]. From the very start, Wittgenstein declared that as the set of coordinates for sensations belongs neither to the *a priori* system of logic nor to the empirically given set of data, it is impossible to articulate. Nimrod Bar-Am follows the reading and observes that this way Wittgenstein's 1929 paper is recognition of total failure of that philosophy. Apologetic commentators deftly dodge it.

As Munz and Hintikka have argued, whatever Wittgenstein had ever said, he never explicitly withdrew it but he repeatedly altered its meaning. His radical change of view was not so much in what he said, as in the approach to it. In particular, he exhibited two new tendencies. First, after the change he tended to ignore science [Musgrave, 1999, 259]. Second, he considered logic as an increasingly broad category, where he placed all the metaphysics that he could not dispose of. (For this reason Feyerabend viewed him as Kantian of sorts.) And he repeatedly rejected explanation and advocated description only. This made his later philosophy quite conservative [Pole, 1958, 80-2, 96].

Georg Henrik von Wright corroborates this quite against his intention. While denying that Wittgenstein was conservative he admits that Wittgenstein disliked modernism and looked for the golden age of pure language [Egidi, 1995, 1-19]. Von Wright also asserts that anti-modernist Wittgenstein redirected the charges of metaphysical confusions against those who exhibit a "craving for general theories" and who "constantly see the methods of science before their eyes", such as the "Vienna Circle". Yet he was reticent about it. Von Wright does not say why. He also discusses the allegation that Wittgenstein fell under the spell of mysticism of a Tolstoyan sort. (The source of the allegation is Russell [Russell, 1951, 298], whom von Wright ignores.) He explains this spell as moral, as hostility to humbug. He likewise explains Wittgenstein's dislike for science as hostility to excess respect for it. This is funny, because it is the respect for science that is at issue.

Consider the disagreement (reported above in the Prologue) between Carnap and Popper about the sentence "this stone thinks about Vienna": is it meaningless? The disagreement was verbal not factual: it concerned grammar, not facts. It took place in 1932. In 1934 Carnap answered Popper's objection, I surmise, if "answer" is the right word, though only obliquely. (Not only did he not mention the objection of Popper or of anyone else; he used a more obvious example: an unproven mathematical conjecture. One may respond to this by admitting that a conjecture has questionable meaning, to elucidate by proof: prior to proof how to understand it is open. (This is Wittgenstein's oblique answer to Carnap, if "answer" is the right word [Wittgenstein, 1951, §§334, 517]). This is why Popper said [*LScD*, §4], the loose use of the word "meaningless" is merely unpleasant but not too objectionable, since it is devoid of clear meaning.) Carnap said then [Carnap, 1937, 162], still as a part of his oblique reply to Popper, I surmise, about the thoughtful stone: meaning grows with the growth of knowledge. *Ergo*, when we know more about the mechanism of thinking, then we will all agree about the celebrated thoughtful stone. This is still not serious enough, as it merely amounts to the admission of the lack of assurance about the thinking mechanism that stones admittedly lack. For, it has nothing to do with language, much less with confusion: once we understand thinking as a certain disposition that, as Turing informs us, computers possess but stones do not, then Popper's assertion is verified and Carnap's refuted.

The discussion just reconstructed is both oblique and frivolous, yet it regrettably still is high on the agenda of the followers of Wittgenstein [Bar-Hillel, 1964, 34]. These days discussions abound of differences of opinion between members of the "Circle" — regarding science, not yet regarding politics. They are all apologetic and so self-defeating, as they pertain to debates in the style of the "Vienna-circle" about the question, why are debates objectionable. This is faintly comic. For the clearest recognition of the dissolution of this discussion with no profit whatsoever, see the recent presentation by Soames [Soames, 2003, 1, 258]:

"logical positivism combined Wittgenstein's emphasis on an explicit test for meaningfulness with Russell's logical techniques and his emphasis on sense experience and observation. The result was an ambitious, logicized [?] version of traditional empiricism, put forward as a theory about the scope and limits of meaning."

To clinch matters the passage continues thus:

"The central doctrine of logical positivism was its analysis of the meaning of empirical sentences in terms of verification."
Forty pages later discussion moves to (297)
"Lessons of the Positivist Failed Attempt to Vindicate Verificationism".
He offers only one lesson here: one should not be too ambitious (299). Does it merit the two generations of intensive debate of the "Circle" and its heirs? His early fans rightly viewed him as revolutionary and wrongly appreciated him. Carnap's enthusiastic early publications (in *Erkenntnis*) report that Wittgenstein had killed metaphysics. Allan Janik and Stephen Toulmin say [Janik and Toulmin, 1973, 257],
"Looking back ... we can see how far the 'revolution' ... was sociological rather than intellectual."
They stop short of saying that intellectually it was a failure, that its success was merely social. They also do not say why it was successful despite its emptiness. Friedrich Stadler admits: the verification theory of meaning was never serious. Proof: the members of the "Circle" did not agree even on basics and even among themselves [Stadler, 2003, xvii ff.]. This total collapse appears in an essay that is mostly an attack on Popper. This makes no sense: the members of the "Circle" advertised the verification principle as a great revelation, as is hard to contest. Yet Stadler uses this as evidence that Popper was in error: he attacked an idea of theirs that was marginal and neglected their valuable ones. Sadly, Stadler neglected to report what these valuable ideas are. Instead, he reports Schlick's claim that Popper's assault on the "Circle" was not intellectual but personal [Stadler, 2003, xvii]:
"I would like to make some similar claims with regard to the perhaps more complex and emotion-laden relationship between Karl Popper and the Vienna Circle ..."
Stadler point is this: if Popper and his Viennese peers were in intrinsic intellectual agreement, then what divided them was not intellectual. *Ergo*, it was emotional. Hence, since they are bankrupt, let us take comfort that they took Popper down with them.
The semi-official record of the "Circle" [Kraft, (1950), 1953] explains this. After a survey of the remains of the "Circle" it comes to the point (page 9): although Popper "never participated in the meetings of the Vienna Circle", he agreed with them "in essentials", namely, hostility to metaphysics (15-16). Thus, the view that metaphysics is meaningless was not "essential". Thus, the "Circle" had nothing new to say: Kraft presents them as positivist, not as "logical" positivist. Kraft's distortion does the job: it does imply that "in essentials" Popper agreed with the "Circle". Thus, since Popper

refused to follow Wittgenstein, so did the "Vienna Circle". This is absurd: possibly the "Circle" agreed "in essentials" with Wittgenstein and possibly it agreed with Popper, but definitely not with both. Admittedly, already in 1936 Carnap made Popper a follower of Wittgenstein. Two documents refute this. Wittgenstein referred to Popper only once; he called him an ass. Russell wrote to Popper regarding the encounter the three had once (*The Wittgenstein Poker*). He said, Popper was right and Wittgenstein was not. (See above page 154, note 49.)

Kraft stuck to his excess consistently. For example, he found it odd that Schlick held that the meaning of terms and proper grammar insure the meaning of statements and that at the same time he held the verification theory of meaning [Kraft, (1950), 1953, 198, n46]. This is an oversight of the novelty of Wittgenstein's message, his identification of proper grammar and verification: they are identical extensionally, one might say, not intensionally. This thesis "in essential" is the new in "logical" positivism as opposed to traditional positivism. Of course, this thesis could not be subject to critical discussion; it is extravagant and patently false. This, however, is no reason for Kraft to fail to recognize its role as the central doctrine of the "Circle", the one that they were proud of as the great philosophical innovation of the century. An old member himself, he knew all that. He found it politic to distort the information that is on easily available record.

Popper's dissent from the "Circle" comprises two items. First, (following Wittgenstein) they demarcated proper language; (following tradition) he demarcated proper science –considering science a system of statements; and, (following tradition) they considered empirical research efforts to validate; (making a revolution) he considered empirical research efforts to detect error. The concern with language has proved misplaced. "The linguistic turn" was into a blind alley. In the name of the "Circle" Hempel admitted defeat on this cardinal point. The disagreement about validity still holds. Some repeatedly ascribe to Popper the view of the "Circle". Hempel has ascribed to Popper first the idea that his demarcation is of the language of science. Without taking this back, he also offered another distortion: he ascribed to Popper the view that only corroborated theories are scientific. The paradigm case of a significant scientific theory that has only refutation and no corroboration to its record is the Bohr-Kramers-Slater theory (that describes the law of conservation of energy as statistical).

Appendix Five: The Heritage of "Vienna Circle"
The biggest compliment a scientist gave Popper thus far is Steven Weinberg's [Weinberg, 1992, 165], who has crowned him "the dean of modern philosophers of science". But, just to keep a sense of proportion, let me mention as a token example to the contrary the essay by the leading historian of science Don A. Howard, "Einstein's Philosophy of Science", published on the internet in the prestigious *Stanford Encyclopaedia of Philosophy* (Spring 2004 Edition). It mentions Carnap a few times and Popper not once, despite Einstein's cavalier dismissal of Carnap's output [Schilpp, 1959-60, 491] and his support of Popper's [Popper, 1959, 461].

This is just one small piece of evidence for the robust popularity of "logical" positivism. What does it rest on? The "logical" positivist manifesto on which it once rested is not detailed (this is no criticism) and the details of its doctrines underwent repeated patching up and no overhaul. Its advocates did not spell out their general views: they hardly had the time. They considered its chief thesis impervious to all past criticism. That thesis is unspoken; it should be explicit and open to discussion. What is it, then?

This question has engaged a few thinkers and it still does. Outstanding among them is Wolfgang Stegmüller. He said, the "Vienna Circle" doctrine was a hope, and hopes are irrefutable. This is too thin [Agassi and Wettersten, 1980]; was the high reputation of the "Vienna Circle" that of a hope?

Friedman's book [Friedman, 2000] is much better. Laudably, he begins with the problem-situation that gave rise to "logical" positivism. He also rightly praises the early works of Schlick (23-5). They belong to a variant of critical realism: science posits putatively true but not certain realist principles. Einstein expressed gratitude to Schlick for his having defended his theory in the light of this philosophy [Howard, 1994, 51]. Schlick moved then away from his critical realism towards "logical" positivism. Friedman explains: Schlick found it necessary to prove realism in order to develop a theory of meaning, yet he could not, and he found this frustrating. Friedman cites here Wittgenstein's explanation for this frustration: we may grasp the right metaphysics yet be unable to articulate it properly. This has a remarkable advantage: the metaphysics that is impossible to articulate properly is in no need for proof. The frustration is thus gone. (What is that metaphysics? Wittgenstein said it is solipsism [Wittgenstein, 1922, 564]; [Coffa and Wessels, 1991, 245]; Schlick dissented; yet his dissent is not open to critical scrutiny. This brings the frustration back.)

The basic contribution of the "Circle" was thus its use of ineffability Wittgenstein-style to dodge the duty to prove their metaphysics. This is his or their celebrated verification principle. Things did not stay put, however. Carnap deviated from Wittgenstein's view to allow for Russell's criticism. Neurath deviated from Wittgenstein to allow for Marxism and for the history of science. They thus jettisoned the doctrine of ineffability but clung to the verification principle all the same. This should have invited reconsideration. It did not. Deviating from Wittgenstein, Neurath and Carnap expressed loyalty to him. (Carnap had a promising career as a logician before he slipped into "logical" positivism [Awodey and Klein, 2004]. His great fame was due to his defense of the new movement by developing a doctrine that nowadays his fans find embarrassing [Agassi, 1988, 86-8]. Neurath was an organizer; his ideas were not taken seriously by other members of the "Circle" [Agassi, 1998]; Popper said, he was the most original among them [Popper, 1973].)

Effort to smooth the transition from Wittgenstein to Popper began early. This tallied with the wish of members of the "Circle" to ameliorate the loss due to Wittgenstein's dismissal of metaphysics. They should have conceded openly that that they were deviating from Wittgenstein. Their initial view was that science is good as it is meaningful and metaphysics is bad as it is not. Their recognition of the impossibility of verification of theories raised for them the problem of the demarcation of science anew: what meaningless theory is scientific and what is metaphysical. It also raises the subsequent question, why is metaphysics bad? They said, it is it is in principle not confirmable: theories given to empirical confirmation are scientific. What then is confirmation? Hempel said, it is instantiation; Carnap said, it is probability. Had they taken this seriously, they would have conceded to Russell (see above, Appendix One). They did not.

The members of the "Circle" declared themselves progressive, and they hoped that clarity suffices as a tool for the defense of the right cause. It does not. A clear example surfaced long after the demise of the "Circle", in the restless late 'sixties, Hilary Putnam, Reichenbach's lead disciple, shifted his rather traditional position to membership in a group that swore by Chairman Mao. He shifted later to membership in a conservative Jewish congregation. This led to no alteration in his output in the philosophy of science and of language. This is suspect: even the wish to stay within the confines of the philosophy of science and of language raises at least one broader problem, that of the demarcation of the diverse fields of their activity. They know this, as they have their diverse solutions to it: some of them are radicals who deem Marxism scientific; others are conservatives of the Chicago school of economics.

Thus, Friedman's presentation of the "Vienna Circle", though agreeable to the historically minded, does not leave much to appreciate in its output. The mood has changed. Increasing numbers of commentators find it hard to make sense of it all: no one says any longer that one and only one claim of the "Circle" makes them peculiar in any way. Even their claim for excellence in clarity is lost. The test of clarity is not only regarding views of the nature of things but also regarding disputes, especially with Popper. On this it is much easier to adjudicate.

In [Stadler, 2003, xviii] we read, "Even Schlick who — apparently for personal reasons — did not invite Popper" to his home "paid tribute to Popper's" book "by including it in the series" of publications of the "Vienna Circle". Not so: Schlick invited Popper to his home to scold him for having submitted the book directly to the publisher. And he included Popper's book in his series to minimize the damage. His first public reference to Popper was a nasty misrepresentation [Ayer, 1959, 213]: he presented him as an idealist, no less. He said Popper's view is idealist, as he denied certitude to observation reports. But then Schlick should have added that Popper was here in agreement with Neurath [*LdF*, §26]. Here is a point of similarity between Popper and Neurath that Schlick missed despite his insistence that they were all advocating similar views. As Hacohen says [Hacohen, 2000, 210, 223], this was not an honest mistake: in a private letter Schlick said, Popper got "almost everything right".

We should backtrack, perhaps. Why did Schlick switch from Einstein to Wittgenstein?

Hempel vindicated the switch [Stadler, 1993, 1-10, 5]; [Jeffrey, 2000, 295-304, 298]: The attraction of Wittgenstein was in his promise that with sufficient effort each problem will meet with a solution or dissolution. As Hempel knew, this passage comes after the one in which Hempel says that Popper exaggerated his differences with the "Vienna Circle".[10]

Stadler admits that the major difference between Popper and the "Circle" was in his rejection of their Wittgensteinian equations of scientific character with grammar; he admits that this was their specific characteristic and their debt to Wittgenstein; yet he insists that this disagreement was minor. Common sense says, a disagreement is major if the differing parties view it as a matter of principle. What is a principle may be under dispute, of course. Here, however, there is no dispute: the change that Wittgenstein advocated concerned principles. After all, nothing is more fundamental than the principle that philosophical principles are in principle ungrammatical. Popper

---

10 Gödel has refuted this even for mathematics. Hempel knew that this refutes both Hilbert and Wittgenstein but he was diplomatic.

could not possibly differ from them more as he denied this principle.

Bar-Hillel praised Popper in print twice. First is the joint review [Bar-Hillel and Sambursky, 1960] of *The Logic of Scientific Discovery* that praises its ideas "as landmarks in the history of scientific methodology and of philosophy in general". The review also notes that the original of the book "did not exert the amount of influence it deserves" despite the praise lavished on it by Carnap, Ayer, and others. The review mentions with faint praise one of Popper's great and unquestionable achievements, his autonomous axiom system for the calculus of probability. It plays this down: "Popper's system is, nevertheless, not essentially stronger than Kolmogorov's, for instance, and is hence incomparably weaker than Carnap's." This is distressing, as the superiority of Popper's system over Kolmogorov's is proven [*LScD*, 346n]; [Schilpp, 1974, 1131-7] and is not under dispute [Spohn, 1986, opening]. The claim that Popper's system is "incomparably weaker" than that of Carnap rests on the claim that the latter system postulates some initial probabilities and the former does not: Popper naturally allowed for all sorts of hypotheses concerning initial probabilities. Carnap postulated his *ad hoc*, and then allowed for variations just like Popper. Carnap invited researchers to make hypotheses that determine initial probabilities [Carnap, 1952, ending]. Popper observed that they do so anyway, but in the object language, not in the meta-language as Carnap had suggested [Popper, 1963, 291-2]. Here Carnap fell prey to the confusion that he was dedicated to fight against, between the object-language and the meta-language.[11]

Bar-Hillel's second printed praise of Popper is in a posthumous publication, his contribution to *The Philosophy of Karl Popper* [Schilpp, 1974, 332-48]. In the penultimate note there (347, note 11) he reported the excitement that had accompanied his reading of Popper's *magnum opus* in one go. It took him 12 hours, including a five-minute break for a meal. (Before I left for England I read his doctoral dissertation and some of his early essays; no reference to Popper there.) Had he said this in his debate with Popper of two decades earlier, things might have been different. He could not be honest with Popper without capitulating, yet capitulate he could not — not in good faith. He was thus on a sticky wicket. He was to the last an enemy of metaphysics, as the last note of that sorry last paper of his makes amply clear. And he knew that capitulation to Popper was the endorsement of the metaphysical realism that he always abhorred. This has its irony: traditionally anti-metaphysics rested on hostility to dogmatism; it then became a dogma.

---

[11] In my last conversation with Bar-Hillel, you remember, he confessed he could not forgive himself having allowed Carnap to mislead him on this very point. Quine, by contrast, was never misled, and he admired Carnap all the way [Quine & Carnap, 1990, last page].

Popper's reply to Bar-Hillel is unfriendly — justly so but all the same regrettably so — brushing aside the compliments and responding to the insults. It ends with his response to Bar-Hillel's charge that Popper had wasted time on controversy. Popper's response should have been, good controversy contains efforts at criticism of good ideas and is thus most constructive. It was not. Instead, he rejected Bar-Hillel's charge, saying [Schilpp, 1974, 1048], throughout his life he had spent more time on constructive work than on controversy. I find this answer quite regrettable.

Bar-Hillel's greatest affront in that paper is his praise of Stopes-Roe's review of Popper in *The Journal of Symbolic Logic* [Stopes-Roe, 1968, 144]. On top of its patent incompetence, that review is too insolent:

"Popper is embarrassed by the recognition that no general laws are really 'highly confirmed' in the common sense, and tries to find another word for his interpretation ('Corroboration' is now favored)."

Stopes-Roe then repeats Bar-Hillel's and Kemeny's censures of Popper for his persistent "failing to appreciate the distinction between incremental and final confirmation" — a distinction that Popper was the first to make, as an exposure[12] of an inconsistency rooted in Carnap's "failing to appreciate the distinction" [Agassi, 1996c, 252-4]. What is the use of logic, asked Popper bitterly, if its advocates high-handedly ignore proofs that they dislike? It is high time for the editors of the esteemed *Journal of Symbolic Logic* to make amends for its complicity in this flagrant transgression. This could hopefully make a difference.

In summer 2000 I surprised a speaker — it was Kurt Rudolf Fischer (professor of philosophy and psychology in the University of Vienna for whom Friedrich Stadler edited a *Festschrift*) — when, after his lecture in the first conference on the history of the philosophy of science that took place in Vienna: I observed that he was echoing Popper. You mention Popper, he responded in obvious consternation; I did not think one has to mention him at all. I know of people, he explained, who got chairs in American universities because they had attacked him. After the meeting, I asked him if he deemed

---

12 Carnap had celebrated fans who threw much mud in the public eye in efforts to rescue his work from Popper's criticism. The distinction that they made in these efforts is no response to the criticism. The distinction is between the view of confirmation as probability and the view of it as increased probability. The criticism is that Carnap confused the two. It also shows a lacuna in Carnap's rules of explication between the classificatory concept (hot, cold), the graded one (hotter, colder), and the quantitative (temperature). The rules should correlate between each two of the three, and Carnap did not complete this simple task. The reason is simple: he tried hard to ignore an obvious fact that exposes the folly of the traditional theory of induction as probability: tautologies are most probable and least given to confirmation. Hence, confirmation is no probability. I said to Popper repeatedly that this profundity makes the whole discussion of inductive probability worthless. He refused to discuss this with me as he found refuting Carnap a great challenge.

this appropriate. I never thought about this, he replied. I wondered who the professors of whom he spoke were. It did not occur to me that he was referring to Paul Feyerabend, until I read a paper of his about him. Never mind.

Hilary Putnam criticized Popper, saying, "general relativity was accepted before there were decisive experiments in its favor" [Putnam, 2002, 180]. This, he explains, "of course contradicts completely the whole Popperian account, which can be characterized as mythological." Here the expressions "of course" and "completely" do not suffice: such severe charges call for corroborating evidence. Putnam should supply a quotation of some text (chapter and verse) reporting Popper's endorsement of the assertion, "no scientist has ever accepts a theory before there is decisive experiment in its favor". As it happens, Popper has stated unequivocally the opposite: he said, the initial degree of corroboration of a theory like that of Einstein is high without any new test, and this initially high degree of corroboration recommends it for acceptance. In other words, contrary to Putnam's misrepresentation, Popper deemed general relativity acceptable with no further ado simply because it explains both the empirical corroborations and the empirical refutations of Newton's theory. This may astonish inductivists, and so they should notice the following: Putnam the inductivist considers acceptability credibility; Popper the critical realist considered it testability: for a theory to undergo a test it should be accepted as an object for test, and for this it should exhibit initial merit. Putnam adds, "Really solid experimental confirmation of general relativity came only in the 1960s". This may suggest that Putnam claims that Popper assumed in flagrant error that already in 1919 Eddington had provided "really solid experimental confirmation". Yet Popper said, "Science does not rest on solid bedrock" [*LScD*, end of §30]. (Here I deem synonymous "really solid experience" and "solid bedrock".)

Putnam implies that whereas his own inductivism allows for the early endorsement of general relativity, Popper's view does not. The boot is on the other foot. Putnam does not discuss the status of the immediate acceptance in 1919 of Eddington's not "really solid experimental confirmation". By the inductivist canon it is a gross error to endorse a theory with no "really solid experience" to back it up. Popper, by contrast, could and did consider Eddington's 1919 information acceptable when he said, information is scientific if and only if it is refutable, thus allowing for the process of improving on it. Putnam makes a big fuss about the fact that astronomers improved upon Eddington's evidence, as if only inductivism, not hypothetico-deductivism, allows for this process. The opposite is the truth. Popper admired Eddington's evidence as it is the outcome of a serious test [Schilpp, 1974, 1035] – this regardless of its outcome, since for all that Eddington knew he could refute general relativity [Schilpp, 1974, 993-8]. Einstein calculated the result for Eddington's observation once on Newton's theory of gravity plus the assumption that light waves carry mass, and once on his own theory. As Eddington's observation, however inaccurate it was (as all observations are), agreed better with Einstein than with Newton, it refuted Newton and failed to refute Einstein. Einstein went further: his theory is still unsatisfactory, he said, despite the empirical backing that it has.

A court would declare Putnam a hostile witness. He cites Quine with qualified approval and Popper with unqualified disapproval [Putnam, 2002, 141-2, 145, 180] — contrary to Quine's presentation of his position. He said [Quine, 1987, 8] of the researcher, "Popper well depicts him as inventing hypotheses and then making every effort to falsify them by cunningly devised experiments." Elsewhere he says [Edwin and Schilpp, 1986, 621],

"Vuillemin reports Popper as arguing, contrary to holism, that scientific hypotheses are separately falsifiable. The reasoning is akin to what I have been reviewing above, so I am in substantial agreement [with Popper, his] slogans aside",

says Quine, gently refraining from correcting explicitly Vuillemin's report on the philosophy of Popper. In his contribution to *The Philosophy of Karl Popper* [Schilpp, 1974, 220], Quine sides with Popper explicitly.

Putnam's ascription to Popper of "the fantasy of doing science using only deductive logic" [Putnam, 2002, 145] is somewhat awkward, since, at the very least, Popper's "fantasy" is, science is fantasy tested by evidence. Deductive logic serves here as a mere tool: the logical relations between theory and evidence are deducibility, independence, and contradiction. What Putnam alludes to is that Popper said only logical relations between theory and any evidence count. The only way to view theoretical learning from evidence with no inductive logic, he added, is the idea that empirical refutations of theories are theoretically instructive. Putnam considers this an error, and this is his right. His calling it a myth or a fantasy, however, is his rhetorical surrogate for rational argument.

My rejection of Putnam's assertion that Popper's view is a myth and a piece of fantasy is in no way an endorsement of his deprecation of myth and of fantasy: all that the advocacy of a myth or of a piece of fantasy requires is its complementation by a critical attitude. It behooves us to try to see what a thinker tries to do; the demand for success is excessive. As it happens, Popper succeeded in surprisingly much of his work, and it took him decades to realize just how very successful he was. Carnap was less fortunate. In particular, his effort to stick to Wittgenstein's vision while endorsing Russell's criticism of it is valiant, even though a failure. His rejection of naturalism made it impossible for him to endorse the thesis that philosophy is inherently (i.e., naturally) meaningless. He suggested instead that the distinction between the object-language and the meta-language [Carnap, 1937, 186, 295-6] is sufficient for boosting Wittgenstein's thesis: he declared all philosophical assertions confusions between the two languages. This, he hoped, allowed him to relax Wittgenstein's strict limitations on the language of science. He attempted to develop a system [Carnap, 1937] that allowed the three items that Russell found necessary for mathematics and wanting in Wittgenstein's system (identity (49-50, 139, 303); existence (140-1, 61); and infinity (46, 101, 321). This made it easy to refute his view of all philosophy as confusion. To block this, he read infinity and existence in an extremely formalist mood. His formalism was far more extreme than that of Hilbert. Once we consider formalism as seriously anti-ontology, as Carnap did, we are bound to notice that his ink-spots (or chalk-marks) are recognizable by their shapes, and that shapes require geometry of sorts, and that geometry comes with its Platonic ontology.

This is an objection not to Carnap's fantasies but to Putnam's clinging to them (with some contorted apologetics, we remember). Carnap has a share in it, though. No matter how often he changed his mind, his assertions are all definitive and unqualified. Thus, in 1937 he still said of anti-philosophy (280) that the "Vienna Circle" had "established in detail and investigated in all its consequences", even though he noted that Wittgenstein's (and Schlick's) exclusion of universal statements from the language of science is an error (102).

All this is much ado about very little indeed. The role that this little discussion plays in this memoir is inflated due to the poverty of my ability to put things in proper proportion. I hope that future historians of philosophy will be rationalists in some sense, and I expect them to pay little attention to the painful episodes expounded here and expand instead on the great marvel of the rise of a new science and a new philosophy that celebrate the freedom of thought, an course of events that occurred in an epoch that saw the greatest human suffering and the greatest crimes against humanity. They will wonder, I suppose, and perhaps try to explain, how this strange conjunction was at all possible.

Appendix Six: Ordinary Language Philosophy

When I was Popper's student, Wittgenstein's influence was at its peak, especially in England. It centered there on ordinary language rather than on artificial ones, and went by the names of ordinary language philosophy, language analysis, and Oxford philosophy. Since Wittgenstein was Cambridge, I wondered about this name. The reason was simple. In the Cambridge philosophy department logic and the history of philosophy were high on the agenda, and Richard Braithwaite was a closet Popper fan. This was for me an indication of the pressure put on philosophy professors to comply. Another was the case of David Pole, who found Popper more attractive than Ayer but wrote a dissertation under the latter for pragmatic reasons.[13]

Russell and Popper judged unserious the kind of occupation that engaged Oxford philosophy then. Any appeal to language — any language, ordinary, artificial, or ideal — comprises efforts to relate metaphysics to grammar; all of them were doomed to fail. In efforts to oust metaphysical theories, Wittgenstein fans dismissed them at times as (grammatically) odd, at times as nothing more than commonsense, and at times as conflicting with commonsense. The distinction between these cases cannot be grammatical. This is no minor matter: Wittgenstein and his fans exposed magical expressions as mere metaphors ("symbolic"); some did the same with mental expressions or moral ones or political ones or religious ones; others took them literally. In particular, views on magic are false, as are many other superstitions. Are views on the human soul scientific or superstitious? Is Marxism scientific? Wittgenstein fans differed about these matters, and they supported their views in the same Wittgenstein-style discourse, thus showing that they followed no strict rule and no clear idea. The presuppositions behind language, Wittgenstein rightly asserted, are the existence of communities of speakers. But he had no more than that to offer. As no language is free of magic, religion, metaphysics, language is obviously very permissive. (It should be permissive, said Popper, since we should use it to decide what opinions we put on our agenda for examination.) Thus, the Wittgenstein set of ideas is woolly, especially his assertion that all language is precise, that we cannot think illogically, etc. [Wittgenstein, 1922, 5.156; 5.473; 5.4731; 5.61; 6.1251]. Not surprisingly, the attack on wooly talk in general is general wooly talk. Moreover, it is advisable to ignore wooly talk as much as reasonably possible, as wooly talk is easier to produce than to disparage.

Superfluous as Popper's critique of ordinary language analysis was, it

---

13 I do not know if these two pieces of gossip are true. I have them from the horses' mouths.

had a positive outcome: it made him assert his own metaphysics more boldly, especially his realism, his indeterminism, and his mind-body dualism. This led him to develop his propensity interpretation of probability as grand-scale metaphysics of which he was very proud. His major critique of ordinary language analysis appears in the preface to his *Logic of Scientific Discovery* (1959), the English translation of his first vintage, his *Logik der Forschung* (1935). He then let it rest. His attacks on "logical" positivism, however, went on, centering on their efforts to develop a theory of meaning and more so on their theory of inductive probability that is a part of the creed of most of its fans yet was older than any version of positivism.

There never was a canonic version of ordinary language philosophy. Adherents could barely say what it is without transcending ordinary discourses, of course. So when forced to articulate some general idea about it, they often said it is the best tools we have for philosophizing.

English Philosophy Since 1900 by G. J. Warnock, 1958, takes it for granted that all English philosophy since 1900 is in the style of the ordinary language school. He does not refer to Popper, although comparison of what he says in that book on Berkeley with the contents of his earlier book that is devoted to Berkeley shows that he has managed to learn something from Popper. Warnock had a thesis: the originator of ordinary-language philosophy was not Wittgenstein but G. E. Moore. Now Moore denied that all metaphysics is meaningless. In his book on that topic he acknowledges the existence of diverse metaphysical systems and he opts for common sense metaphysics. His reason is this: he was deeply convinced that commonsense ideas are true. There remains the question, are those who opt for other alternatives fools? If not, why do they do not see the light? This question troubled Moore. He seems to have followed Wittgenstein, although only partly, in order to answer it in a reasonable manner.

This is how I saw things when I was a philosophy student in England. For more detail, let me offer here only one example. It is the once famous *The Vocabulary of Politics* of T. D. Weldon, 1953, that naturally interested many people in the London School of Economics.

Consider Weldon's discussion of the justification of democracy and its foundations [Weldon, 1953, 100-01]. Of these he says,
"their rationality or advisability is a matter of discussion and controversy in which empirical arguments are highly relevant. There is nothing then sacrosanct about them and it is inappropriate to describe them as foundations."
This is a tacit rejection of all justification of democracy. It is the closest that Weldon came to endorse Popper's ideas in public. The book is mainly verbal

analyses Wittgenstein-style. (It does not mention Wittgenstein, but refers approvingly of some of his famous followers.) Popper obtains here a single remark (82) that insults him with faint praise:
"Provided that this is done consciously and the limitations of it are recognized (as they are, for instance, by Professor Popper in his advocacy of 'piecemeal' as distinct from 'utopian' engineering) there is much to be said for it. Unfortunately, however, matters do not always stop there. If theoretical puzzle-solving fails to deal with difficulties …"
(The term "Puzzle-solving" is a Wittgenstein-crowd buzzword.)

We should not judge Weldon harshly: he discussed Popper when it was easier to ignore him. And in private, we remember the testimony of Sir Peter Medawar, he encouraged his students to read Popper. This should serve as a reflection on the state of affairs in English philosophy in the mid-twentieth-century.

I know of nothing else said on the political power of puzzle-solving, except, perhaps, the brief preface of John T. D. Wisdom to his collection of essays, where he admits that they are boring while correcting the famous observation that the charge of the light brigade was magnificent but not war: obviously it was war and not magnificent in the least. This is presumably his expression of a preference for boring puzzle solving over killing. No one would object, but one may ask, how can anything to do with boredom ever prevent war?

Appendix Seven: Nails on the Socratic Problem

The hostility to Popper's pen-portrait of Plato rests on two frivolous arguments: the views that Plato expressed that are unpalatable to the modern democratic taste were common in his day; and he presented his utopia as a thought-experiment, not as a proposal for implementation. No text supports all this. Texts to the contrary abound. Popper mentions the texts that Gomperz and Field cite that expose Plato's view as unpopular even in his own day [*OS*, ii, 281] and he adds Plato's expressions of sharp disapproval against the increase of the popularity of the advocacy of the abolition of slavery [*OS*, i, 221-2, 224-5, 236, ii, 282]. Plato's dream of an orderly society won popularity in due course. Few commentators disliked it and they all took it as realistic, even as practicable [*OS*, ii, 302, 310]. So far for the excuse that Plato's prejudices were popular; as to the excuse that his republic is utopian, it is clearly not as inviting a utopia as that of the prophet Isaiah. Whereas Isaiah dreamt of no more war, Plato dreamt of internal harmony, not of world peace [*OS*, i, 43, 198, 259], and he discussed it in practical details such as the duty of the authorities to educate human watchdogs and to deceive their subjects. Nevertheless, as a thought experiment, Plato's utopia won Popper's admiration despite his revulsion against its totalitarianism. Popper's critics present Plato as an average Athenian; this is an insult, at least in contrast with Popper's image of him as politically amazingly astute. And, Popper added, even as a utopia not intended for implementation, Plato had success in aiming his utopia to have reactionary influence.

In addition to these two popular excuses against Popper's charges, another popular one is the allegation that his translations of central passages from Plato are too biased to count. This hurt him deeply. He repeatedly referred to the review of the authoritative Oxford classicist Richard Robinson [Robinson, 1951], who agreed with both sides as to the evidence that is most relevant to the dispute: final judgment on Popper's charges against Plato, he adjudicated, depends chiefly on the decision as to whether his translations of pertinent passages are less partial than the received ones. He (reluctantly) admitted that they are systematically better than those of the defense, although not always the very best available. Received opinion nevertheless still goes with Ronald Levinson's *In Defense of Plato*. For example, influential Kraut declared that one can find there "the fullest defense" of Plato against Popper [Kraut, 1992, 489]. He insultingly passed in silence over Popper's response to Levinson [*OS*, i, Appendix]. (Incidentally, the reprint of Levinson's book too makes no mention of Popper's response, and has no correction, not even of the slips of Levinson's pen that Popper has noted.)

Amir Meital invited me to work with him on an essay, "Slaves in Plato's Laws". That paper shows the standard of the commentaries on Popper by Levinson, Kraut, and others, as considerably below the standard received among classicists. We failed to find a classicist periodical that would accept it for publication. Eventually it appeared in *The Philosophy of the Social Sciences* (2007).

Let me discuss here Debra Nails on the Socratic Problem, as this problem was dear to Popper. It is because she is one of the best and most fair that I have decided to add here a review of one of her fascinating books, *Agora, Academy, and the conduct of Philosophy* (1995). It is perhaps the most thorough study of the problem to date. Let me sum up her argument through the eyes of Popper.

Nails mentions Popper in her text (not in her index) only once, in passing, in an insignificant passage (p. 131). She also mentions him only once in her bibliography (*The open society and its enemies*, vol. 1). She ignores Richard Robinson's review of that book, even though he discusses translations and she devotes much space in this book to philological studies. She ignores Richard Kraut's highly pertinent "Socrates and Democracy" in *Popper and the Human Sciences* [Currie and Musgrave, 1985]. Yet she agrees with Popper in a few significant ways. First and foremost, she is skeptical all the way: she considers both Socratic dialogues and science inherently open-ended: no presupposition is immune to challenge, she says. She differs here from almost all other Plato commentators, as they judge dialectics as ending in affirmations, and scientific research even more so. Second, her approach to the Socratic problem is nearest to Popper's, except that she is very careful not to refer to what makes the problem so hot, namely, Plato's political philosophy. She notices that the Socratic problem fills a literature that began in all seriousness only after World War II, but she does not explain: doing so would involve her in explaining Popper's role in it, and this she evidently declined to do for reasons that she did not even hint at. Her approach is nonetheless fresh and very attractive: it is valuable (even from the narrow viewpoint adopted here): all presuppositions are open to challenge, she says. Popper ascribes this utter openness to Socrates and to the young Plato, not to the mature Plato. By contrast, Nails ascribes it to both Socrates and Plato all the way. She does not refer to this disagreement, although in the brief passing remark about Popper (p. 131) she shows full awareness of it.

There is a caveat here: to declare that Socrates was open-minded about the views he held is not to say that he had no opinion. Unfortunately, tradition goes with the Pyrrhonists here and tacitly equates openness with having no opinion. The contributions of Popper and of Nails are both significant just because they do not make this mistake.

Nails splits the Socratic problem into two: first doctrinal and second methodological. She notices doctrinal discrepancies between dialogues. (Socrates abides once by the law of the land and once by his own conscience; he once supports hedonism and once opposes it, and he does the same with the doctrine of ideas.) She finds in this no problem at all, since Plato was a teacher who taught independent thinking, not any specific doctrine, she says. She then shows that efforts to arrange the works of Plato chronologically fail since Plato constantly rewrote them, and efforts to use some background information fail too: to make it work, the young and the old Plato should have different backgrounds, but they hardly do. Even Socrates and Plato had almost the same background, as the one was only 40 years older than the other. And so, as far as doctrines are concerned, the Socratic problem seems to vanish. There is then the matter of method, and here Nails distinguishes the dialectical part of the writings of Plato from their didactic parts, linking them to the fact that Socrates did not write and Plato did. She explains this by reference to the low efficiency of the oral practice unaided by written texts. Finally, she also notices that Socrates conversed with simple people and with slaves in a highly democratic style.

Nails sees in the foundation of the Academy a reaction to the inefficiency of Socrates' purely oral practice. In this context, she does not refer to the teaching of autonomy, although it is highly relevant, since writing is a powerful tool that educators may use for developing autonomy as well as for developing submission. Popper restated the Socratic problem forcefully as he declared Socrates a democrat devoted to autonomy and Plato the opposite: Popper refused to see as mere exercises in dialectic both expressions of readiness to die in defense of free speech and expressions of readiness to kill its practitioners. If his ascription of the first to Socrates and the second to Plato is an error, then it deserves criticism. So far, almost none has surfaced. So his solution to the Socratic problem still stands as an option: these two attitudes belong to two individuals. This division, as it happens, fits extremely well Nails' own division of Plato's texts to dialectical and didactic, and not by accident but as a strong corroboration of Popper's view, the view to which she devotes just one word ("betrayal", p. 131) in her whole meticulous, detailed text.

What brought about the founding of the academy? Before discussing this question we may ask, how did a playboy come to be a Plato? For, success in any activity tends to increase devotion to it, but for this to happen some motive to begin doing so is necessary. How does one who has no intent to partake in some activity come to do so anyway? The most obvious narrative is this. Plato bumped into Socrates, found him fascinating, and joined his entourage. He began writing by taking notes of Socrates speech in his own defense during his trial.[14] When he published them, he caused sufficient embarrassment to raise the rumor that Socrates was to blame for his death as he had the opportunity to escape and live in exile. This led him to write his second work, his *Crito*. He was sick when the execution of Socrates took place, probably because of it, so he asked about Socrates' last hours and wrote down the record as he had heard it as his *Phaedo*. Then came the other Socratic dialogues quite naturally. Millennia later, Popper wished to explain how faithful Plato came to betray the ideals of his beloved teacher. Popper spent much effort in his *The Open Society and Its Enemies* on this in as friendly a manner as he could. Most classicists did not take this friendliness seriously, as it did not soften Popper's harsh verdict that Plato was a reactionary and a totalitarian.

In the first part of her book Nails takes pains to show that no solution to the Socratic problem is comfortable. This is right and very valuable. She concludes that the problem does not really exist, at least not doctrinally, but even methodologically too, since despite the great difference between Socrates' dialectics and Plato's didactics, neither was a dogmatist, she says. All this is forceful, but a nagging problem stays: why did so many recent scholars devote efforts to a problem that she dismisses? Because Popper's arguments will not go away, I say, even if they are pooh-poohed as no more than unscholarly propaganda or remain unmentioned. Popper stressed and endorsed here a received view: Plato's argument for his granting unchecked power to the philosopher-king is valid: the possession of absolutely true knowledge permits wielding absolute power. And as this claim for the possibility of true knowledge is invalid, Popper added, Plato's case collapses. Moreover, Popper noticed, Plato's dogmatism (or his didactics, as Nails has it) is at odds with Plato's use and defense of the dialectic method. How could he defend it yet demand that its teaching be limited to reliable people past their prime? The enigma lingers.

The answer that critics repeatedly pitched against Popper and that Nails perfects is that the betrayal is merely a dialectic exercise, a thought-experiment meant to teach the opposite, to show that even the very best tyranny is worse than democracy. This is possible, but it meets with a wealth of arguments against it that Popper has paraded and that his critics should

---

14 Momigliano doubted that Socrates had ever existed. He viewed the reference in the *Apology* to Plato's taking notes as an instance of a known ploy for raising verisimilitude. It seems that later on the chatty gossip by Aristoxenus made him change his mind.

respond to but thus far did not. In particular, one should note that many of Plato's anti-democratic arguments are widely accepted in current democracies. Chief among these are two: first, his view that democracy leads to disorder and thus to tyranny and, second, his view that the sophists were not serious as they stayed skeptical to the last. Nails, at least, takes side with Popper on this issue, even though only tacitly. And so, she has something to explain. She does not.

The most pertinent point here, however, is the stress that Nails puts on the question, who was Plato? She falls back, alas, on the traditional answer: we can never answer the question, as he hid behind his interlocutors. Popper cites passages that show Plato clearly as power hungry. Nails ignores them.

Thus far I spoke in support of Popper. Let me end this by speaking for myself. The crux of the argument has to do with Plato's advocacy of inquisition in his *Laws*. His liberal-minded fans must consider this not a proposal for a viable option but a thought experiment meant to convince Plato's readers that the inquisition is atrocious. There is no room for indifference here. What may decide the choice between these two extreme options? Consider the way people argue against a cause even though they do not find it challenging, simply because many people do find it attractive. They may decide in such cases to start with a defense of that cause in a provocative mood, even though their hearts are not in the least in that defense. This is what Nails deems didactic: it is not quite dialectic. Alternatively, such people may argue against the objectionable cause because they find this discussion challenging just because they deem it a possible cure for contemporary ailments. This is what Nails deems dialectic proper. And she is right, and very significantly so. Discussing the use of inquisition or of torture is no fun. Suppose that Plato's *Laws* is didactic, as Nails says it is. She assumes, then, that the proposal to use an inquisition that Plato advocated provocatively was popular in Greece at the time. This is why Popper's quotation of Plato's protest against those Greeks who opposed slavery is so pertinent. Again, Nails offers no comment although she praises Socrates for his readiness to converse with slaves. Nails' careful study highlights the great importance of Plato's emphatic protest against the movement to free slaves and Popper's equally emphatic tribute to it. (For more, see [Meital and Agassi, 2007].)

Appendix Eight: Popper on Xenophanes

Popper's posthumous work [Popper, 1998] is clearly incomplete: understandably, it has much repetition, and includes passages that are clearly less informative than they would be, had Popper lived longer. It includes three previously published essays, (1) "Back to the Presocratics", (2) a chapter from *The Self and Its Brain* about the antiquity of the mind-body problem, and (3) a new appendix to *Logik der Forschung* that is what Popper (rather naïvely) considered his *coup de grace* to the traditional Aristotelian doctrine — inductivism-essentialism-probability. It is bitter, much in the classical German style of academic disputes; it is, indeed, a translation from the German. Had he found the time to polish it, most likely he would have translated it into a friendlier style, in line with his other late works. The book also includes some miscellaneous items, such as a correction of a traditional reading of a text of Aristotle, regrettably with no explanation: he obviously meant to add this later and to find a proper home for it. There is one moving essay on Xenophanes, on which more soon. Most of the book is on Parmenides, including some predecessors, especially Heraclitus, and some successors, especially Democritus and even a little on Plato and on Aristotle. Somehow, even Schopenhauer enters the text, and his influence on some contemporary leading physicists. Again, its relevance would have turned out later on had Popper lived to complete the book.

With the exception of the one essay against inductivism, the book is in a relaxed, witty style. Popper cites here in a friendly manner scholars who have neglected the duty to refer to him, and he does not chastise them for this neglect. He mentions Lakatos with gratitude and with no reservation. The book includes very little morals, and then not as homiletics but as a part of the praise for Xenophanes for his intellectual humility as a fallibilist. Popper hardly snubs the Jews here. (He snubs them once briefly (54): the monotheism of Xenophanes is better than the Christian or the Jewish, as Jehovah is a jealous god (*Exodus* 20:5). No mention of possible affinity or of possible influence. (I sense here the early nineteenth-century classicist J. G. Droysen, who tried to divorce the Hebrew ethos as much as possible from the Greek ethos and even Christianity.) The only people on whom Popper pours some scorn here are the language analysts, and this, too, he does gently. The chief thesis here is that Parmenides was a cosmologist, not an ontologist and not a meaning philosopher. It is odd: the word "ontology" is used here not in the traditional sense (as a part of metaphysics) but in a sense that analytic philosophers employ. (Is it an allusion to the book on Parmenides by G. E. M. Anscombe?) Parmenides had the denotation theory of meaning that Popper views as that of a blind child. Oddly, he ignores the fact that it was older, and the only one extant until Frege refuted it in the late nineteenth century. Nor did Popper need this argument to beat the analysts with. To that end suffice it that Parmenides had addressed a genuine philosophical problem. There are dazzling passages here that belong to classical studies and especially to philology that I cannot judge, having no Greek and no Latin. They contain much criticism, but here it is couched in respectful and gentle language.

All this indicates how changed Popper was in the twilight of his life.

The book reads to me as a letdown, though, despite its lovely tone and many exciting passages. It has one incredible flaw of omission and it rests on another flaw, one that is the book's agenda that is not explicit. (To say this of Popper is dreadful, I know. But as the book is incomplete, this is no censure.) The flaw is this: the book overlooks completely the transition from magic to science that Gellner calls "the big divide": the only mention of magic here is in the reproduced essay on the mind-body problem. Popper proves there that the mind-body problem is ancient (although tradition ascribes it to Descartes) by reference to the prevalence of dualism. (This seems to me irrelevant to the modern version of the problem that rests on the theory of substance that Russell and Popper rightly rejected.)[15] Popper also mentions that most cultures postulate the existence of ghosts. He mentions animism here only once, in oversight of its postulates that the world is full with (abstract) meanings, even though he refers to Evans-Pritchard. Nor does he mention the story of Thucydides about Pericles describing eclipses as devoid of meaning since they are natural phenomena. The idea that meanings are abstract and always involve anthropomorphism is totally missing here, even as the evil eye is thrown a passing glance. The view of magic as it transpires here seems to be the one that prevailed in the science-oriented circles of Popper's youth: it is the same as that of Neurath. (Whatever they had in common must be non-specific.) And so the great discovery of universalism (monotheistic at first) and its relation to the distinction between truth by nature and truth by convention is missing despite its centrality in the magnificent Chapter Five of *The Open Society and Its Enemies*. Perhaps this is so because here Popper overlooks the centrality of the idea of substance in Greek thought, preferring to it the new and exciting idea that the idea of ve-

---

15 In the 'fifties Wisdom tried to replace the mind-body problem with the mind-phantom-body problem (how do people view their mind's interaction with their phantom bodies, including their phantom limbs?). Popper objected that this is not the traditional problem. But then, without substance, his problem is not the traditional one either. This is a very rare example: Popper's research here has hardly helped break new ground.

risimilitude was common among the Presocratics.[16] He even says that Parmenides had held it. How this links with his idea of proof is unclear, and Popper agrees with Arpad Szabo that Parmenides is the originator of the demand for proof. A comment is sorely needed at this point. Popper praises Kirk, Raven and Schofield with no mention of his early quarrel with Kirk and Raven in "Back to the Presocratics". He also praises ("my late friend") Guthrie here (42) with no reference to his (Guthrie's) erroneous rejection of the Greek origin of the distinction between truth by nature and truth by convention. (He identified it with the distinction between nature and culture that Claude Lévi-Strauss rightly declared ubiquitous. Popper ignores Lévi-Strauss.)

It is not clear how "Back to the Presocratics" fits in with *The World of Parmenides*. The Ionians were critical, but the criticisms (say Anaximander's criticism of the doctrine of Thales) were not empirical; Democritus' criticism of Parmenides (and of Zeno) is. (This is doubtful: motion is all too obvious, and so its observation differs greatly from observations like that of deviations of planets from circular orbits.) No reference to the crisis due to the assertion that as appearances are truths by convention they are false and so inexplicable by truths by nature (as truths yield only truths) and the fact that Popper overcomes it (to the extent that he does) by his doctrine of verisimilitude (as the verisimilar is mostly false). Yet Heraclitus and Parmenides overcame this in the two possible extreme ways — Heraclitus by the denial of the truth by nature and Parmenides by the denial of the truth by convention, namely of the very existence of appearance. This is his view of appearances as sheer delusion. Popper rejected the traditional ascription of this view to Parmenides, it being a solution to no particular problem that he knew of. This is a weak argument [*PoH*, §28]. Also, he knew the problem here stated: it is the famous problem of the one and the many.

Finally, Popper emulated here Wisdom's psychological explanation for origins of ideas [Wisdom, 1975, 187-92] that he had earlier severely denounced; he offered the beautiful speculation that Parmenides was raised by a blind sister. (For the blind, the sense of touch is the most basic, perhaps the only trustworthy one. Also, Popper shows that some puzzling texts of Parmenides make sense on the assumption that in his view not darkness but light is privation. This resembles the suggestion of Szabo that puzzling texts on acoustics make sense on the assumption that the words for high and low

---

16 Without expressing an opinion on Popper's conjecture, let me observe that it is hard to avoid identifying a new elaborate idea with its old vague predecessor. Students of these texts should beware of this pitfall. How much the ancient use of verisimilitude that Popper speaks of relates to approximation to the truth I cannot judge for want of Greek.

notes are different in ancient and in modern texts.)

Popper presents here early Greek thought with two innovations: with no reference to metaphysics and with reference to verisimilitude, evidently in an (incomplete) effort to have the latter replace the former as a theory of intellectual frameworks. He identified with the Greek Enlightenment, especially with Xenophanes, "the founder of epistemology" and "the founder of the Greek Enlightenment" (33), who advocated a "critical empiricism" (46), a "critical theory of knowledge" (46), and *"a theory of objective knowledge"* (47), no less. He describes him as belonging to a population of greatly learned and enormously talented refugees. He cites (p. 54) Xenophanes describing himself as a weary old man, a guest comfortably inclined by the fireside, sipping sweet wine and nibbling nuts, and answering the queries of his host by narrating about his long life of travails.

Perhaps the strongest identification that Popper had with Xenophanes is his repeated assertion that the reason for the neglect of the tremendous contribution of Xenophanes to the Western way of life was that he did not impress historians just because he did not boast as a great wise man having invented the humble view that all human thinking is fallible.

The new and impressive aspects of this last book of Popper's are two. First is his systematic identification with Xenophanes, a person resigned to his sad fate, whose reputation he was rescuing in the twilight of his life. Second is the sparseness of harsh tone and in the mellowness of this book, inundated with a beautiful, wistful fragrance of extraordinary character, sparkling with insight and wit and irony. Both aspects testify to the conjecture that he died resigned to his fate and relatively happy.

May he rest in peace.

LETTERS

THE LONDON SCHOOL OF ECONOMICS AND POLITICAL SCIENCE.
(UNIVERSITY OF LONDON)

HOUGHTON STREET,
ALDWYCH,
LONDON W.C.2.

Telephone: Holborn 7686 (7 lines).
Telegrams: "Poleconics, Estrand," London.

TO WHOM IT MAY CONCERN.   January 22nd, 1958.

Dr. Joseph Agassi has been working in close co-operation with me in the field of the philosophy and history of physics since 1952, and I have never had a student in this field who was his equal in keenness and intelligence.

Agassi's thesis, which is a voluminous work, has two roots: his interest in the problem of the interpretation of physical theories, and his admiration for Einstein. (Relativity theory was his special subject in his studies for his M.Sc. in physics.) Einstein attributed to Faraday the fundamental idea of his own favourite programme - the unified field theory. This remark of Einstein's led to the suggestion to study Faraday as a theoretician - in opposition to the usual view which sees Faraday's significance mainly in his epoch making experiments.

Dr. Agassi has read, first as my research student, and later as my research assistant, almost everything I have written since 1953 in manuscript, and his criticism has been very valuable to me. We have worked in daily contact, discussing in detail every aspect of his and my work, and these discussions have been very fruitful. I have taken the opportunity of acknowledging his contributions in several of my publications; and Mr. J. W. N. Watkins, who is now a member of my department, has also acknowledged in two of his publications his indebtedness to Dr. Agassi.

Dr. Agassi was invited to accompany me to the Palo Alto 'Center for Advanced Study in the Behavioural Sciences', first as a Research Associate, and later as a Resident Fellow. This was in 1956-1957. Since our return, he has been a temporary lecturer in my department, replacing Dr. J. O. Wisdom who is on sabbatical leave. Owing to the fact that there is no permanent vacancy in my department, he is at this moment looking for a position to be taken up, if possible, in the autumn of 1958.

Sir Karl Popper, CH, FRS
136 Welcomes Road,
Kenley, Surrey CR8 5HH
ENGLAND

Herzlia, 1992.07.28

Dear Karl,

Nathaniel Laor has shown me a recent letter from Melitta Mew, from which I surmise that a letter of apology from me to you will not meet with displeasure.

I request your indulgence. I do not know what form becomes this letter and it suits me to write different versions in the hope that one of them will meet with your acceptance. This is, of course, time consuming, and I am never oblivious to your being so very much pressed for time. I hope that this letter will at least help to terminate a standing cause of irritation to you, so that you will not find utter waste the distraction it incurs.

2. I apologize for all the occasions, remembered as well as forgotten, of my expression of disrespect for you, for all the situations in which I was overbearing, arrogant and haughty and insolent and supercilious in your presence and/or regarding yourself, and for all the times I took liberties with your kindness. I apologize for all I did to cause you the annoyance and the disappointment and the displeasure and the waste of precious time. As I have made some statements about you in public, allow me to add that, for each and every disrespectful personal remark about you in public I now humbly apologize too.

3. We are both past our prime, and we have the choice of letting things stand as they are or conclude in peace. My intent is to conclude in peace. I have no other intent. Since you are so overworked, I do not ask for a reply, though if Mrs. Mew will acknowledge for you the receipt of this letter it will be a great kindness. Nor do I see myself troubling you further in the near future, except on your instruction, of course. So, unless you say something to the contrary, this is it. My sole request is that you consider this apology as most sincere, that you accept it graciously, and that you accept this as a peace offering.

May I wish you long life in good health and happiness, my old, revered teacher.
Be well.

Sir Karl Popper, CH, FRS

136 Welcomes Road,
Kenley, Surrey
CR8 5HH

1-8-92

Dear Joske,

Thank you for your Fax dated "1992.07.28". I accept it gladly, as "a peace offering", as you put it.

I also have a telegram, signed by Judith and yourself. Please give my love to Judith! And thanks for the telegram.

Karl.

P.S. As it seems to be your intention to include everything – including cases of plagiarism and of distortion – I also wish to make my acceptance of your apology very general, and I include cases coming in the future.

Agassi 1993.02.16
TEL AVIV UNIVERSITY

Dear Karl,

last night I was happy to receive your letter, informing me kindly, good-humouredly, of your acceptance of my peace offering. Thank you very much indeed. This settles matters, finally and satisfactorily. I will inform some of those whom the news will please. If there is anything you wish me to do, you will do me a great favour to inform me of it. Meanwhile I am,

as ever,

yours,

Joske

P.S. Judith thanks you for your kind words.

BIBLIOGRAPHY

N. B. The following titles of works mentioned in this volume appear here merely for convenience; they differ in range and level, as well as in value.

Adler, S., and O. Theodor, 1931. Investigations on Mediterranean Kala Azar. II. Leishmania Infantum. *Proceedings of the Royal Society B*, 108, 447-463.

Agassi, Joseph, 1959. Wittgenstein the Elusive, two letters to the editor. *Times Literary Supplement*, 22 and 29 May 1959.

———, 1959b. Corroboration versus Induction. *The British Journal for the Philosophy of Science*, 9, 311-17.

———, (1963) 1967. *Towards an Historiography of Science*.

———, 1963b. Between Micro and Macro. *The British Journal for the Philosophy of Science*, 14, 26-31.

———, 1971. *Faraday as a Natural Philosopher*.

———, 1972. The Twisting of the IQ Test. *Philosophical Forum*, 3, 260-72.

———, 1974. Conventions of Knowledge in Talmudic Law. In B. Jackson, editor, *Studies in Jewish Legal History, in Honour of David Daube*, 16-34. (Published also as *Journal of Jewish Studies*, 25, 16-34).

———, 1974b. The Last Refuge of the Scoundrel. *Philosophia*, 4, 315-17.

———, 1975. *Science in Flux*.

———, 1976. The Lakatosian Revolution. In Cohen *et al.*, 1976, 9-22.

———, 1977. *Towards a rational philosophical Anthropology*.

———, 1977b. Who Discovered Boyle's Law? *Studies in the History and Philosophy of Science*, 8, 189-250.

———, 1981. *Science and Society: Studies in the Sociology of Science*.

———, 1982. In Search of Rationality: A Personal Report. In Levinson, 1982, 237-248.

———, 1985. Hegel's Scientific Mythopoiesis in Historical Perspective. *In* Cohen, R. S., R. M. Martin, and Merold Westfal, editors, 1985. *Studies in the Philosophy of John N. Findlay*, 445-458.

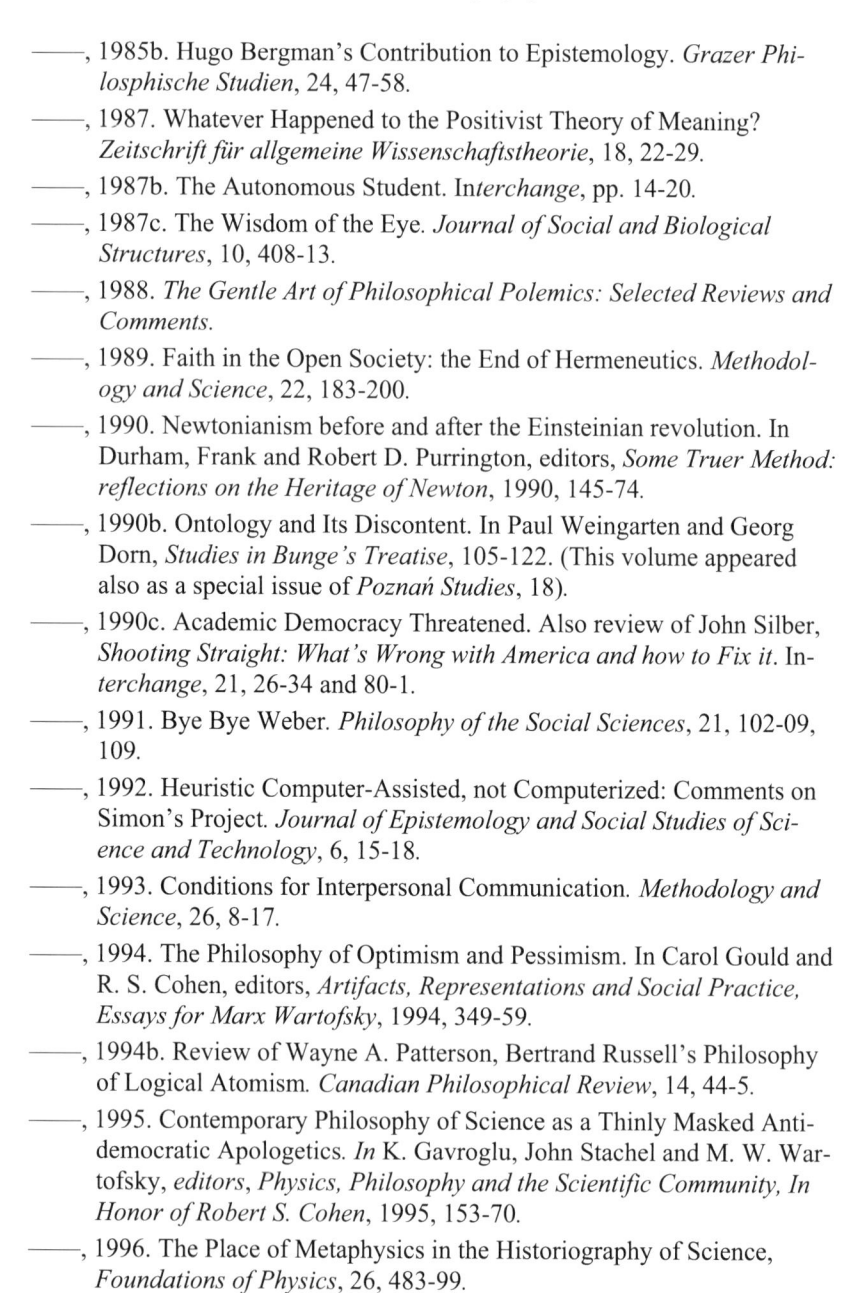

——, 1985b. Hugo Bergman's Contribution to Epistemology. *Grazer Philosphische Studien*, 24, 47-58.
——, 1987. Whatever Happened to the Positivist Theory of Meaning? *Zeitschrift für allgemeine Wissenschaftstheorie*, 18, 22-29.
——, 1987b. The Autonomous Student. In*terchange*, pp. 14-20.
——, 1987c. The Wisdom of the Eye. *Journal of Social and Biological Structures*, 10, 408-13.
——, 1988. *The Gentle Art of Philosophical Polemics: Selected Reviews and Comments*.
——, 1989. Faith in the Open Society: the End of Hermeneutics. *Methodology and Science*, 22, 183-200.
——, 1990. Newtonianism before and after the Einsteinian revolution. In Durham, Frank and Robert D. Purrington, editors, *Some Truer Method: reflections on the Heritage of Newton*, 1990, 145-74.
——, 1990b. Ontology and Its Discontent. In Paul Weingarten and Georg Dorn, *Studies in Bunge's Treatise*, 105-122. (This volume appeared also as a special issue of *Poznań Studies*, 18).
——, 1990c. Academic Democracy Threatened. Also review of John Silber, *Shooting Straight: What's Wrong with America and how to Fix it*. In*terchange*, 21, 26-34 and 80-1.
——, 1991. Bye Bye Weber. *Philosophy of the Social Sciences*, 21, 102-09, 109.
——, 1992. Heuristic Computer-Assisted, not Computerized: Comments on Simon's Project. *Journal of Epistemology and Social Studies of Science and Technology*, 6, 15-18.
——, 1993. Conditions for Interpersonal Communication. *Methodology and Science*, 26, 8-17.
——, 1994. The Philosophy of Optimism and Pessimism. In Carol Gould and R. S. Cohen, editors, *Artifacts, Representations and Social Practice, Essays for Marx Wartofsky*, 1994, 349-59.
——, 1994b. Review of Wayne A. Patterson, Bertrand Russell's Philosophy of Logical Atomism. *Canadian Philosophical Review*, 14, 44-5.
——, 1995. Contemporary Philosophy of Science as a Thinly Masked Antidemocratic Apologetics. *In* K. Gavroglu, John Stachel and M. W. Wartofsky, *editors, Physics, Philosophy and the Scientific Community, In Honor of Robert S. Cohen*, 1995, 153-70.
——, 1996. The Place of Metaphysics in the Historiography of Science, *Foundations of Physics*, 26, 483-99.

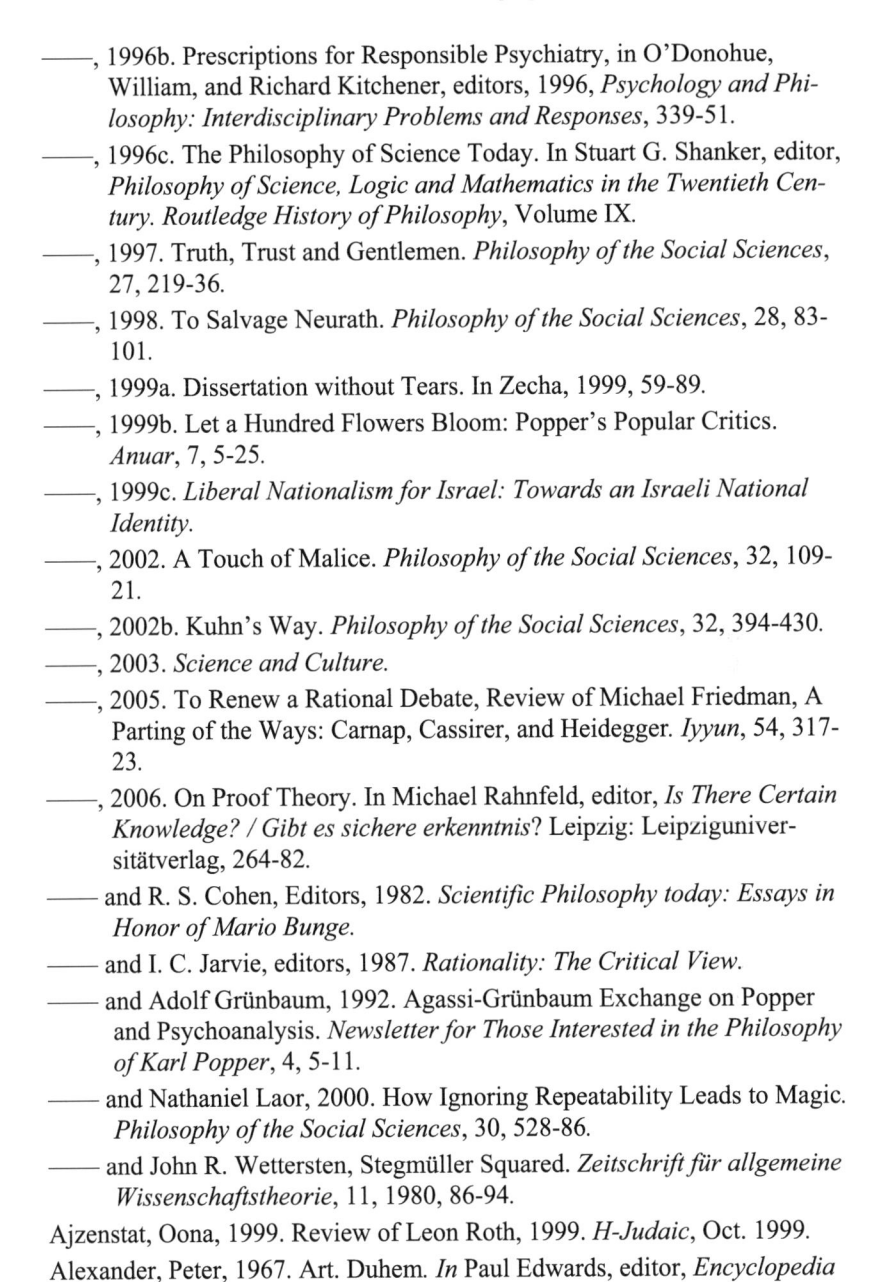

―, 1996b. Prescriptions for Responsible Psychiatry, in O'Donohue, William, and Richard Kitchener, editors, 1996, *Psychology and Philosophy: Interdisciplinary Problems and Responses*, 339-51.

―, 1996c. The Philosophy of Science Today. In Stuart G. Shanker, editor, *Philosophy of Science, Logic and Mathematics in the Twentieth Century. Routledge History of Philosophy*, Volume IX.

―, 1997. Truth, Trust and Gentlemen. *Philosophy of the Social Sciences*, 27, 219-36.

―, 1998. To Salvage Neurath. *Philosophy of the Social Sciences*, 28, 83-101.

―, 1999a. Dissertation without Tears. In Zecha, 1999, 59-89.

―, 1999b. Let a Hundred Flowers Bloom: Popper's Popular Critics. *Anuar*, 7, 5-25.

―, 1999c. *Liberal Nationalism for Israel: Towards an Israeli National Identity*.

―, 2002. A Touch of Malice. *Philosophy of the Social Sciences*, 32, 109-21.

―, 2002b. Kuhn's Way. *Philosophy of the Social Sciences*, 32, 394-430.

―, 2003. *Science and Culture*.

―, 2005. To Renew a Rational Debate, Review of Michael Friedman, A Parting of the Ways: Carnap, Cassirer, and Heidegger. *Iyyun*, 54, 317-23.

―, 2006. On Proof Theory. In Michael Rahnfeld, editor, *Is There Certain Knowledge? / Gibt es sichere erkenntnis*? Leipzig: Leipziguniversitätverlag, 264-82.

― and R. S. Cohen, Editors, 1982. *Scientific Philosophy today: Essays in Honor of Mario Bunge*.

― and I. C. Jarvie, editors, 1987. *Rationality: The Critical View*.

― and Adolf Grünbaum, 1992. Agassi-Grünbaum Exchange on Popper and Psychoanalysis. *Newsletter for Those Interested in the Philosophy of Karl Popper*, 4, 5-11.

― and Nathaniel Laor, 2000. How Ignoring Repeatability Leads to Magic. *Philosophy of the Social Sciences*, 30, 528-86.

― and John R. Wettersten, Stegmüller Squared. *Zeitschrift für allgemeine Wissenschaftstheorie*, 11, 1980, 86-94.

Ajzenstat, Oona, 1999. Review of Leon Roth, 1999. *H-Judaic*, Oct. 1999.

Alexander, Peter, 1967. Art. Duhem. *In* Paul Edwards, editor, *Encyclopedia of Philosophy*.

Amsterdamski, Stefan, 1996. *The Significance of Popper's Thought: Proceedings of the Conference Karl Popper: 1902-1994, March 10-12, 1995, Graduate School for Social Research, Warsaw. Poznań Studies,* 49.
Annan, Noel, 1990. *Our Age.*
Austin, John L., 1962. *How to Do Things with Words.*
Awodey, Steve and Carsten Klein, editors, 2004. *Carnap Brought Home: The View from Jena.*
Ayer, A. J., 1957. *The Problem of Knowledge.*
——, 1977. *Part of My Life: The Memoirs of a Philosopher.*
——, editor, 1959. *Logical Positivism.*

Bambrough, Renford, editor, 1967. *Plato. Popper and Politics: some contributions to a modern controversy.*
Banton, Michael, 1965. *Roles: An Introduction to the Study of Social Relations.*
Bar-Am, Nimrod, 2003. A Framework for a Critical History of Logic. *Sudhoffs Archiv,* 87, 80-9.
Bar-Hillel, Yehoshua, 1964. *Language and Information: selected essays on their theory and application.*
——, 1964. The Present Status of Automatic Translation of Languages, *Advances in Computers,* 1, 91-163.
——, 1974. Popper's Theory of Corroboration. In Schilpp, 1974, 332-48.
Bar-Hillel, Yehoshua and Shmuel Sambursky, 1960. Review of Popper, 1959, *Isis,* 51, 91-4.
Bartley, W. W. III, 1962, 1984. *The Retreat to Commitment.*
1976. On Imre Lakatos, in Cohen, Feyerabend, and Wartofsky, 1976, 37-8.
——, 1990. *Unfathomed Knowledge, Unmeasured Wealth: On Universities and the Wealth of Nations a collection of essays.*
Baum, Wilhelm, editor, 1997. *Briefwechsel, Paul Feyerabend and Hans Albert.*
Baumgardt, Carola, 1952. *Johannes Kepler: Life and Letters.* With an Introduction by Albert Einstein.
Berkson, William and John Wettersten, 1984. *Learning from Error: Karl Popper's Psychology of Learning.*
Berlin, Isaiah, 1953. *Historical Inevitability.*
——, 1969. *Four Essays on Liberty.*
——, 1997. *The Proper Study of Mankind.*

Bernard, Harvey Russell, 2005. *Research Methods in Anthropology: Qualitative and Quantitative Approaches.*

Bernays, Paul, 1959. Comments on Ludwig Wittgenstein's Remarks on the foundations of mathematics. *Bernays Project*: Text No. 23.

Berry, D., 1977. Buber's View of Jesus as Brother. *Journal of Ecumenical Studies* 14, 203-218.

Bischof, Norbert 1993. *Gescheiter als alle die Laffen. Ein Psychogramm von Konrad Lorenz.*.

Black, Max, 1962. *Models and Metaphor: studies in language and philosophy.*

———, 1947. Review: Bertrand Russell, *Logical Positivism*. *Journal of Symbolic Logic*, 12, 24.

Blaug, Mark, 2002. Review of Hacohen's life of Popper. *History of Political Thought*, 23, 545-564.

Blaukopf, Kurt, 1994. *Logik der Musikforschung: Karl Poppers methodologische Beitrag.* In Seiler and Stadler, 1994, 181-200.

Bloom, Harold, 2000. *How To Read and Why.*

Boas, George, 1948. Fact and Legend in the Biography of Plato. *Philosophical Review*, 57, 439-57.

Bok, Sissela, 1974. The Ethics of Giving Placebos. *Scientific American*, 2331, 17-23.

Bonk, Thomas, 2003. *Language, truth and Knowledge: Contributions to the Philosophy of Rudolf Carnap.*

Born, Max, 1949. *Natural history of Cause and Chance.*

———, 1953. The Interpretation of Quantum Mechanics. *The British Journal for the Philosophy of Science*, 4, 95-106.

———, 1969. *Physics in my Generation.*

Born, Max and Hedwig, and Albert Einstein, 1971. *The Correspondence between Albert Einstein and Max and Hedwig Born, 1916-1955.*

Bosetti, Giancarlo, 1997. See Popper, 1997.

Bouwsma, O. K., 1961. Review of Wittgenstein, *The Blue Book*. In *Journal of Philosophy*, 58, 141-162.

Bridgman, P. W. 1955. *Reflections of a Physicist*, 2nd edition.

———, 1964. The Mach Principle. In Bunge, 1964, 224-33.

Broad, William and Nicholas Wade, 1982. *Betrayers of the Truth: Fraud and Deceit in the Halls of Science.*

Bromberger, Sylvain, 1992. *On What We Know We Don't Know: Explanation, Theory, Linguistics, and How Questions Shape Them.*

Bronowski Jacob, 1971. The Disestablishment of Science. *Encounter*, 37, 8-16.
Buber Agassi, Judith, 1985. Gender Discrimination through Recruitment. *Organizational Studies*, 13, 472-5.
—— and Joseph Agassi, 1985. The Ethics and Politics of Autonomy: Walter Kaufmann's Contribution. *Methodology and Science*, 18, 165-185.
Bunge, Mario, editor, 1964. *The Critical Approach to Science and Philosophy: In Honor of Karl Popper.*
Burtt, E. A., 1926. *The Metaphysical Foundations of Modern Physical Science.*

Calaprice, Alice, 2000. *The Expanded Quotable Einstein.*
Campbell, Ian and David Hutchinson, A Question of Priorities: Forbes, Agassiz and Their Disputes on Glacier Observations. *Isis*, 69, 1978, 388-99.
Campbell, Norman, 1921. *What is Science.*
Caplow, Theodore, and Reece McGee, 1958 (1965, 2001). *The Academic Marketplace.*
Carnap, Rudolf, (1928), 1967. *The Logical Structure of the World; Pseudo Problems of philosophy.*
——, 1932. Überwindung der Metaphysik durch logische Analyse der Sprache. *Erkenntnis*, 2, 219 - 241.
——, (1934) 1937. *The Logical Syntax of Language.*
——, 1936-7. Testability and Meaning. *Philosophy of Science*, 3, 419-71, and 4, 1-40. (Appeared as a book in 1950.).
——, 1942. *Introduction to Semantics.*
——, (1950) 1962. *The Logical Foundations of Probability.*
——, 1952. *The Continuum of Inductive Methods.*
——, 1956. The Methodological Character of Theoretical Concepts. In Herbert Feigl and Michael Scriven, editors, *The Foundations of Science and the Concepts of Science and Psychology*, pp. 38-76.
——, 1963. Replies. In Schilpp, 1963.
Cartwright, Nancy, 1989. *Nature's Capacities and their Measurement.*
Cartwright, Nancy, Jordi Cat, Lola Fleck, and Thomas E. Uebel, 1996. *Otto Neurath: Philosophy between Science and Politics.*
Chiariello, Michael, 1996. A review of Agassi, 1993. *The Review of Metaphysics.*
Cobban, Alfred, 1954. The Open Society: A Reconsideration. *Political Science Q*uarterly, 69, 119-26.

Coffa, J. Alberto and Linda Wessels, 1991. *The Semantic Tradition from Kant to Carnap: To the Vienna Station.*
Cohen, Percy S., 1968. *Modern Social theory.*
Cohen, R. S., P. K. Feyerabend, and M. W. Wartofsky, 1976. *Essays in memory of Imre Lakatos.*
Colodny, R. G., editor, 1966. *Mind and Cosmos.*
Crick, Francis, 1990. *What Mad Pursuit: a Personal View of Scientific Discovery.*
Crombie, Alistair C., 1952. *Augustine to Galileo: A history of Science A. D. 400-1650.*
Crossman, Richard, 1937, 1959. *Plato Today.*
Currie, Gregory and Alan Musgrave, 1985. *Popper and the Human Sciences.*

Dawidowicz, Lucy S., 1975. *The War against the Jews, 1944-1945.*
Davies, Phillip and Reuben Hersh, 1980. *The Mathematical Experience.*
Davies, Robertson, 1982. A Giant of the Stage. *Maclean's Magazine*, July, 24, 48-49.
Dirac, P. A. M., 1978. *Directions in Physics.*
Drake, Stillman, editor and translator, (1953) 1967. Galileo, *Dialogue concerning the Two Chief World Systems.*
Dukas, Helen and Banesh Hoffmann, editors, 1979. *Albert Einstein, the Human Side.*
Duerr, Hans Peter, editor, 1995. *Paul Feyerabend: Briefe an einen Freund.*
Dynes, Wayne R. and Stephen Donaldson, 1992. *Ethnographic studies of homosexuality.*

Eccles, John, 1970. *Facing Reality: Adventures of a Brain Scientist.*
Edmond, Gary, and David Mercer, 2002. Conjectures and Exhumations: Citations of History, Philosophy and Sociology of Science in US Federal Courts. *Law and Literature*, 14, 309-366.
Edmonds, David, and John Eidinow, 2001. *Wittgenstein's Poker: the Story of a Ten-minute Argument between Two Great Philosophers.*
Egidi, Rosaria, 1995. *Wittgenstein : Mind and Language.*
Eidlin, Fred, 1999. Matching Popper's Theory to practice. In Jarvie and Pralong, 1999, 203-7.
Einstein, Albert, 1920, 1946, 1950, 1955. *The Meaning of Relativity.*
——, 1920b. *Relativity: The Special and General Theory.*
——, 1949. Autobiographic Notes and Replies in Schilpp, 1949.

——, 1949b. *The World as I See It.*
——, 1954. *Ideas and Opinions.*
——, 1959-60. Letter to P. A. Schilpp. In Schilpp [1959-60].
——, 1981. *Briefe. Aus dem Nachlass* herausgegeben von Helen Dukas und Banesh Hoffmann.
——, 1953. Preface to Stillman Drake, 1953.
——, 1998. The Collected Papers of Albert Einstein, vol. 8A, edited by R. Schulmann, A. J. Fox, and J. Illy.
Engelmann, Paul, 1967. *Letters from Wittgenstein.*
Ettinger, Shmuel, 1976. The Origins of Modern Anti-Semitism. Yisrael Gutman and Livia Rethkirchen, editors, *The Catastrophe of European Jewry: Antecedents, History, Reflections*, 3-39.

Fadiman, Clifton, editor, 1939. *I Believe. The Personal Philosophies of Certain Eminent Men and Women of Our Time.*
Farrell, Robert P., 2000. Will the Popperian Feyerabend please step forward: pluralistic, Popperian themes in the philosophy of Paul Feyerabend. In *Intl. Stud. Phil. Sci.*, 14, 257-266.
Feferman, Solomon, 2005. The Gödel editorial project: a synopsis. *Bulletine of Symbolic Logic*, 11, 132-149.
Festinger, Leon, 1957. *Theory of Cognitive Dissonance.*
Feuer, Lewis, 1987. Introduction: John Stuart Mill as a Sociologist: The Unwritten Ethology, in Mill, 1987.
Feyerabend Paul K., 1955. Review of Wittgenstein, *Philosophical Investigations. Philosophical Review*, 64, 449-483.
——, 1958. Reichenbach's Interpretation of Quantum-Mechanics. *Phil. Stud.*, 9, 49-59.
——, 1962. *Knowledge without Foundations*. Also in Feyerabend, 1999, 50-77.
——, 1975. *Against Method.*
——, 1978. *Science in a Free Society.*
——, 1979. *Erkenntnis für freie Menschen.*
——, 1980. *Erkenntinis für freie Menschen Veränderte Ausgabe.*
——, 1987. *Farewell to Reason.*
——, 1991. *Philosophical Papers*, Vols. 1 and 2. Edited by John Preston.
——, 1991b. *Three Dialogues on Knowledge.*
——, 1995. *Killing Time: the Autobiography of Paul Feyerabend.*
——, 1999. *Philosophical Papers*, Vol. 3. Edited by John Preston.

——, 2001. *Conquest of Abundance: A Tale of Abstraction Versus the Richness of Being*.
Feyerabend, Paul K., and G. Maxwell, editors, 1966. *Mind, Matter and Method. Essays in Philosophy and Science in Honor of Herbert Feigl*.
Field, G. C., 1925. Socrates and Plato in Post-Aristotelian Tradition-II. *Classical Quarterly*, 19, 1-13.
Findlay, John N., 1966. Review of Kaufmann, 1965. *Philosophical Quarterly*, 16, 366-368.
Finn, B. S., 1972. Review of Agassi and Williams. *Science*, 176, 665-7.
Fisch M. and S. Schaffer, 1991. *William Whewell: A Composite Portrait*.
Flugel, John Carl, with supplements by Donald John West, 1935, 1970. *A Hundred Years of Psychology*.
Flusser, David. 1988. *Judaism and the Origins of Christianity*.
Fösling, Albrecht, 1997. *Albert Einstein: A Biography*, translated from the German by Ewald Osers.
Ford, Kenneth M., Clark Glymour and Patrick J. Hayes, 1996. *Android Epistemology*.
Forster, E. M., (1951) 1965. *Two Cheers for Democracy*.
Fraassen, Baas C. van, 1976. Representations of Conditional Probabilities, *Journal for Philosophical Logic*, 5, 417-30.
Fraenkel, Abraham Adolf, (1953), 1968. *Abstract Set Theory*.
Fraenkel, Abraham Adolf, Yehoshua Bar-Hillel, and Azriel Levy, with the collaboration of Dirk van Dalen, 1973. *Foundations of Set Theory*.
Frank, Philipp, 1949. Einstein's philosophy of Science, *Rev. Mod. Phys.*, 21, 349-55.
Franklin, James, 2000. Last bastion of Reason. *The New Criterion*, 74-8.
Freyne, Sean, 2002. *Texts Contexts and Cultures: Essays on Biblical Topics*.
Freud, Sigmund, 1923. *The Ego and the Id*. Also in Freud et al., 1953-74, vol. 9, 3-66.
Freud et al., 1953-74. *The Standard Edition of the Complete Psychological Works of Sigmund Freud*.
Frey, Gerhard, Schlick's Konstitutierungen etc., in Eugene Gadol, 1983. 145-159.
Fried, Yehuda and Joseph Agassi, 1976. *Paranoia: A Study in Diagnosis*.
——, 1983. *Psychiatry as Medicine*.
Friedman, Michael, 1999. *Reconsidering Logical Positivism*.
——, 2000. *A Parting of the Ways: Carnap, Cassirer, and Heidegger*.

Gadol, Eugene, editor, 1983. *Rationality and Science: Memorial Volume for Moritz Schlick in Celebration of the Centennial of his Birth.*

Gattegno, Caleb, 1970. *What We Owe Children: the Subordination of Teaching to Learning.*

Gellner, Ernest, 1959. *Words and Things.*

——, 1994. Karl Popper — the Thinker and the Man. In Amsterdamski, 1996, 75-85.

——, 1998. *Language and Solitude: Wittgenstein, Malinowski and the Habsburg Dilemma.*

Gladstone, John H., 1872. *Faraday.*

Golde, Peggy, editor, 1986. Women in the Field: Anthropological Experiences. Second edition.

Goldstein, Laurence, 1999. *Wittgenstein's development and His relevance to Modern Thought.*

Gombrich, Ernst H., 1970. *Aby Warburg: An Intellectual Biography.*

——, 1974. The Logic of Vanity Fair. In Schilpp, 1974.

——, 2001. See Kiesewetter.

Gombrich, Ernst H. and Ernst Kris, 1940. *Caricature.*

Goodman, Edward, 1951. *Forms of Public Control and ownership.*

——, 1969. *The Impact of size: A study of economic and human values in modern industrial society.*

Goodman, Nelson, 1955. *Fact, Fiction and Forecast.*

Gopnik, Adam, 2002. A Critic at large: The Porcupine: A pilgrimage to Popper. *The New Yorker for April 1.*

Grant, G. P., 1954. Plato and Popper, *Canadian Journal for Economics and Political Science*, 20, 185-194.

Grattan-Gueness, Ivor, 1992. Russell and Karl Popper: Their Personal Contacts. *Russell*, 12, 3 -18.

Gregory, Richard, 2004. A Student's view of Cambridge philosophy Post-Wittgenstein, 1947-49, *Philosophy at Cambridge, Newsletter of the Faculty of philosophy*, May, 2004.

Gross, Berl, 1992. *Before Democracy.*

Grünbaum, Adolf, see Agassi and Grünbaum.

Habermas, Jürgen, (1999) 2003. *Truth and Justification.*

Hacohen, Malachi Haim, 1999. Dilemmas of Cosmopolitanism: Karl Popper, Jewish Identity, and "Central European Culture", *Journal for Modern History*, 71, 105-49.

Bibliography 363

——, 2000. *Popper: The Formative Years, 1902-1945.*

——, 2002. Critical Rationalism, Logical Positivism, and the Poststructuralist Conundrum: Reconstructing the Neurath-Popper Debate. In M. Heidelberger and F. Stadler, editors, *History and Philosophy of Science,* 307-24.

Hahn, Lewis Edwin, and Paul Arthur Schilpp, editors, 1986. *The Philosophy of W. V. Quine.*

Hajek, Petr, 1998. *Metamathematics of fuzzy logic.*

Hall, John A., 1999. The Sociological Defect of The Open Society analyzed and remedied. In Jarvie and Pralong, 1999, 83-96.

Harré, Rom, 1970. *The Principles of Scientific Thinking.*

——, 1986. Persons and Power. In Shanker, 1986, 135-153.

Harel, David, (1987) 2004. *Algorithmics: The Spirit of Computing.*

Hattiangadi, Jagdish N., 1985. The Realism of Popper and Russell. *Philosophy of the Social Sciences,* 15, 461-86.

Hay, S. I., M. John Packer, and D. J. Rogers. 1997. The impact of remote sensing on the study and control of invertebrate intermediate hosts and vectors for disease. *International Journal for Remote Sensing,* 18, 2899-2930.

Hayek, F. A. von, 1974. Wittgenstein. Letter to the editor of T. L. S., 8 Feb. 1974.

——, 1977. Remembering My Cousin, Ludwig Wittgenstein. *Encounter,* August 1977, 176-81.

Heal, Jane, 1995. Wittgenstein and Dialogue. In Timothy Smiley, editor, *Philosophical Dialogues: Plato, Hume, Wittgenstein,* 63-83.

Heaney, Barbara, 1990. The Assessment of Educational Outcomes. *ERIC Digest, ERIC Clearinghouse for Junior Colleges Los Angeles CA.*

Hebb, Donald O., 1958. Alice in Wonderland, or Psychology among the biological sciences. In H. F. Harlow and C. N. Woolsey, editors, *Biological and Biochemical Basis of Behavior.*

Hempel, Carl Gustav, 1950. Changes and Shifts in the Criteria of Meaning, *Rev. Intl. Phil.,* 11, 41-63.

——, 1951. The Concept of Cognitive Significance: A Reconsideration, *Proc. Am. Acad. Arts and Sciences,* 80, 61-77.

——, 1965. *Aspects of Scientific Explanation and Other Essays.*

——, 1965. Empiricist Criteria of Cognitive Significance: Problems and Changes, in his 1965.

——, 2000. Richard C. Jeffrey, editor, *Selected Philosophical essays.*

———, 2001. James H. Fetzer, editor, *The Philosophy of Carl G. Hempel: Studies in Science, Explanation, and Rationality.*
Hendrickson, Linnea, 1987. *Children's Literature: A Guide to the Criticism.*
Herder, Johann Gottfried (1773) 2004. *Another Philosophy of History and Selected Political Writings.* Translated, with introduction and notes by Ioannis D. Evrigenis and Daniel Pellerin.
Hintikka, Jaakko, 1975. Editor, *Rudolf Carnap Logical Empiricist: Materials* and *Perspectives.*
———, 1993. On Proper (Popper) and Improper Uses of Information Epistemology. *Theoria,* 59, 158-65.
———, 1996. *Ludwig Wittgenstein: Half-Truths and One-and-a-Half-Truths: Selected Papers,* vol. 1.
Hollinger, David A., 1996. *Science, Jews, and Secular Culture: Studies in Mid-Twentieth-Century American Intellectual History.*
Hook, Sidney, 1987. *Out of Step: An Unquiet Life in the 20th Century.*
Howard, Don (1984). Realism and Conventionalism in Einstein's Philosophy of Science: The Einstein-Schlick Correspondence. *Philosophia Naturalis,* 21: 618-629.
———, 1994. Einstein, Kant, and the origins of Logical Empiricism. In Salmon and Wolters, 1994, 45-106.
Howson, Colin, editor, 1976. *Method and Appraisal in the Physical Science.*
Husserl, Edmund, (1929) 1969. *Formal and Transcendental Logic.*
Hutchins, John, 2000. *Early Years of Machine Translation.*

Jaki, Stanley, 1978. *The Road to Science and the Ways to God.*
———, 1989. *God and the Cosmologists.*
James, William, 1909. *A Pluralist Universe*: Hibbert Lectures on "the Present Situation in Philosophy".
Janik, Alan, and Stephen Toulmin, 1973. *Wittgenstein's Vienna.*
Jarvie, Ian C., 1964. *The Revolution in Anthropology.*
———, 1982. Popper on the Difference between the Natural and the Social Sciences. In Levinson, 1982, 83-107.
———, 2001. *The republic of Science: The Emergence of Popper's Social Views of Science 1935-1945.*
———, 1998. Popper, Sir Karl. In *Routledge Encyclopedia of Philosophy,* vol. 7, 533-40.
———, 2003. Review of David Edmonds and John Eidinow, *Wittgenstein's Poker. Journal of the History of the Behavioral Sciences,* 39, 205-06.

Jarvie, Ian C., and Joseph Agassi, 1987. The Rationality of Magic. In Agassi and Jarvie, 1987, 363-83.
Jarvie, Ian C., and N. Laor, 2001. *Critical Rationalism Metaphysics and Science: essays in honor of Joseph Agassi*, Volumes 1 and 2.
Jarvie, Ian C., and Sandra Pralong, 1999. *Popper's Open Society After 50 Years.*
Jeffrey, Richard, editor, 2000. *Carl G. Hempel: Selected Philosophical essays.*
Jensen, Arthur, 1969. How Much Can We Boost I. Q. and Scholastic Achievement? *Harvard Educational Review.*
Joad, C. E. M., 1950. *A Critique of Logical Positivism.*
Jones, Bence, 1870. *The Life and Letters of Faraday.*
Jowett, Garth S., Ian C. Jarvie and Kathryn H. Fuller, 2007. *Children and the Movies: Media Influence and the Payne Fund Controversy.*
Julesz, Béla, 1960. Binocular Depth Perception of computer-generated patterns. *Bell System Technical Journal*, 39, 1125-62.
——, 1994. *Dialogues on Perception.*

Kamerschen, D. R., editor, 1967. *Readings in Microeconomics.*
Katz, Elihu and Paul F. Lazarsfeld, 1955. *Personal Influence.*
Kaufmann, Walter, 1951. The Hegel Myth and Its Method. Phil. Rev., 60, 459-86.
——, 1955. Review of Levinson, 1953. *Journal of Politics*, 17, 126-8.
——, 1959. *The Owl and the Nightingale.*
——, 1961. Review of John N. Findlay, *Hegel: A Re-examination. Mind*, 70, 264-269.
——, 1965. *Hegel: Reinterpretation, Texts and Commentary.*
——, 1973. *Without Guilt and justice: from Decidophobia to Autonomy.*
——, 1980. *The Discovery of the Mind.* Three volumes.
Kemeny, John, 1959. *A Philosopher Looks at Science.*
Kiesewetter, Hubert, 2001. *Karl Popper — Leben und Werk.* Interview mit Sir Ernst und Lady Gombrich über ihre Freundschaft mit Popper, p. 105 ff.
Klappholz, Kurt, and Joseph Agassi, 1959. Methodological Prescriptions in Economics, *Economica*, 26, 60-74; reprinted in Kamerschen, 1967.
Klett, Ronald, 1988. George Bernard Shaw's Letter to the Editor, May 1945. *The Journal of Historical Review*, 8, 509-511.
Knorr-Cetina, Karin D., 1981. *The Manufacture of Knowledge: An Essay on the Constructivist and Contextual Nature of Science.*

Koelbl, Herlinde, 1989. *Jüdische Portraits: Photographien und Interviews.*
Korczak, Janusz, 1923; English translation, 1986. *King Matt the First.*
Koyré, Alexandre, 1939. *Galilean Studies.*
———, 1968. *Metaphysics and Measurement.*
Kraft, Julius, 1932. Von Husserl zu Heidegger. Kritik der phänomenologischen Philosophie.
Kraft, Victor, (1950) 1953. *The Vienna Circle; the origin of neo-positivism, a chapter in the history of recent philosophy.*
Kraut, Richard, 1984. *Socrates and the State.*
———, 1985. Socrates and Democracy. In Currie and Musgrave, 1985, 185-204.
———, editor, 1992. *The Cambridge Companion to Plato.*
Kuhn, Thomas S., 1962. *The Structure of Scientific Revolutions.*
Kulka, Tomas. 1989. Art and science: An outline of a Popperian aesthetics. *British Journal of Aesthetics*, 29, 197- 212.
Kulick Don and Margaret Willson, 1995. *Sex, Identity and Erotic Subjectivity in Anthropological Fieldwork.*

Lakatos, Imre, 1976. *Proofs and Refutations.*
——— and Alan Musgrave, editors, 1968. *Problems in the Philosophy of Science, Proceedings of the International Colloquium in the Philosophy of Science London 1965*, vol. 3.
———, 1970. *Criticism and the Growth of Knowledge.*
Landé, Alfred, 1973. *Quantum Mechanics in a New Key.*
Lane, Homer, 1928. *Talks to Parents and Teachers.*
Langley, P., 1996. *Elements of Machine Learning.*
Langley, P., and H. A. Simon, 1995. Applications of machine learning and rule induction, *Communications of the Association for Computing Machinery*, 38, 55-64.
Langley, P., G. L. Bradshaw, and H. A. Simon, Rediscovering chemistry with the BACON system, Chapter 10 of Michalski et al., 1983.
Langley, Pat, Herbert A. Simon, Gary L. Bradshaw, and Jan M. Zytkow, 1987. *Scientific Discovery: Computational Explorations of the Creative Processes.*
Laor, Nathaniel and Joseph Agassi, 1990. *Diagnosis: Philosophical and Medical Perspectives.*
Latour, Bruno and Steve Woolgar, 1986. *Laboratory life: the construction of scientific facts*; introduction by Jonas Salk.

Latsis, Spiro, editor, 1976. *Method and Appraisal in Economics.*
Lehmann, Winfred P. 1982. Linguistics at Wisconsin (1937-41) and at Texas (1949-): A retrospective view. *Notes on Linguistics*, 23, 22-25.
Leblanc, Hughes, 1960. On requirements for conditional probability functions. *The Journal of Symbolic Logic*, 25, 238-242.
———, 1981. What Price Substitutivity? A Note on Probability Theory. Philosophy of Science, 48, 317-322.
———, 1989. The Autonomy of Probability Theory (Notes on Kolmogorov, Renyi, and Popper). *The British Journal for the Philosophy of Science*, 40, 167-181.
Leblanc, Hughes and Bas van Fraassen, 1979. On Carnap and Popper Probability Functions. *The Journal of Symbolic Logic*, 44, 369-373.
Levinson, Paul, editor, 1982. In *Pursuit of Truth: Essays in Honor of Karl Popper's 80th Birthday.*
Levinson, Ronald Bartlett, 1953, 1970, 1987. In *Defense of Plato.*
Lewis, Roger, 1996. *The Real Life of Laurence Olivier.*
Linsky, Leonard, 1967. *Referring.*
Long, Jancis, 1998. Lakatos in Hungary. *Philosophy of the Social Sciences*, 28, 244-311.
Lorenz, Konrad, 1992. Analogy as a Source of Knowledge, in Jan Lindsten, Editor, *Nobel Lectures, Physiology or Medicine 1971-1980.* .
Lukasiewicz, Jan, 1951. *Aristotle's Syllogistic from the Standpoint of Modern Formal Logic.*

Mach, Ernst, (1905) 1976. *Knowledge and Error.*
MacIntyre, Alasdair, editor, 1972. *Hegel: A Collection of Critical Essays.*
Magee, Bryan, 1997. *Confessions of a Philosopher.*
Malcolm, Norman, 1958, 1984. *Ludwig Wittgenstein, A Memoir* With a Biographical Text by Georg Henrik von Wright.
———, 1986. *Wittgenstein: Nothing is Hidden.*
Manuel, Frank, 1968. *A Portrait of Isaac Newton.*
Marchi, Neil De, 1988. The *Popperian legacy in Economics: Papers presented at a symposium in Amsterdam, December 1985.*
Maxwell, Nicholas, 2002. Popper. In P. Dematteis, P. Fosl and L. McHenry, editors, *British Philosophers 1800-2000, Dictionary of Literary Biography*, volume 262, pp. 176-194.
McCarthy, Timothy, and Sean C. Stidd, 2001. *Wittgenstein in America.*
McClure, Alexander K., 1904. *"Abe" Lincoln's Yarns and Stories.*

McGuinness, Brian, 1988. *Wittgenstein: A Life: Young Ludwig, 1889-1921.*
McVittie, George C., 1956. *General relativity and cosmology.*
Mead, Margaret, 1928. *Coming of Age in Samoa.*
Medawar, Peter, 1961. *The Future of Man.*
——, 1986. *Memoirs of a Thinking Radish: An Autobiography.*
Meital, Amir, and Joseph Agassi, 2007. Slaves in Plato's Laws. *Philosophy of the Social Sciences*, 37, 315-47.
Merriman, Brian, 1976. *The Man who Talked Babytalk.*
Merton, Robert K., 1968. The Matthew Effect in Science: The reward and communication systems of science are considered. *Science*, 159, 56-63, January 5, 1968.
——, 1972. Insiders and Outsiders: A Chapter in the Sociology of Knowledge, *Am. Journal of Sociology*, 78, 9-47.
——, 1976. The Ambivalence of Scientists, in R. S. Cohen, P. K. Feyerabend, and M. W. Wartofsky, 1976, 433-455.
Michalos, Alex C., editor, 2003. *The Best teacher I ever Had: Personal reports from Highly Productive Scholars.*
Michalski, R. S., J. G. Carbonell and T. M. Mitchell Editors, 1983. *Machine learning, an artificial intelligence approach.*
Michotte, Edmond, (1858, 1860) 1968. *Richard Wagner's Visit to Rossini and An Evening at Rossini's in Beau-Sejour.* Translated from the French and annotated by Herbert Weinstock.
Mill, John Stuart, 1843, 1987. *On Socialism.*
——, 1843. *A System of Logic.*
Miller, David, 1978. Review of Agassi, 1975. *Phil. QUARTERLY*, 28, 308-9.
——, 1997. Sir Karl Raimund Popper. *Biographical Memoirs of Fellows of The Royal Society*, 43, 367-409.
Mitchell, Basil and J. R. Lucas, 2003. *An Engagement with Plato's Republic: A Companion to the Republic.*
Momigliano, Arnaldo D., 1990. *The Classical Foundations of Modern Historiography*, with a foreword by Riccardo Di Donato.
——, 1994. *Studies on Modern Scholarship.*
Monk, Ray, 1996. *Bertrand Russell: The Spirit of Solitude.*
——, 2000. *Bertrand Russell 1921-70: The Ghost of madness.*
Morgenstern, Martin and Robert Zimmer (editors), 2005. *Karl R. Popper, Hans Albert, Briefwechsel* 1958–1994.

Motterlini, Matteo, editor, 1999. Imre Lakatos and Paul Feyerabend, *For and Against Method; Including Lakatos's Lectures on Scientific Method and the Lakatos-Feyerabend Correspondence*. Edited with an Introduction by Matteo Motterlini.
Moulinas, Carlos Ulises, 1991. Making Sense of Carnap's *Aufbau*, in Spohn, 1991, 263-86.
Munz, Peter, 2004. *Beyond Wittgenstein's Poker: New Light on Popper and Wittgenstein*.
Murray, Gilbert, 1925. *Five Stages of Greek religion*.
Musgrave, Alan, 1999. *Essays in Realism and Rationalism*.
Myers, Gerald E., 1986. *William James: His Life and Thought*.

Naess, Arne, 1968. *Four Modern Philosophers*.
Nagel, Ernest, and James Newman, 1958. *Gödel's Proof*.
Nails, Debra, 1995. *Agora, Academy, and the conduct of Philosophy*.
Nelson, Leonard, 1965. *Socratic Method and critical philosophy: Selected Essays*.
Neurath, Otto, 1973. *Empiricism and Sociology*.
——, 1983. *Philosophical Papers 1913-1946*.
Nicod, Jean (1924) 1970. *Geometry and Induction*, with an Introduction by Bertrand Russell.
Nietzsche, Friedrich, (1886) 1996. *Birth of Tragedy*.
——, (1889) 1969. *Ecce Homo*.
Nino, Carlos Santiago, 1996. *The Constitution of Deliberative Democracy*.
Notturno, Mark Amadeus, 2003. *On Popper*.

Oberdorfer, Don, 1995. *Princeton: The First 250 Years*.
O'Hear, Anthony, Editor, 2004. *Karl Popper: Critical Assessments of Leading Philosophers*.
Oldham, John M., and Lois B. Morris, 1995. *The New Personality Self-Portrait: Why You Think, Work, Love, and Act the Way You Do*.
Orwell, George, (1946) 1968. *Why I Write. In* his *Collected Essays, Journalism and Letters*, Vol. I, 1-7.

Pap, Arthur, 1957. Once more: colors and the synthetic *a priori*. *Phil. Rev.*, 66, 94-9.
Parrini, Paolo, 2002. *Popper e Carnap su marxismo e socialismo, Nuova Civiltà delle Macchine*, 20, 90-8.

Passmore, John, 1975. The Poverty of Historicism Revisited. *History and Theory*, 14, *Beiheft* 14: *Essays on Historicism*, 30-47.
Perls, Fritz, 1969. *Gestalt Therapy Verbatim.*
Pieper, Annemarie, 2003. Das Individuum, in *Handbuch philosophischer Grundbegriffe.*
Pinder, Craig C., and L. F. Moore, 1980. *Middle Range Theory and the Study of Organizations.*
Pitcher, George, 1964. *The Philosophy of Wittgenstein.*
Pitte, Frederick P. van der, 1971. *Kant as a Philosophical Anthropologist.*
Polanyi, Michael, 1958. *Personal Knowledge.*
———, 1969. *Knowing and Being.*
Pole, David, 1958. *The later philosophy of Wittgenstein.*
Popper, Karl R., 1925. Über die Stellung des Lehrers zu Schule und Schüler, *Schulreform*, 4, 204-8.
———, 1927. Zur Philosophie des Heimatgedankens, *Die Quelle*, 77, 899-908.
———, (1935) 1994. *Logik der Forschung.*
———, (1945) 1966. *The Open Society and Its Enemies.*
———, 1956. Adequacy and Consistency: A Second Reply to Bar-Hillel. *The British Journal for the Philosophy of Science*, 57, 249-56.
———, 1957. *The Poverty of Historicism.*
———, (1959) 1968. *The Logic of Scientific Discovery.*
———, 1962. Julius Kraft, *Ratio*, 4, 2-10.
———, 1963. *Conjectures and Refutations.*
———, 1964. The Demarcation between Science and Metaphysics, in Popper, 1963 and in Schilpp, 1964.
———, 1966. 'A Theorem on Truth-Content', in Feyerabend and Maxwell, 1966, 343-353.
———, 1969. The Moral Responsibility of the Scientist. *Encounter*, March, 52-57.
———, 1970. Normal Science and Its Dangers, in Lakatos and Musgrave, 51-8.
———, 1972, 1979. *Objective Knowledge: An Evolutionary Approach.*
———, 1973. Memories of Otto Neurath. In *Empiricism and sociology*, edited by Robert S. Cohen and Marie Neurath, 51-6.
———, 1974. *Unended Quest.*
———,1976. The Myth of the Framework. In Eugene Freeman, editor, *The Abdication of Philosophy: Philosophy and the Public Good. Essays in Honor of Paul Arthur Schilpp*, pp. 23-48.
———, 1979. *Die beiden Grundprobleme der Erkenntnistheorie.*

——, 1981. Joseph Henry Woodger, an obituary notice. In *The British Journal for the Philosophy of Science*, 32, 328-330.
——, 1982. *Quantum Theory and the Schism in Physics*, edited by W. W. Bartley, III.
——, 1983. *Realism and the Aim of Science*.
——, 1990. *A World of Propensities*.
——, 1994a. *The Myth of the Framework: In defence of Science and Rationality*. Edited by Mark Notturno.
——, 1994b. In *Search of a Better World: Lectures and Essays from Thirty years*.
——, 1997. *The Lesson of this Century; With Two Talks on Freedom and the Democratic state*. Karl Popper Interviewed by Giancarlo Bosetti.
——, 1998. *The World of Parmenides*. Edited by Arne F. Petersen.
——, 1999. *All Life is Problem Solving*.
Popper, Karl R., and Lorenz, Konrad, 1985. *Die Zukunft ist offen: Das Altenberger Gespräch*.
Potok, Chaim, 1967. *The Chosen*.
Preston, John M., 1997. *Feyerabend: Philosophy, Science and Society*.
Pryke, Jo, 1999. Wales and the Welsh in Gaskell's fiction: Sex, Sorrow and Sense. *The Gaskell Society Journal* Vol. 13.
Putnam, Hilary, 1961. Review of Nagel and Newman, 1958, *Phil. Sci.*, 28, 209-211.
——, 2000. To Think with Integrity. *Harvard Rev. Phil.*, Spring 2000, 4-13.
——, 2002. *The Collapse of the Fact/Value Dichotomy*.

Quine, W. van O., 1951. Two Dogmas of Empiricism. *Phil. Rev.*, 60, 20-43. Reprinted in Quine 1953.
——, 1953. *From a Logical Point of View: Nine Logico-Philosophical Essays*.
——, 1977. Review of Gareth Evans and others, Truth and Meaning: Essays in Semantics, *J, Phil.*, 74, 225-241.
——, 1986. Reply to Jules Vuillemin. In Hahn and Schilpp, 1986, 619-22.
——, 1987. *Quiddities: An Intermittently Philosophical Dictionary*.
——, 1988. A Comment on Agassi's Remarks, *Z. allg. Wiss.theorie*, 19, 117-18.
——, 1994. Comments on Tennant, In Tennant, 1994, 345-51.

Quine, W. van O. and Rudolf Carnap, 1990. *Dear Carnap, Dear Quine: The Quine-Carnap Correspondence and Related Work*, edited with Introduction by Richard Creath.

Quinton, Anthony, 1961. Review of Ernest Gellner, *Words and Things*, The British Journal for the Philosophy of Science, 11, 337–344.

Radice, Lisanne, 1984. *Beatrice and Sidney Webb.*

Radnitzky, Gerhard, 1982. *Das Verhältnis von individuellen Freiheitsrechten und Sozialrechten. Zeitgeist im Zeichen des Fetisch der Gleichheit*. In Bossle, L. and Radnitzky, G. editors, 1982, *Selbstgefährdung der offenen Gesellschaft*, 63-126.

———, 1995. Karl R. Popper.

Radnitzky, Gerard, and Gunnar Andersson, editors, 1978. *Progress and Rationality in Science.*

———, 1979. *The Structure and Development of Science.*

Reck, Erich H., 2004. From Frege and Russell to Carnap: Logic and Logicism in the 1920s. In Awodey and Klein, 2004.

Ree, Jonathan, 2007. Review of Mark T. Mitchell, Michael Polanyi, *T. L.S.*, 5444, 2007.

Reichenbach, Hans, 1944. *The Philosophical Foundations of Quantum Mechanics.*

Reid, Constance, 1970, *Hilbert.*

Robbins, Lionel, 1979. On Latsis's *Method and Appraisal in Economics*: A Review Essay, *Journal for Economic Literature*, 17, 996-1004.

Roberts, John M., and C. Ridgeway, 1969. Musical involvement and talking. *Anthropol. Linguist.*, 11, 224-46.

Robinson, Guy, 1998. *Philosophy and Mystification: A Reflection on Nonsense and Clarity.*

Robinson, Joan, 1937. *Essays in the Theory of Employment.*

———, 1960. *Collected Economic Papers*, vol. 2.

———, 1962. *Economic Philosophy.*

———, 1978. *Contributions to Modern Economics.*

Robinson, Richard, 1951. Dr. Popper's Defense of Democracy, *Phil.Rev.*, 60, 487-507. (Reprinted in his 1969 *Essays in Greek Philosophy*.).

———, 1954. Plato's Modern Enemies and the Theory of Natural Law. The Philosophical Review, 63,596-598.

Roeper Peter and Hugues Leblanc, 1999. *Probability Theory and Probability Semantics.*

Röseberg, Ulrich, 1998. The Rise of Scientific Philosophy Revisited. In Dionysios Anapolitanos, Aristides Baltas and Stavroula Tsinorena, editors, *Philosophy and the Many Faces of Science*, 196-205.

Rorty, Richard, 1985. Solidarity or Objectivity? In J. Rajchman and C. West, editors, *Post-Analytic philosophy*, 1985.

Roth, Leon, 1926. *Correspondence of René Descartes and Constantyn Huygens, 1635-1647.*

———, 1929, 1954. *Spinoza.*

———, 1999. *Is There a Jewish Philosophy? Rethinking Fundamentals.* With a Forward by Edward Ullendorf.

Russell, Bertrand, 1896. *German Social Democracy.*

———, 1897. *An Essay on the Foundations of Geometry.*

———, 1912. *The Problems of philosophy.*

———, 1917. *Mysticism and Logic.*

———, 1920. *The Practice and Theory of Bolshevism.*

———, 1922. Introduction to Wittgenstein, 1922.

———, 1927. *An Outline of Philosophy.*

———, 1930. *The Conquest of Happiness.*

———, 1934. *Freedom versus Organization: 1814-1914.*

———, 1938. *Power: A New Social Analysis.*

———, 1945. *A History of Western Philosophy.*

———, 1948. *Human Knowledge: Its Scope and Limits.*

———. 1950. *Unpopular Essays.*

———, 1950a. Logical Positivism. *Logic and Knowledge*, 1956, 367-82.

———, 1951. Ludwig Wittgenstein, *Mind*, 60, 978-9.

———, 1956. *Portraits from Memory.*

———, 1959. *My Philosophical Development.*

———, 1967, 1968, 1969. *The Autobiography of Bertrand Russell.* 3 vols.

———, 1985. *The Collected Papers of Bertrand Russell*, Volume 12: *Contemplation and Action, 1902-14.*

———, 1996. *The Collected Papers of Bertrand Russell*, Volume 10: *A Fresh look at Empiricism.*

———, 1997. *The Collected Papers of Bertrand Russell*, Volume 11: *Last Philosophical Testament, 1943-68.*

———, 1999. *Russell on Religion.*

Ryle, Gilbert, 1937. Taking Sides in Philosophy. In *Philosophy*, 12, 317-32. Reprinted in Ryle, 1990. ii. 153-69.

———, 1990. *Collected Papers.*

——, (1949) 2000. *The Concept of Mind.*
——, 1994. Comments: Reichenbach on Realism, in Salmon and Wolters, 1994, 139-46.
——, 1994b. Carnap, Hempel, and Reichenbach on Scientific realism, in Salmon and Wolters, 1994, 237-54.

Salmon, Wesley and Geron Wolters, 1994. *Logic, Language, and the Structure of Scientific Theories: Proceedings of the Carnap-Reichenbach Centennial, University of Constant, 21-24 May, 1991.*
Sambursky, Samuel, 1971. Kepler in Hegel's Eyes. In *Proc. Israel. Acad. Sci. Hum.*, 5, 92-104.
Santayana, George, 1968. *The German mind: a philosophical diagnosis.*
Santillana, Giorgio, editor, 1953. Galileo's *Dialogue.*
Sassower, Raphael, 1999. Pedagogy as Psychology: A View from Within, in Jarvie and Laor, 1999.
——, 2004. *Confronting Disaster: an Existential Approach to Technoscience.*
Sassower, Raphael and Joseph Agassi, 1994. Avoiding the posts: reply to Friedman. *Crit. Rev.*, 8, 95-111.
Schacter, Daniel L., 1997. *The Seven Sins of Memory.*

Shanker, S., and Shanker, V. A., 1986. *Ludwig Wittgenstein: critical assessments.*
Scharfstein, Ben-Ami, 1980. *The Philosophers. Their Loves and Their Thoughts.*
Scheffler, Israel, 1958. Inductive Inference: A New Approach, *Science*, 127, 177-81.
Schilpp, Paul Arthur, editor, 1949. *Albert Einstein Philosopher-Scientist.*
——, 1959-1960. The Abdication of Philosophy, *Kant-Studien*, 51, 480-95.
——, 1963. *The Philosophy of Rudolf Carnap.*
——, 1974. *The Philosophy of Karl Popper.*
Schilpp, Paul Arthur, and Maurice Friedman, editors, 1967. *The Philosophy of Martin Buber.*
Schlick, Moritz, 1979. *Philosophical Papers.* Vol. 1, 1909-1922; vol. 2, 1925-1936.
——, 1987. *The Problems of Philosophy in their Interconnection. Winter Semester Lectures, 1933-34.*
Scholem, Gershom, 1965. *On the Kabbalah and Its Symbolism.*

Schrödinger, Erwin, 1952. Are there Quantum Jumps? *The British Journal for the Philosophy of Science*, 3, 109-23,233-43. Now also in Schrödinger, 1956.

——, 1956. *What is Life and Other Essays.*

Schwartz, Robert, Israel A. Scheffler and Nelson Goodman, 1970. An Improvement on the Theory of Projectability. *Journal of Philosophy*, 67, 605-8.

Segre, Vitorio Dan, 1974. *The High Road and the Low: A Study in Legitimacy, Authority and Technical Aid.*

——, 2005. *Memoirs of a Failed Diplomat.*

Seiler, Martin, and Friedrich Stadler, editors, 1994. *Heinrich Gomperz, Karl Popper und die 'österreichische Philosophie'.*

Shae, William, 1982. The Young Hegel's Quest for a Philosophy of Science, or Pitting Kepler Against Newton, in J. Agassi and R. S. Cohen, 1982, 381-97.

Shahar E., 1997. Popperian perspective of the term 'evidence-based medicine'. *Journal of Evaluation in Clinical Practice*, 109-116.

Shanker, Stuart G., editor, 1986. *Philosophy in Britain Today.*

Shaw, Bernard, 1965. *Prefaces.*

Shortt, E. H., 1967. Saul Adler, 1895-1966, *Biographical Memoirs of Fellows of the Royal Society*, 13, 1-34.

Simey, T. S. and M. B., 1960. *Charles Booth: Social Scientist.*

Simkin, Colin, 1993. *Popper's Views on Natural and Social Science.*

Sluga, Hans, (1993) 1995. *Heidegger's Crisis: Philosophy and Politics in Nazi Germany.*

Smiley, Tim, 1998. Letter to the editor of the *T.L.S.*, March 13, 1998.

——, 2004. Popper and the Poker, Philosophy at Cambridge, *Newsletter of the Faculty of Philosophy*, May 2004, page 7.

Soames, Scott, 2003. *Philosophical Analysis in the Twentieth Century.* Vol. 1, *The Dawn of Analysis*; Vol. 2, *The Age of Meaning.*

Solomon, Maynard, 1977. *Beethoven.*

——, 1988. *Beethoven Essays.*

Soros, George, 2002. *George Soros on Globalization.*

Spohn, Wolfgang, 1986. The Representation of Popper Measures. *Topoi*, 5, 69-74.

——, 1991. Editor, *Erkenntnis Orientated: A Centennial Volume for Rudolf Carnap and Hans Reichenbach.*

SPSS, Inc. 1986. *Statistical Package for the Social Sciences User's Guide.*

Stadler, Friedrich, 2004. *The Vienna Circle: Studies in the Origins, Development, and Influence of Logical Empiricism.*
Stadler, Friedrich, editor, 2003. *The Vienna Circle and Logical Empiricism: Re-evaluation and Future Perspectives.*
Steiner, George, 2003. *Lessons of the Masters.*
Stern, David, 2007 Wittgenstein, the Vienna Circle, and Physicalism: A Reassessment. In Alan Richardson and Thomas Uebel, *The Cambridge Companion to Logical Empiricism*, 305-321.
Stopes-Roe, Harry V., 1968. Review of the Carnap-Popper Controversy, *Journal of Symbolic Logic*, 33, 142-6.
Stove, David, 1982. *Popper and After: Four Modern Irrationalists.*
Strauss, Leo, 1941. Review of Crossman, Plato Today, *Soc. Research*, 8, 250-1.
———, 1997. *Spinoza's Critique of Religion as the Foundation of his Science of the Bible, Investigations into Spinoza's Theologico-Political Treatise.* (Also in his *Ges. Schrift.* Vol. 1, 1996.).
Strawson, Peter F., 1971. *Logico-Linguistic Papers.*

Tennant, Neil, 1994. Carnap and Quine. In Salmon and Wolters, 1994, 303-344.
Toulmin, Stephen, 1958. *The Uses of Argument.*
Toynbee, Arnold, 1934-61. *A Study of History.*
Tribe, Keith, 1997. *Economic Careers: Economics and Economists in Britain 1930-1970.*

Velody, Irving, Herminio Martins, and Peter Lassman, editors, 1988. *Max Weber's 'Science as a Vocation'.*
Voegelin, Eric, 2001. *The Collected Works*, Volume 13.

Wagner, Richard, 1850. *Das Judentum in der Musik.*
Warnock, Geoffrey J., 1958, 1969. *English Philosophy Since 1900.*
Watkins, John, 1957-8. Epistemology and Politics, *Proceeding of the Aristotelian Society*, 58, 79-102; Agassi and Jarvie, 1987, 151-67.
———, 1958. Confirmable and Influential Metaphysics, *Mind*, 67, 344-65.
———, 1975. Metaphysics and the Advancement of Science, *The British Journal for the Philosophy of Science*, 26, 91-121.
———, 1977. My LSE. In J. Abse, Editor, *My LSE*, 64-82.

——, 1985. On Stove's book, by a fifth 'irrationalist'. *Australasian Journal for Philosophy*, 63, 259 – 268.
——, 1995. Epiphenomenalism and Human Freedom, in Jarvie and Laor, 1995, 33-9.
——, 1997. Karl Raimund Popper, 1902-1994. Proc. Brit. Acad., 94, 645-84.
——, 1997a. Karl Popper: a memoir, *Am. Scholar*, 66, 205-219.
Webb, Beatrice, 1926. *My Apprenticeship*.
Weiler, Gershon, 1988. *Jewish Theocracy*.
Weinberg, Steven, 1992. *Dreams of a Final Theory*.
Weldon. T. D., 1953. *The Vocabulary of Politics*.
Wettersten, John, 1985. Russell and Rationality Today, *Methodology and Science*, 18, 140-163.
——, 1987. On Education and Education for Autonomy, *Interchange*, 18, 4, 21-26.
——, 1987. Selz, Popper, Agassi, On the Unification of Psychology, Methodology and Pedagogy. In*terchange*, 18, 1-13.
——, 1987b. On Education and Education for Autonomy, *Interchange*, 18, 4, 21-26.
——, 1992. *The Roots of Critical Rationalism*.
——, 1999. A Rationalist Ethic for Today: Popper's Theory of Rationality Combined with Russell's Moral Practice. In Zecha, 1999, 99-115.
——, 2005. *Whewell's Critics: Have They Prevented him from Doing Good?*.
——, 2006. *How Do Institutions Steer Events? An Inquiry into the Limits and Possibilities of Rational Thought and Action*.
Wettersten, John and Joseph Agassi, 1991. Whewell's Problematical Heritage. In M. Fisch and S. Schaffer, *William Whewell: A Composite Portrait*, 345-69.
Weyl, Hermann, 1931. *The Theory of Groups and Quantum Mechanics*.
Wheelis, Allen, 1958. *The Quest for Identity*.
Whewell, William, 1847. *History of the Inductive Sciences*.
——, 1860. *Philosophy of Discovery*.
Wills, W. D., 1964. *Homer Lane: a Biography*.
Wilson, Fred Forster Jr., 1995. The Dialectic of the Master-Student Relationship: Critical Notice on J. Agassi's *A Philosopher's Apprentice: In Karl Popper's Workshop, Interchange*, 26 193-203.
Wisdom, John O., 1975. *Philosophy and Its Place in Our Culture*.
——, 1984. What is Left of Psychoanalytic Theory? *Intl. Rev. Psycho-Analysis*, 12, 73-85.

Wittgenstein, Ludwig, 1922. *Tractatus Logico-Philosophicus.*
——, 1929. Some Remarks on Logical Form, *Proceedings of the Aristotelian Society*, Suppl. vol. 9, 162-171.
——, 1953. *Philosophical Investigations.*
——, 1956. *Remarks on the Foundations of Mathematics.*
Woffinden, Bob, 1987. *Miscarriages of Justice.*
Wolniewicz, Boguslaw, 1969. A parallelism between Wittgensteinian and Aristotelian ontologies. In R. S. Cohen and M. Wartofsky, editors, *Boston studies in the philosophy of science*, Vol. IV, 208-217.
Wood, Alan, 1957. *Bertrand Russell: The Passionate Sceptic.*
——, 1959. Russell's Philosophy: A Study in Its Development. In Russell, 1959.
Worral J. and E. G. Zahar, 1976. *Imre Lakatos: Proofs and Refutations.*
Wrigley, Michael, 1977. Wittgenstein's Philosophy of Mathematics. *The Philosophical Quarterly*, 27, 50-59.
Wyman, David, 1985. *The Abandonment of the Jews: America and the Holocaust 1941-1945.*

Zecha, Gerhard, editor, 1999. *Critical Rationalism and Educational Discourse.*
Ziman, John, 1959. Review of Popper, 1959, *New Statesman.*
——, 2000. *Real Science: What It Is, and What It Means.*
Zuckerman, Solly, 1978. *From Apes to Warlords.*
——, 1989. *Monkeys Men and Missiles, an Autobiography, 1946-88.*
Zweig, Stefan, 1942, 1943. *The World of Yesterday.*

GLOSSARY

Ad hoc: the slightest modification of views under the pressure of criticism, meant to save them from criticism; changes made minimally and reluctantly.

Analysis. Literally dissection, taking apart. The opposite of synthesis. Philosophical analysis is the spelling out of ideas in detail.

Analytic philosophy, a school of philosophy that preaches Anti-philosophy. It thus seemingly says nothing about society. In truth it is a reactionary social philosophy thinly disguised as radical to hide its élitist contempt for common people.

Anti-philosophy: the view that philosophy has nothing to say and that philosophers should only perform conceptual Analysis; the view that metaphysical utterances are meaningless; Logical Positivism.

Apriorism, see Intellectualism.

Autonomy (moral and/or intellectual): literally, self-rule; independence.

Collectivism: the theory that societies shape individuals and their aims; the theory that social ends precede individual aims. See Sociologism.

Conservatism: the most widespread theory; it says, to fare well society should follow and guard its own traditions.

Convention: agreement, socially endorsed opinions and rules, arbitrariness.

Conventionalism: the demand that laws should be obeyed though they are conventional, *i.e.*, enacted by some arbitrary agreement; the theory that science is binding only as a convention.

Criterion: a rule for judging.

Critical attitude: the readiness and the wish to engage in criticism. See Criticism.

Critical rationalism: the view that rational conduct involves trial and error, conjectures and criticism.

Criticism: attempting to disprove, refute, locate any error.

Deductivism: the view of science as based on axioms; Intellectualism.

Definition: see "Nominal definition" and "Essential definition".

Dialectical method: the rules of critical debate, of cross-examination, of interrogation in the attempt to refute — a testimony, a theory, a proposal.

Dialectics: critical debate; the rules of critical debate.

Egalitarianism: the theory that we all are (or should be) equal (before the law).

Empiricism: the theory that knowledge is wholly based on observations; the theory that only observations justify theories.

Enlightenment: understanding; the theory that science will save humanity; an intellectual movement ruling the intellectuals in the seventeenth and eighteenth centuries.

Essence: the quality that makes a thing what it is; the opposite of accident, which is inessential.

Essential definition: the description of an essence. See Essence.

Ethics: the rules of proper conduct; the theory of such rules; their justification.

Existentialism. The philosophy that says that life is purposeless and that the realization of this fact is the only salvation of the individual. It thus seemingly says nothing about society. In truth it is a reactionary social philosophy thinly disguised as progressive and scarcely hides its contempt for common people.

Fallibilism: the view that no human endeavor is free of error.

Historicism: the doctrine of historical inevitability or of destiny.

Humanism: faith in humanity; the theory that all humans possess dignity.

Individualism: the theory that only individuals (and not collectives or societies or institutions) are real; that only have purposes. See Psychologism.

Induction: the process by which scientific theories (allegedly) evolved from available scientific factual information; the process by which scientific theories (allegedly) gain scientific justification from available factual information.

Inductivism: the view of empirical science as resting inductively on factual information that justifies it.

Instrumentalism: the theory that (not resting on information) scientific theories are mere instruments for classification of past observations that miraculously help predict future observations.

Intellectual frameworks: metaphysical systems, presuppositions, taken upon faith according to fideists and tentatively according to critical rationalists.

Intellectualism: *a priorism*: the theory that scientific knowledge rests on axioms whose truth is attained by intuition; justification by intuition.

Logical Positivism: The justification of Positivism by the (absurd) claim that metaphysical utterances are inherently meaningless; the conceit that this claim follows from an existing, satisfactory theory of meaning; the conceit that this claim comprises a non-controversial demarcation between the utterances of science and of metaphysics.

Marxism: the theory (originated by Marx) that (the essence of) history is the history of class struggle. A school of thought that assumes a monopoly over radicalism and progressivism and condemn parliamentary democracy sham since in their view only a socialist revolution can bring about social justice.

Metaphysics: the first principles of physics; the metaphysics of any science is the foundations or the intellectual framework of that science. See Intellectual frameworks.

Method: literally, way; scientific, metaphysical or any other method is the (allegedly) proper way to do science, philosophy or anything else.

Methodology: the theory of (proper) method. See Method.

Nationalism: the theory that the proper citizenry of the state comprise a (coherent) nation and the state is its instrument.

Naturalism: the theory that there is nothing supernatural.

Nature: reality; the real; the opposite of appearances.

Nominal definition: any formula that merely replaces a long word (set of words) with a short one.

Nominalism: the theory that all names are proper names; that only individual objects have names; that a general name does not name a generality (as this does not exist); the theory that a general name names the individual objects that share it.

Orthodoxy: literally, right opinion; established or received doctrine.

Paradigm, bi-paradigm and multi-paradigm: literally, chief example, two of them or many of them; a fancy and confusing name for intellectual framework(s); a fancy and confusing name for some given individual scientific theory that is a chief example of some given intellectual framework.

Phenomena: literally, appearances; experience.

Phenomenalism: the theory that only appearances exist; that an experience is not of any object.

Pluralism: the preference for a diversity over unanimity — of opinions, of denominations, of political ideologies, of scientific theories.

Positivism: literally, matter-of-fact-ness; the theory that philosophy has nothing to say. See also anti-philosophy; Logical Positivism.

Presocratics: literally, (those who were active) before Socrates; the earliest Greek thinkers, mostly physicists and sophists.

Progressivism: the theory that (human) history is governed by a law of progress.

Proof: decisive justification; the act of making everyone see the truth of a proposition; showing that a proposition is unquestionably true.

Protestant ethic: see Puritanism.

Psychologism: a form of Reductionism, the theory that since only individuals really exist all social science is actually psychology; the theory that backs individualism and was once progressive.

Puritanism: (slang) the moral view that work is of supreme value; that having fun is a waste of time and so it is evil.

Radicalism: the demand to reject all tradition and all arbitrariness; the theory that unless one starts afresh utterly and thoroughly, all of one's effort is in vain.

Rational belief: (supposedly) scientific belief; (supposedly) belief based on proof or on evidence; obligatory belief; incontestable belief.

Rational disagreement: disagreement where it is not obvious which of the different sides is mistaken, where the different sides have sound arguments for their own views and/or against opposite views.

Rationalism: the theory that one should use one's own brain to select one's lifestyle. Classical rationalism is radical; modern rationalism need not be; critical rationalism is not.

Rationality: the disposition to think and act rationally, *i.e.*, critically, *i.e.*, while arguing, presumably thereby justifying one's views and actions.

Reaction: literally opposite action; backlash; the traditionalist (backlash) response to the radicals.

Realism: the theory that reality exists, that our experience emanates from objects, that there is something other than (behind) the appearances.

Reality: that which is behind the appearances, the solid behind the transient, the substance. See Substance.

Reductionism: the idea that one science is really another, for example, the idea that psychology is really physics because we are computers of sorts. In the social sciences there are two traditional versions of reductionism: the psychologism of the Enlightenment and the sociologism of the reaction. See Psychologism; Sociologism.

Reformism: the theory that the best practical approach (to society, politics, legislation, to science — perhaps to all other matters) is (not to endorse and not to reject but) to try to improve it.

Relativism: the opposite of absolutism, the idea that the truth differs in different times and places or that the true ethic is, etc.; relativism is the denial of absolutism; though absolutists need not reject the relative, relativists are those who reject the absolute.

Religion (revealed and rational): doctrine plus ritual plus tradition plus the community of the faithful and its organizations.

Romanticism: the hankering after the past.

Scientism: the idea that only science counts intellectually, and only the methods of the (natural) sciences leads to valid results. See Positivism; Radicalism.

Skepticism: the demand to suspend judgment on every issue, since proof is impossible plus the view that the suspension of judgment brings peace of mind (*ataraxia*).

Socialism: the idea that justice is impossible to reach unless ownership of the means of productions is centralized; the proposal to abolish (private) ownership of capital.

Sociologism: the theory that only societies (but not individuals) really exist and all social science is actually sociology. See Reductionism.

Socratic doubt: the readiness to examine every view however obvious and/or well established it may be.

Substance: the real; that which is utterly independent of everything else; the simplest and the unanalyzable. See Reality.

Synthetic proposition: a proposition which is not analytic; an informative proposition.

Thomism: the philosophy of St. Thomas of Aquinas; refined Aristotelianism.

Traditionalism: the theory that the best way to run a society is to stick to tradition as much as possible and to preserve it as the best possible.

Ultimate values: the (presumed) axioms of an ethical system.

Universalism: the rejection of everything parochial especially revealed religion, of nationalism, of anything not shared by all humanity.

Utilitarianism: the idea that the highest good is the greatest happiness for the greatest number.

Utopianism: the search for an ideal state.

Values: central norms or criteria by which to determine norms; whatever makes life worth living.

Work ethic: moral views concerning work in general; puritanism; the reference to the work ethic, unspecified, as if there were only one, is to the work ethic that stresses the supreme value of work.

Zionism: the apex of the Jewish national movement that aspired for a normal nation-state, organized mass migration of Jews to Palestine, founded Israel, yet sadly refused to disband.

INDEX OF NAMES

Achilles.....9
Acton, John Dalberg.....165,220
Adams, Walter.....103
Adler, Alfred.....17,74,295,301,308
Adler, Saul.....61-2,353,374
Adler, Sophie.....61
Agassi, Aaron..... 119
Agassi, Aaron..... 5
Agassi, Judah.....5
Agassi, Judith Buber, see Buber
Agassi, Tirzah.....8,310
Agassiz, Louis.....201,358
Ajzenstat, Oona.....67,355
Albert, Hans.....16,118,238,244,356,369
Alexander, Peter.....130,355
Alexander, Samuel.....67
Amsterdamski, Stefan.....355,362
Anapolitanos, Dionysios.....372
Anaximander.....347
Anderson, Michael.....198
Andersson, Gunnar.....39,372
Androcles.....273
Annan, Noel.....153,356
Anscombe, G. E. M.....109,171,319,345
Aquinas, St. Thomas.....384
Archimedes.....250
Aristotle.....32,63,93-4,104,148-9,164,178, 187,221,232,250,289,295,301,309,319-20, 325,343,345,361,377,384
Aristoxenus.....343
Asimov, Isaac.....36
Atwood, George.....151
Aubrey, John.....76
Augustine, St.....172,359
Austen, Jane.....74,197
Austin, John.....44,146,267,356
Awodey, Steve.....21,330,356,372
Ayer, Sir Alfred..... 56,65,148-9,158,185,218, 316,318-19,331-2,337,356

Bacon, Francis.... 19,32,34,48-9,79,814,87, 115,139,147,185,194,223,237,250,262, 271,274,277,280,290-1,296,301,316, 319-20,366
Baltas, Aristides.....372
Bambrough, Renford.....223,356
Banton, Michael.....258,356
Bar-Am, Nimrod.....16,94,325,356

Bar-Hillel, Yehoshua..... 21-2,49,53-4, 71,93,116-17,150-2,188,304,326,332-3, 356,361,370
Bartlett, Josiah.....218
Bartley, William Warren III....22, 24,34,37-9, 51,67,77,105,115,132,151,162,164-6,189, 199-200,202,214,226,228-9,241,243,247, 269,291,295,298,308,356,370
Baum, Wilhelm.....200,356
Baumol, William.....165
Baumgardt, Carola.....87,356
Beethoven.....93,132,179,375
Bellarmino, St. Roberto.....29,157
Belvedere, Lynn.....221
Bence Jones, Henry.....29,365
Bergman, Hugo.....66,140-1,195,354
Bergson, Henri.....43,167-8,260
Berkeley, George..34,215,307,338
Berkson, William.....51,356
Berlin, Isaiah.....152,155-7,168,223,293,356
Bernard, Harvey Russell.....145,357
Bernays, Paul.....170,187,357
Berry, Donald L.......111,357
Bion, Wilfred.....201
Bischof, Norbert.....169,357
Black, Max.....101,217,357
Blaug, Mark.....143,357
Blaukopf, Kurt.....171,357
Bloom, Harold.....74,357
Boas, George.....89,357
Bohm, David.....45,128
Bohr, Niels.....62,79,107,159,242,278,328
Bok, Sissela.....65,357
Bonaparte, Napoleon.....234
Bonk, Thomas.....323,357
Boole, George.....93-4,247,319
Booth, Charles.....81,375
Borges, Jorge Luis.....118,311-12
Born, Hedwig..... 357
Born, Max.....75,83,92,318,357
Bosetti, Giancarlo.....357,371
Bossle, Lothar......372
Boswell, James.....41
Bouwsma, O. K......181,357
Boyle, Robert.....19,52,76,115,167,200,254, 262,273,353
Bradshaw, Gary L.......82,366
Brahe, Tycho.....82
Braithwaite, Richard Bevan....154,337
Brewster, David.....30

Bridgman, Percy W......50,287,357
Broad, William.......37,357
Bromberger, Sylvain.....18,357
Bronowski, Jacob.....248,358
Bruckner, Anton.....305
Bruno, Giordano.....29
Buber Agassi, Judith.....16,23,249,255-6, 358
Buber, Martin.....36-7,53,111,234,247,259, 319,357,375
Buddha, Gautama, the.....230
Bunge, Mario.....50,83,92,112,190,254,290, 297,354-5,357-8
Buonamico, Francisco.....250
Burke, Edmund.....234
Burnet, John.....149
Burtt, E. A......75,321,358
Butler, Samuel.....51,197,231

Calaprice, Alice.....30,358
Campbell, Ian.....201,358
Campbell, Norman.....14,358
Cantor, Georg.....235
Caplow, Theodore.....160,358
Caravaggio.....189
Carbonell, Jaime G......368
Carnap, Rudolf..... 7,17,21-2,36,48-9,53-6,73, 116-17,130,136-9,141,150-2,157-8,171, 184-5,187,189-200,204,211,216-18,225, 228-9,267,293,300,303-4,315-16,322-3, 326-30,332-3,335-6,355-8,362,364,367, 369-70,372,374-6
Carroll, Lewis.....98,241
Cartwright, Nancy...34,185,316,320,358
Cassirer, Ernst.....355,361
Cavell, Stanley.....177
Chaplin, Charlie.....175
Chiariello, Michael.....11,358
Chomsky, Noam.....203,214,283
Christ, Jesus.....111,357
Churchill, Winston.....249,253
Cobban, Alfred..... 222,234,358
Coffa, Alberto..... 316,318-19,329, 359
Cohen, Hermann.....51
Cohen, Paul..... 39
Cohen, Percy..... 129,359
Cohen, Robert S...... 51,353-6,359, 368. 371,375,377
Collingwood, R. G.....75,84,191,233
Colodny, Robert G.....84,359
Columbus, Cristopher......34
Comte, Auguste.....152

Conant, James.....185
Copernicus, Nicolaus.....19,32,60,157,162, 250,311
Cormier, Robert.....74
Cornelius Nepos.....73
Cornelius, Henry.....284
Coulomb, Charles-Augustin.....115
Creath, Richard.....371
Cressey, Paul G......74
Crick, Francis.....75,359
Crito.....343
Croesus.....224
Crombie, Alistair C......32,131,359
Crossman, Richard...221,261,359,376
Currie, Gregory.....222,341,359,366

Daniel..... 238
Darwin,Charles.....14,18,79,100,213,234,255, 294,301
Daube,David.....353
David..... 232
Davidson, Donald.....271
Davies, Phillip..... 39,359
Davies, Robertson William..... 202,359
Davis, Bette......104
Davis, Louis E......104
Davy, Humphry.....29,107,163
Dawidowicz, Lucy S......234-5,359
Dematteis, P.......367
Democritus.....32,246,345,347
Descartes, René.....32,46,56,66,73,76,242, 250,262,274,320,346,372
Dingle, Herbert.....64,140
Dirac, P. A. M.....66,359
Donaldson, Stephen.....145,359
Donato, Riccardo Di.....368
Donovan, Charles.....61
Doolittle, Dr. John.....241
Dorn, Georg.....324,354
Dostoevsky, Fyodor.....41,304
Dover, Kenneth.....228
Drake, Stillman.....32,359-60
Droysen, Johann Gustav......345
Duerr, Hans Peter.....16,21,109-10,118,120, 200,241,359
Duhem, Pierre.....32,45-6,52,130,138,140, 151,295-6,300,323,355
Dukas, Helen.....30,33,359-60
Dummett, Michael.....170
Durham, Frank.....354
Dyke, Jerry van.....53

Dynes, Wayne.....145,359
Eccles, John.....30,92,134,359
Eddington, Arthur Stanley.....44,48,139,334-5
Eden, Martin..... 104
Edmond, Gary.....291,359
Edmonds, David.....153,185,298,303,305,359, 364
Edwards, Paul.....355
Edwards, Ronald.....203
Edwin, Lewis.....335,363
Egidi, Rosaria.....53,325,359
Eidinow, John..153,185,298,303,305,364
Eidlin, Fred.....282,359
Einstein,Albert....9,11,17,19,22,29-33,43,45, 49-52,58-9,62,75,81-3,100,115,117,143, 147,150-1,159,1623,175,188,192,202,204, 213,230,232,240,248,253,255,264,278, 287,303,307,312-13,317-18,324,329,331, 334-5,354,356-61,364,374
Eisenhower,Dwight.....265
Eisenstein,Sergei Mikhailovich..106
Eliot, George.....198
Eliot, T. S.......245
Ellis, Robert Leslie..... 81
Engelmann, Paul.....319,360
Engels, Friedrich.....44
Ettinger, Shmuel.....45-7,67,189,193,360
Euclid.....322
Euler, Leonhard.....46
Evans, Timothy.....91
Evans-Pritchard, Edward.....19,145,346,371
Ezioni, Amitai.....259

Fadiman, Clifton.....253,360
Farabi, Abū Nasr Muhammad ibn al-Farakh al-.....221
Faraday, Michael.....11,29,45,52,61,76-7,80, 113-15,132,148,159-60,225,248,250-1, 280,353,362,365
Farrell, Robert P......37,360
Feferman, Solomon.....170,360
Feigl, Herbert....17,136,140,225,240,271,358, 361
Festinger, Leon.....194,360
Fetzer, James H......84,364
Feuer, Lewis.....80,360

Feyerabend, Paul K....14,16,21-2,32,34, 37,94,101,108-10,118,128,136, 141,149,157,162,170,176,187,191,200, 203,209,245-6,258,267,269,275,280, 283-4,294,299,308,310-11,324-5,334, 356,359-61,368,370-1
Findlay, John.....233,235-6,354,361,365
Finn, Bernard S...... 250,361
Finn, Huckleberry......74
Fisch, Menachem.....118,361,377
Fischer, Kurt Rudolf.....333
Fitzgerald, F. Scott.....125
Fleck, Lola.....358
Flexner, Abraham.....95
Floyd, Juliet.....139
Flugel, John Carl.....44,361
Flusser, David.....240,361
Foldes, Lucien.....16
Forbes, J. D. 358
Forster, E. M......253,361,377
Fosl, P......367
Fösling, Albrecht.....30,361
Fraassen, Bas van...130,313,361,367
Fraenkel, Adolf.....47,53,57-8,63,93-4,195-6, 251,361
Frank, Philipp.....34,49,134,361
Franklin, Benjamin.....39,79,361
Frazer, James.....19,238-9,273
Freedman, Maurice.....130
Freeman, Eugene.....370
Frege, Gottlob.....90,122,200,296,321-2,346, 372
Freud, Sigmund.....18,43,74,83,96,100, 132,171-4,178,180,188,205,237,246, 272,275,297,307,300.307.311-12,361
Frey, Gerhard..... 318,361
Freyne, Sean.....59,361
Fried, Yehuda..86,100,113, 174,361
Friedman, Alexander..... 45
Friedman, Jeffrey.....374
Friedman, Maurice..... 204, 375
Friedman, Michael.....117,185,329-30, 355,361
Fries, Jakob Friedrich.....233,309
Fulbright, J. William.....164
Fuller, Kathryn H...... 365

Gadamer, Hans-Georg.....270
Gadol, Eugene.....361-2

Galilei, Galileo..... 19,29-30,32,76,91,130, 140,143,157,161-2,240,250,255,262,274, 291,296,311,359,374
Garbus, Morton.....69
Gaskell, Elizabeth.....172,198,371
Gattegno, Caleb.....251,362
Gattei, Stefano.....16
Gautama the Buddha.....230
Gavroglu, Kostas.....354
Gellner, Ernest..... 109,129-30,141,145,148, 153,160,177,185,193,233,239,293-4,303, 305-6,346,362,372
Gentzen, Gerhard.....94,187
Ghiberti, Lorenzo.....37
Gilbert, William.....250
Golde, Peggy.....145,362
Gombrich, Ernst..... 7,12-14,24-8,37,67,69, 120-1,189,223,229,231,252,285,293-4, 362,365
Gombrich, Ilse.....365
Gombrich, Richard.....229
Gomperz, Heinrich.....340,375
Goodman, Edward.....165,362,389
Goodman, Nelson.....117,165,180-1,183, 362,374
Gopnik, Adam.....172,362
Grant, G. P......223,362
Grattan-Gueness, Ivor......153-4,362
Greenberg, Daniel.....144
Gregory, Dick......282
Gregory, Richard......153,362
Grey Walter, William.....67,186
Gross, Berl......245,362
Grote, George.....221
Grünbaum, Adolf.....12,81-3,84,237, 271,313,355,362
Guthrie,William K. C......347
Gutman,Yisrael.....360

Habermas, Jürgen.....234,253,362
Hacohen, Malachi..11,16-17,21,132,216,316,331,357,362
Hahn, Hans.....317
Hahn, Lewis Edwin..... 363,371
Hajek, Petr.....236,363
Hall, John A......282,363
Halley, Edmund.....29
Hanson, Norwood Russell..140,291
Harel, David.....139,363
Harlow, H. F......363
Harré, Rom......146,363

Hartshorne, Charles.....66
Harvey, William.....250
Hattiangadi, Jagdish.....155,203,363
Hay, S. I.....61,363
Hayek,Friedrich A. von.....256-8,264,270,363
Hayes,Patrick J......83,361
Heal, Jane.....122,363
Heaney, Barbara.....97,363
Hebb, Donald O......92,363
Hegel, Georg Wilhelm Friedrich...51,156, 220,226,232-5,237-8,253,288,353,356, 365,367,374-5
Heidegger, Martin.....29,133,177-8,193-4, 220,355,361,366,375
Heine, Heinrich.....11,30,60,110,166,169,194, 233,240
Heisenberg, Werner.....175,194,298
Hempel, Carl Gustav..... 12,22,34,54, 56,84,136-9,212-13,218,271,293,302, 315,317,320,328,330-1,363,365,373
Henderson, Archibald.....202
Hendrickson, Linnea.....74,364
Heraclitus.....345,347
Herbst, David.....104
Herder, Johann Gottfried von.....364
Herodotus.....224
Herschel, John.....111,274
Hersh, Reuben.....39,359
Hesse,Hermann.....179
Hibbert, Robert.....364
Higgins, Henry.....198
Hilbert, David..306,322,331,336,372
Hill, Oliver..... 76
Hillel the Elder..... 65
Hintikka, Jaakko.....12,130,157,218,306,313, 325,364
Hippel, Paul T......359
Hippocrates.....279
Hobbes, Thomas.....222
Hoffmann, Banesh.....30,33,359-60
Hollinger, David A......55,364
Holloway, Stanley.....284
Holyoake, George Jacob.....81
Homer..... 240,307
Hook, Sidney.....43,364
Horace.....41,247
Howard, Don A......204,213, 329,364
Howard, Leslie.....198
Howson, Colin.....39,364

Index of Names

Hume, David.....12,34,151,212,222,232,276, 295,302,363
Husserl, Edmund.....205,364,366
Hutchins, John.....54,364
Hutchinson, David.....201,358
Huxley, Aldous.....175
Huygens, Constantyn.....66,373

Iago.....105
Isaiah..... 340
Ivan the Terrible.....106
Ivory, James.....74

Jackson, B.....353
Jaki, Stanley.....57,364
Jakobson, Roman.....203,297
James, William..... 124,233,369
Janik, Allan.....327,364
Jarvie, Ian Charles.......6,16,22-3,56,67,82, 89,92-3,107,129-30,145,160,166,170,189, 235,239,246-7,255-7,280,306,308,355, 359,363-5,374,376
Jarvis Thompson, Judith.....12
Jeffrey, Richard..... 218,331,365
Jensen, Arthur.....36,365
Jhabvala, Ruth Prawer.....74,201
Joad, C. E. M....158,290,315,365
Johnson, Samuel.....124
Jowett, Garth S......74,365
Judas Iscariot.....200
Julesz, Béla.....89,365
Jung, Carl Gustav.....36-7,50

Kafka, Franz.....173,180,312
Kahlo, Frida.....179
Kamerschen, D. R......129,365
Kant, Immanuel..... 11,51,56,65,89,92,99,110, 122,132,140,194,201,212,215,234-5,240, 247,250,270,273-4,276,296,303,317,321, 325,359,364,370,375
Kastor, Robert.....205
Katz, Elihu.....264,365
Kaufmann, Walter.....23,194,225,233-5,358, 361,365
Kemeny, John.....150-2,333,365
Kennedy, John F......269-70
Kepler, Johannes.....60,82-3,87,107,118,161, 250,356,374-5
Keynes, John Maynard..207,256,264
Kierkegaard, Søren.....167
Kiesewetter, Hubert.....13,362,365
King, Martin Luther.....282

King, Preston.....13-14,69-71,263
Kirk, Geoffrey Stephen.....149,347
Kitchener, Richard.....355
Klappholz, Kurt.....129,191,365
Klein,Carsten.....21,330,356,372
Klett, Ronald.....176,366
Klibansky, Raymond.....134
Knight, Frank.....11
Knorr-Cetina, Karin D......146,365
Koelbl, Herlinde.....105,285,365
Kohn, Hans.....134
Kolmogorov, Andrey.....332,367
Kook, Hillel.....43,66,260,286
Korczak, Janusz..55,74,104,264,365
Körner, Stephen.....128
Kotarbinski, Tadeusz.....134
Koyré, Alexandre.....115,366
Kraft, Juilius..... 133-4,370
Kraft, Victor...109,225,327-8,365
Kramers, Hendrik.....242,328
Kraut, Richard..... 37,222,340-1,366
Krautheimer, Richard.....37
Kresge,Stephen.....24
Kreutzer,Rodolphe.....96
Kripke,Saul.....155,306,313,321
Kris, Ernst.....27,362
Krylov, Ivan.....258
Kuhn, Thomas S...... 22,32,46,52,55,62,64, 138,141,144,146,166,188,210,250,278, 282-3,355,366
Kulick, Don.....145,366
Kulka, Tomas.....171,366
Külpe, Oswald.....51,89
Kundera, Milan.....19

Lafargue, Paul.....104
Lagerlöf, Selma.....231
Lagrange, Joseph-Louis.....19
Lakatos, Imre..... 16,20-2,32,34,37-40,46, 51-2,55-6,67,78,83,91,97,105-6,118,120, 131-2,149,151,157,164,166,169,189,194, 199-200,244,250,280-3,291,293-4,298-9, 306-9,345,353,356,359,366-7,368,370,378
Lamarck, Jean-Baptiste.....294
Landauer, Gustav.....259
Landé, Alfred.....52,366
Lane, Homer.....55,74,104,366,377
Langley, Pat.....82-4,366
Laor, Nathaniel.....18,23,56,124,246,308, 350,355,365,366,374,376
Laplace, Pierre-Simon..18,49,83,274

Index of Names

Lassman, Peter.....376
Latour, Bruno.....146,366
Latsis, Spiro.....39,366,372
Lavoisier, Antoine-Laurent.....81,107,290
Lazarsfeld, Paul.....264,365
Leblanc, Hugh.....130,367,372
Lehmann,Winfred P......54,366
Leibniz, Gottfried Wilhelm..212,297
Leishman, William.....61-2,353
Lejewski, Cesław.....112,165
Lenin, Vladimir Illich Ulyanov.....44
Lermontov, Mikhail.....43
Lerner, Aba.....134
Lessing, Gotthold Ephraim.....75,110,322
Levinson, Paul.....353,364,366
Levinson, Ronald Bartlett.....17,221-2,225-8, 340-1,353,364-5,367
Lévi-Strauss, Claude.....289,347
Levy, Azriel.....361
Lewis, Roger..... 202,367
Ligeti, György.....179
Lincoln, Abraham.....285,367
Lindsten, Jan.....367
Linsky, Leonard.....321,367
Lipsey, Richard.....129
Locke, John.....34,88, 317
Lofting, Hugh.....241
London, Jack.....104,178
Long, Jancis.....21, 281,294,367,391
Lorentz, Hendrik Antoon.....163,278
Lorenz, Konrad..... 37,83,86,169-70, 194,247,297,357,367,371
Lucas, J. R......368
Lukasiewicz, Jan.....94,367
Lycophron.....222
Lynkeus (Josef Popper).....256

Macbeth.....304
Mach, Ernst.....49,92,99,167,183,215,232, 296,307,320-1,357,367
Machiavelli, Niccolò.....222
Macintyre, Alasdair.....233,367
Magee, Bryan.....134,162,367
Maimon, Solomon.....11,100,204,239,247, 257,296
Maimonides, Moses.....100,204,239,257,296
Makarenko, Anton.....55
Malcolm, Norman.....53,233,367
Malinowski, Bronislaw.145,189,362
Manuel, Frank.....29,367

Mao Tsetung.....144,330
Marchi, Neil De.....129, 367
Martin, R. M.......354
Martin, R. N. D......196
Martins, Herminio.......376
Marx, Karl..... 11,22,37,41,44-5,51,80,111, 125,160,213,219-220,232,234-8,241,261, 265-6,275,283,330,337,354,369,381
Maslow, Abraham.....256
Massaryk, Tomáš.....288
Matthew, St.....248-9,368
Maugham, William Somerset.....104,169,253
Maxwell, Grover..... 361,370
Maxwell, James Clerk..... 115
Maxwell, Nicholas.....67,367
McCarthy, Joseph.....54
McCarthy, Timothy.....185,367
McClintock, Barbara.....161
McClure, Alexander K......253,367
McGee, Reece.....160,358
McGuinness, Brian.....184,367
McHenry, Leemon.......367
McTaggart, John McTaggart Ellis.....184
McVittie, George C......50,36y
Mead, Margaret......74,367
Medawar, Peter.....152,177,186,339,368
Meegeren, Han van.....69,189
Meital, Amir.....15,341,344,368
Mercer, David.....291,359
Merchant, Ismail.....74
Merriman, Brian.....172,368
Merton, Robert K...... 50,92,142-4,146-7,150, 188,209,248-9,268,301,368
Methuselah.....275
Mew, Melitta.....16, 28,350
Meyerson, Émile.....75
Michalos, Alex..... 368
Michalski, R. S......366,368
Michotte, Edmond......127,368
Mill, John Stuart..... 80-2,360,368
Miller, David.....6,15-16,105,150,280,305,368
Millhone, Kinsey.....132
Mises, Richard von.....67,94,106-7,118-19,124,236,248,281,289,307
Mitchell, Basil.....368
Mitchell, Mark T..... 372
Mitchell, T. M.....368
Momigliano, Arnaldo D......159,293,343,368
Monk, Ray...18,180,190,253,368,378
Montessori, Maria.....104

Index of Names

Moore, G. E.....53,184,259,338
Moore, L. F..... 207,370
Morgan, Lewis Henry.....145
Morgenstern, Martin..... 118,136,238,244,368
Morgenstern,Oskar.....82
Morris, Lois B..... 132,369
Morton, George..... 130
Moses.....247
Motterlini, Matteo....108-10,200,294,368
Moulinas, Carlos Ulises.....190,323,368
Mozart.....90,280,309
Munz, Peter..... 16-17,82,87,106-7,110,149, 153,155,169,179,184,280,325,369
Murray, Gilbert.....289,369
Musgrave, Allan.... 6,34,37,129,169,222,250, 283,325,341,359,366,369,370
Myers, Gerald E.....124,369
Myrdal, Gunnar.....248

Naess, Arne.....54,369
Nagel, Ernest.....94,368,371
Nails, Debra.....313,340-4,369
Neill, A. S.....55,97
Nelson, Leonard..... 117,128,133-4,180,187, 233,362,369,374
Neumann, John von.....82
Neurath, Marie.....370
Neurath, Otto..... 19,34,54,117,136,138,185, 211,213,216-17,233,267,296,300,304, 316,320,330-1,346,355,358,363,369,370
Newman, James.....94,368,371
Newton, Isaac..... 18-19,29-31,80-1,83,115, 118,151,183,236,247,255,274,278,334-5, 354,367,375
Nicod, Jean.....311,369
Nietzsche, Friedrich..... 41,60,103,132, 168,178,235,369
Niewski, Alexander.....112
Niiniluoto, Ilkka.....323
Nino, Carlos Santiago.....253,369
Niven, David.....198
Nobel, Alfred.....82-3,91,161,176,231,367
Notturno, Mark A.....35,369,371
Nugent, Elliot.....221
O'Hear, Anthony.....18,213,217,220,369

Oakeshott, Michael.....303
Oberdorfer, Don.....30,369
Oldham, John M......132,369
Olivier, Laurence.....202,369
Orwell, George.....43,125,127,369

Osers, Ewald.....361
Oswald, Wilhelm.....51
Pap, Arthur..... 109,369
Parmenides.....247,345-7,371
Parrini, Paolo.....,304,369
Pascal, Blaise.....182
Passmore, John.....218,369
Pasteur, Louis.....19,62
Patterson, Wayne A......354
Pecksniff, Seth.....221
Pellerin, Daniel.....364
Pericles.....247,346
Perls, Fritz.....29,369
Petersen, Arne.....6,371
Philo Judaeus..... 167
Pickwick, Samuel.....92
Pieper, Annemarie.....27,369
Pilate, Pontius.....27
Pinder, Craig C......207,370
Pitcher, George.....200,370
Pitte, Frederick P. van der...276,370
Plamenatz, John.....223
Planck, Max.....50,82,92,248
Plato.....11,31-2,75,78,87-90,92,99,104,107, 134,164,166,172,196,220-2,225-8,237-8, 252-4,261,289-90,303,310, 321-2,336, 340-7,349,361-63,366-8,373,376
Poe, Edgar Allan..... 179,
Polanyi, Michael.....32,143-4,146-7,179, 248-50,278,282,370,372
Pole, David.....25,157,171,177,226,226,234, 325,337,354,370
Popper, Josef.....256-7
Popper, Lady Josephine (Hennie).....10,111, 113,121,132,135,197,284,309
Popper, Simon Carl Siegmund....294
Potok, Haim.....74,371
Pralong, Sandra.....306,360,363,365
Preston, John M......361,371
Priestley, Joseph.....163
Prior, Arthur..... 112,132,134
Protagoras.....88
Pryke, Jo.....172,371
Putnam, Hilary..... 12,81,94,246,251,284,313, 323,330,334-6,371
Pyrrho.....341

Quine, Willard Van Orman..... 49,55-6,90, 137,139,151-2,158,181,187,189-90,271, 295,303,321,323,332,335,363,371-2,376

Index of Names

Quinton, Anthony.....153,372
Quixote.....41
Radice, Lisanne.....80-1,372
Radnitzky, Gerard.....39,257,264,372
Rajchman, John......372
Raven, John Earle......149,347
Reck, Erich H...... 322,372
Ree, Jonathan......143,372
Reichenbach, Hans.... 7,117,158,188,213,217, 236,251,317,322,324,330,360,372-4,375
Reichmann, Yehuda.....195
Reid, Constance.....306,372
Reininger, Robert. 296
Rényi, Alfred.....367
Richards, Lynn.....83
Richardson, Alan.....375
Ridgeway, C......290,372
Robbins, Lionel.....39,129,190-1,283,372
Roberts, John M.....290,359,372
Robespierre, Maximilien.....265
Robinson, Abraham.....39
Robinson, Guy.....200,372
Robinson, Joan.....236,261,372
Robinson, Richard.....30,35,221-2,227,340-1, 372
Rockefeller, John D.....103,122
Roeper, Peter.....130,372
Rolland, Romain.....190
Rorty, Richard.....234,303,323,372
Röseberg, Ulrich.....324,373
Rosen, Edward.....284
Roth, Josef.....21
Roth, Leon..... 13,66-7,73,355,372
Rousseau, Jean-Jacque..222,254,258
Rubens, Bernice..... 74,202
Russell,Bertrand..... 7,11,18-19,22,29,49-53, 55-6,59,65,72,81,84,90,97-8,104,117-18, 134,140,151,153-6,160,167-9,176,178-9, 182,184-5,188-90,198,200,202-3,208,211, 214-16,220-1,233,235,240,247,249,252-5, 257,259,261,264,271,290-1,295,297, 301-2,306,311-16,319,321-2,325-6,328, 330,335-7,346,354,357,362-3,369,372-3, 377-8
Ryle,Gilbert..... 35,132,183,185,196,220,244, 249,271,308-9,359,373

Sabra, Abdelhamid Ibrahim..67,189
Salinger, Jerome D.....172
Salk, Jonas.....366

Salmon, Wesley.....324,364,373-4,376
Sambursky, Samuel..83,332,356,374
Samuel, Herbert.....5,191
Sándor, György.....203
Santayana, George.....235,374
Santillana, Giorgio.....19,374
Sargeant, Winthrop.....305
Sassower, Raphael..118,147,255,374
Savage, C. Wade.....323,
Savoye, Jeffrey.....179
Sawyer, Tom.....74
Schacter, Daniel L......14,374
Schaffer, Simon.....118,361,377
Scharfstein, Ben-Ami.....132,374
Scheffler,Israel.....180-3,374
Schiffer,Menahem Max.....47,94-5,115
Schiller, Friedrich.....273,276
Schilpp, Paul Arthur..... 19,21,35,50-1,56,82, 117,149-50,188,204,211,225,324,329,332-6, 358-9,360,362-3,370-1,374
Schlesinger, John.....201
Schlick, Moritz..... 34,36,158,170,185, 210-13,216-18,225,267,276,316-19,322-3, 327-9,331,336,361-2,364,374
Schoenberg, Arnold.....179,305
Schofield, Malcolm......347
Scholem, Gershom.....60,374
Schopenhauer, Arthur.....178,233,294,345
Schrödinger, Erwin......11,49,75.108,241,374
Schulmann, Robert......360
Schwartz, Robert.....181,374
Scriven, Michael.....358
Segre, Dan Vittorio.....116,258,375
Seiler, Martin.....357,375
Selz, Otto.....51,89,116,377
Semmelweis, Ignaz Philipp.....161
Serkin, Rudolf.....133
Shae, William.....83,375
Shahar, Eyal.....375
Shakespeare, William.....74
Shanker, Stuart G...... 18,355,363,374-5
Shanker, V. A...... 374
Shannon, Claude Elwood.....89
Shaw, Bernard..... 11,38,41,51,70,74,82,104, 125,136,176,178-9,197-8,202,220,231, 233,253,255,273,275,287,301,366,375
Shearmur, Jeremy.....6,16,118,120
Shearmur, Pamela.....6
Shimony, Abner.....196,284
Sholokhov, Mikhail.....44

Index of Names

Shortt, E. H......62,375
Silber, John.....354
Simey, T. S......81,375
Simkin, Colin.....16,375
Simmel, Georg.....202,238,258,265
Simon, Herbert A.....82, 87,234,354,366
Slater, John C......242,328
Sluga, Hans.....133,375
Smiley, Tim.....153,306,363,375
Snell, Willebrord.....76
Snow, Charles P......176,309
Soames, Scott.....314,326,375
Socrates..... 7,9,11,19,87-9,104,125,133,
    149,209,221-2,226,230,239,252,262,
    313,340-5,347,361,369,366,382-3
Solomon, Maynard.....132,376
Solomon.....232
Solon.....224
Soros, George.....73,270,375
Sousatzka, Yuvline.....201-2
Spinoza, Benedict... 9,67,92,134,143,160,167,
    176,179,222,230,235,247,253-4,256-7,
    262,266,296-7,373,376
Spohn, Wolfgang....130,332,368,375
Stachel, John.....354
Stadler, Friedrich.....158,170,185,213,218,
    316,318,320,324,327,331,333,357,363,
    374-5
Stalin, Joseph.....44-5,105,176,199,208,261,
    265,281,283
Steif, Jacob.....16
Stein, Gertrude..... 241
Steiner, George.....11,375
Stern, David.....316,375
Stidd, Sean C......185,367
Stoller, Paul.....145
Stopes-Roe, Harry V......333,376
Stove, David.....241-2,376
Strauss, Leo.....247,347,376
Stravinsky, Igor.....179
Szabo, Arpad.....347
Szasz, Thomas S......110,168

Tarski, Alfred.....11,93,112,190,216,300
Temkin, Owsei.....234
Tennant, Neil.....90,371,376
Tennyson, Alfred.....70
Thales.....347
Theodor, O......61,353
Thompson, Judith Jarvis.....12
Thorsrud, Einar.....104

Thucydides.....247,346
Thurman, Wallace.....74
Todd, Michael.....198
Toklas, Alice B......241
Tolstoy, Leo.....96,182,325
Toulmin, Stephen.....146,153,327,364,376
Toynbee, Arnold.....221,376
Tribe, Keith.....118,376
Trollop, Anthony.....198
Trotsky, León.....109, 125
Trudeau, Pierre Elliot.....264
Tsinorena, Stavroula.....372
Turing, Alan..... 326
Twain, Mark (Samuel Clemens).....74

Uebel, Thomas E......358,376
Uexküll, Jakob von.....247
Ullendorf, Edward.....373

Velody, Irving.....204,376
Vermeer, Johannes.....69,189
Verne, Jules.....198
Voegelin, Eric.....376
Vonnegut, Kurt.....201
Vuillemin, Jules.....335,371

Wade, Nicholas.....37,357
Wagner, Richard.....305,368,376
Waismann, Friedrich.....18,132
Waite, Arthur Edward.....75
Warburg, Aby.....362
Warnock, Geoffrey J......43,177,313,338,376
Wartofsky, Alice......284
Wartofsky, Marx W...... 354,356,359,368,377
Washington, George..... 265
Watkins, John.....11,15-16,38-40,56,67,76,
    78-80,82,86,106,118,127,132,145,188-9,
    200,242-3,249,253,259,280,283,293-4,
    297-8,307,376
Webb, Beatrice.....80-1,202,371,376
Webb, Sidney.....80-1,202,371
Weber, Max.....59-60,92,166,178-9,204,
    209,239,264-5,268,354,376
Webern,Anton.....178-9
Weiler,Gershon.....60,376
Weinberg,Steven.....204,329,376
Weingarten,Paul.....354
Weininger,Otto.....49
Weldon,T. D......177,338-9,377
Wells, H. G...... 176
Wendel, Hans Jürgen.....213

Index of Names

Wessels,Linda.....316,319,329,359
West, Cornel.....373
Westfal, Merold.....353
Wettersten, John R.....51,81,116,118,151,203, 295,308,311,317,329,355,356,377
Weyl, Hermann.....31,377
Wheelis, Allen.....74,377
Whewell, William.....30,46,51,81,83,92,118, 140,144,146,194,233,274-5,296,302,361, 377
Whitehead, Alfred North.....52,134
Whittaker, Edmund.....64,83,143
Wilde, Oscar..... 172
Williams, L. Pearce.....250-1,361
Wills, W. D.....74,377
Willson, Margaret.....145,366
Wilson, Fred Forster Jr......11,377
Wisdom, John O......41,44,56,64,67-9,71,78, 94,101,114,170,173,189,205-6,215,241, 249,280,298,306-7,339,346-7,353,377
Wisdom,John T. D.....339
Wittgenstein, Ludwig.....11,14-16,18- 19,22,29,34,36-7,49,54,56,81,87-8,109, 122,131-2,143,151-7,166,170-2,174,177- 8,180-5,190,200,211,214-15,218,232-4, 241,247,259,267,270-1,290,293,296,299- 306,313-331,335-9,353,357,359,360, 362-4,367,369-70,373-5,377-8

Woffinden,Bob.....91,377
Wolfson,Harry A......250
Wolniewicz, Boguslaw.....219,377
Wolters, Geron.....364,373,375
Wood, Alan.....18,190,202,378
Woodger, Joseph Henry..94,134,370
Woolgar, Steve.....366
Woolsey, C. N......363
Worral, John.....378
Wright, Georg Henrik von..325,367
Wrigley, Michael.....170,377
Wyman, David.....260,378

Xenophanes.....309,345,348

Yates, Frances.....19
Yehezkely, Chen.....16

Zahar, Eli G......378
Zecha, Gerhard.....355,377-8
Zeno.....347
Zermelo, Ernst.....57
Ziman, John.....30,377
Zimmer, Robert.....118,136,238,244,368
Zin,Howard..... 283
Zollschan,George.....109,190-1
Zuckerman, Solly.....56,378
Zweig, Stefan.....21,378
Zytkow, Jan M.....82,366

INDEX OF SUBJECTS

Academism, Academy.....11-12,16-17,20, 47,54,65,67,77,78,97,130,144-5,177, 192,195-6,206,209-10,241,268,279, 291,304-5,310,313,341-2,345,354,357
Acknowledgement.....9,17,34,50-1,76,87, 106-8,140,142,149-50,169,187,190, 192,194,209,217,222,226,228,240,242, 280,295,303,306,308,320,338,350
Aesthetics..... 79,171,177,235,252,366
Agenda, public.....98,141,166,248,252, 257,269,274,278-9,283,286,291,299, 326,337,346
Anarchism.....91,136, 284
Anarchy.....220
Anthropology.....89,129,144-6,189
Anti-metaphysics.....29.48-9,57-8,98-9, 115,154-6,171,185,212-13,215-16,233, 247,272-4,296,298,301,303,314,320-3, 327,330,332,336
Anti-Semitism.....44-5,59,176,360
Apology.....106,121-2,126,159,169,350
Apologetics....61,73,91-2,137,158,326,336
Approximation.....30,32,37,45,57,61,89, 159,162,196,255,307,347
Astronomy.....60,81,162
Asymmetry.....139,190,302
Ataraxia.....383
Atomism.....100,354
Autobiography, see Biography
Autonomy, individual.....23,29-31,41,45, 108,148,160-1,193,209,218,235,240, 248,252-3,257,262,264,266,342,354, 358,379
Autonomy, scientific...62,139,160,195,210
Autonomy of the social sciences.....277
Avant-garde...74,160-1,205-6,221,250,252

Axiom.....19,94,130,150,315,332, 380-1, 384
Bad faith.....54,95,138,148,183
Behaviorism.....49,90,300
Bootstrap operation.....50,300
Bureaucracy.....65,209,279,284
Cabbala.....19,60,81,147,276
Cargo-cult.....147,289
Catharsis.....178
Chauvinism.....60,175-6,189
Christianity.....58-9,169,180,345,361
Citation index.....188
Collectivism.....209,234,256,263,266,286, 379
Commonsense.....13,31,34,51,56,65,73,98, 107,110,167,190-200,215.221,244,255, 257,270,295,301-2,313-14,320,331, 333,337-8
Communication.....25,28,82,89-90,106, 134,161-2,171,225,228,265,270-5,280, 287,297-8,354,366,368
Communism.....44-5,103,125,281
Concept, Concept-formation, Conceptual analysis.....46-7,53,80,92,94,141,147, 156,185,189,200,209,216,240,259,261, 265-6,297,319,333,358,363,373,379
Confirmation.....47,62,117,130,139,141, 150-1,157,185,189,272,274-5,291,311, 330,333-4,376
Conformism....50,90,141,147,159, 177-8,258-9
Conscription.....257
Consensus.....222,234,250,253,258-9,265, 283
Conservatism...23,99,173,179-80,186,243, 264,284,288-9,311,325,328,330,379

Controls, democratic, see Democracy
Convention, Conventionalism.....36,62,73,
 81,93,100,212,214,222,234,253-5,300,
 307,346-7,364,379
Coordination.....255,257,262
Corroboration, see Empirical support
Cosmos, Cosmology.....239-40,345
Cosmopolitanism.....21
Counter-culture.....160
Credibility, see Confirmation
Credulity.....19175
Cryptomnesia.....50
Curiosity.....58,97,172,174,250

Data.....81-3,89,143,162,193,289,323,325
Decidophobia.....23,365
Deducibility.....141,150,335
Deduction, Deductivism.....94,112,118,
 162,185,187,190,289-90,311,334-5,379
Defensiveness...20,31-2,43,45-6,60,86-7,
 89,96,101,152,176,179,193,198,214,
 227,301-2
Definition.... 46,78,93-4,96, 232, 297,
 379-81
Democratic controls see Democracy
Democracy,.....9,11,14,23,31,43,55,63,73,
 96-7,107,127, 169,176,180,190,202,
 208-9,218,220-3,245,247,252-4,255-9,
 260-1,263-6,269-70,275-6,278-9,282-3,
 286,288,338,340-4,354,361-2,366,
 369-70,372
Desire.....92,95-6,173-4,256,270,278
Despair.....93,159,178,238,260,311
Destiny..98,132,152,155,175,218-21,
 233-5,237,258,260,278,380
Determinism, Indeterminism.....167,338
Diagnosis.....18,61-2,361,366
Dialectics.....46,107,149,185,196-7,228-9,
 234-5,240,289-91,341-4,377,380

Dialogue.....9,31-2,71,87,104,122,185,196,
 250,272,290,316,341-3,359,360,374
Didactics in Plato.....342-4
Disagreement.....56,94,106-7,116,156-7,
 217,220,227,283,328,331
Disbelief, suspension of.....19
Discipline.....45-7,67,90-1,116-17,173,
 271,283
Discipline, scholastic.....283,299-300
Disconfirmation.....139
Dissent.....107,205,249-50,261,283,301,
 307,316,329
Dogmatism.....56,58,79-80,109,125,137-8,
 143,163,185,197,205,234,236,247,254,
 265,275,323,332,343,371

Economics.....91,129-30,165,186-7,207,
 218,236,248,262,264,268-9,279,285,
 330
Education.....36,41,55,80,86,90,92,96-7,
 104,107,116,119,129,160,173-4,207,
 213,220,231-2,248,252,257,264,276,
 289-91,342
Élite, Élitism.....23,147,155,176,210,220,
 249,253,300
Empirical evidence, Empirical test...14,31,
 45,49,59,65,76,81,91-2,99,107,129,141,
 156,160,212,214,247,276-7,290,296,
Empirical learning....29-31,45,49-52,59,
 62, 76,81-3,89,100,145-7,154,160,163,
 166-7,174,179,193,210,216,224,249-
 50,259,274,277,289,299,320,328,341
Empirical support.....45,62,151,159,162,
 165,212-13,242-4,247,254-5,274-5,
 290-1,302-3,307,314,328,333-4,342
Empiricism, critical.....308,328,335,347-8
Empiricism, traditional & "logical"....22,
 37,62,100,137,141,185,203,212,239,
 244,296,300,303,307,314,325-6,330

Index of Subjects

Enlightenment.....22,37,43,49,58,60,125, 169,175-6,213,222,230,234,236, 251-3, 257-8,262-3,265-6,278,286,348,380
Entanglement, quantum.....159,160
Equality.....14,29,30,41,59,61,65,75,81,95, 106-7,115,119,121,139,150,153,156, 168,175,178-9,240,282,312,344
Equanimity.....167
Escapism.....74
Eschatology.....238-9
Essence, essentialism.....51,82,96,99,104, 156,184,200-1,232,242-3,262,265,294, 307, 318-19,325,327-8, 345
Ethics.....9,30,66,87,103-4,123-4,126,132, 154,209,257,273,287,305
Expertise.....30,37,55,67-9,83,121,128, 130,140,177,189,221,253,259,278
Explication.....150, 200,333
Extremism.....96,145,169,302,322,336

Fabians.....202,220
Fads.....20,22,150,170,177,210,253,360
Fallibilism.....9,91,125-6,139,190,221, 231-2,254,276,345,348,380
Fanaticism.....236,256
Fideism.....22,125,166,189,381
Fieldwork.....145,366
Friction.....91-2

Gambling.....141,145,273,304
Gossip.....22-3,38,76,120,141,147,153, 230,232,317,337,343
Government, see Democracy

Harmony.....18-19,29,238-9,340
Hedonism.....124,126,342
Heresy, Heretics.....61,66,90,143,159, 171,248,313
Heteronomy.....41,161,251
Heuristic.....92,100,131,184,273-4,354

Historicism.....99,124,152,178,207,218, 220-1,235,237-40,256,259,369-70,380
Historiography.....64,115,241,284,291
History of science.....62-4,153,188,330
Holism.....190,235,335
Holocaust.....220,234,260,378
Humor.....16,80,87-9,134,167-8,189,201
Hypochondria.....113
Hypocrisy.....149,173,205
Hypothetico-deductivism, see Deduction

Ideals, Idealism.....45,56,71,74,89,91,98, 137,147,175-7,180,194,208,215,222, 234,266-70,331,343
Ideology.....11,46,79,175-6,178,193,219, 234,258,286,382
Idleness.....104
Impeachment.....283
Imperialism.....45,270
Incommensurability.....138,195
Indeterminism, see Determinism
Inductivism.....12,17,32,59,68,81-3,212, 242,254,290-2,295,307-8,311,334-5, 345,380
Inquisition.....29,74,200,344
Institutions, Institutionalism..... 99,107, 130,166,193,223,248,250,252-3,255, 258,262-6,268-9,275,277-9,283,285-6, 290,296-7,377,380
Instrumentalism.....45,49,52,200,212,254, 296,300,307,313,380
Internet.....161,324,329
Interpretation.....27,90,106-7,168-9,180, 191,222,227,239,244,320,333,338
Irrational, Irrationality, Irrationalism.....60, 125,136,147,160,170,175-7,182,234, 242,283,289,303,376
Irrefutability.....96,236,261,272,329
Irresponsibility, see Responsibility

Joie de vivre......178

Justice.....13,15,30,66,77,87,107,137,190, 226-8,239,258,279,365,377
Justification.....39,52,56,87,95,151,165, 178,180,182,202,214,227,238,248-9, 253-4,300,325-6,338,362

Kangaroo court.....21,38,243-4
Kibbutz.....261

Language, Linguistics.....90,93-4,200,203, 214,234,296-7,300, 330,332
Law-likeness.....181
Leadership.....20,23,104,117,140,143,148, 157,160-1,164,208-10,217,220,223, 247-91
Lectures.....11,14,19,35,42,50,52,61,63,67, 69-71,75,77-80,82-4,88,90-1,93,97, 103,109,111,116,124-6,128-9, 132,143,148,152,155-6,165,168,175, 179,186-7,189,194,197-8,221,235,243, 245,248,258,281-3,288,304,309, 317-18,333,364,367-8,370,374
Legislation.....187,222,257,264,383
Legitimation.....34,45,100,119,120,132, 146-8,155-6,158-9,181,183,185,236, 252,253,265,283,286,316,374
Liberalism.....19,29,34,43,66,74,86,96-7, 125,130,156,165,168,177,180,220-2, 225,227,245,249,253,256-8,264, 268-70,282,285-6,344,355
Logic course, Popper's.....93-5
Loyalty.....25-6,39,45,64,74,109-10,170-2, 176,258,277,288,293-4,305,319,330

Machine translation.....54
Magic.....19,31,60,73,100,124,130,145, 200,213,273,283,337,346
Magic Rites.....145-6,195,205,229, 239,275,288-9
Manners,.....11,67,70,153,192-3,195-8

Marxism.....11,22,41,44-5,51,111,125,160, 219,236,241,261,266,275,283,330,337
Mathematics.....18-19, 39,66,91,131,137, 139,141, 289,322, 336
Medicine.....62,142,161
Medieval thought.....175,215,221
Memory.....134,158,191,247
Meritocracy.....249
Meta-biology.....300
Metapsychology.....300
Metaphor.....81,147,184,273,337,
Migration.....74,207,310
Miracles.....239
Monotheism.....239,345-6
Movies..74,86,104,157,198,201,221,284
Music.....171,173,175,178-9,221,287,291, 305,309
Mysticism.....18-19,60,184,190,250,273, 325,373
Myth.....81,125-6,141-3,147,149,158,162, 172,193,209,240,248,283,289,308

Nationalism.....9,43,45,169,258-60,284-6, 288,380-1,384
Nazism.....44,112,133,169,176-7,193-4, 207-8,220,233-4,309,375
Neurosis.....74,97,172-5
Nominalism.....200-1,381
Non-justificationism.....165,202,214

Obscurantism.....33,57,205,210,272
Operationalism.....49-50,71,130,159
Opportunism.....52,58
Overwork.....103,173,350

Paradox.....19,50,53-4,93,159,172,178
Paternalism.....110,120,168
Perfectionism.....114,116,175
Philosopher King.....220,237,261,343
Physicalism.....201,300,375
Physiologoi, see Presocratics

Plagiary.....76,152,187-8,280,316,320
Platonism.....90,92,172,321,336
Pluralism...37,44,218,254,259,282-3,382
Positivism, see Empiricism
Populism.....218
Power base.....209-11,216,282,310
Pragmatic paradox.....53,
Pragmatism, American.....22,234
Prediction.....234,255,294,323
Pre-science.....238
Presocratics.....11,125,149,226,
　239,345,347,382
Problem-shift.....210,299
Problem-situations.....18,47,85,297,329
Procedure.....68,253,271
Programs, computer.....82,84,93,
Programs, research, see Reserch program
Psychoanalysis.....100,174,275,308
Psychologism.....258,263,3800,382-3
Puritanism..58,173,180,246,268,273,382-4

Racism.....36,176,213
Radicalism.....22,235,330,382
Reaction.....59,160,176,179,202,234,245,
　251,258,284,291,340,342-3,379
Realism....77,81,98,109,208,213,215-16,
　267-8,283,294-6,298,307,317-19,329,
　332,338,362,364,368,370,373,382
Realpolitik.....286
Recognition.....23,36,67,77,91,93,101,112,
　122,124,140-4,146,158-61,177,186,
　189-90,199,205,208-10,213,219,223-5,
　244,248-9,250,264,278-80,291
Reduction, Reductionism.....83,124,190,
　263-4,266,277,285,381,381-3
Reform..36,45,55,99,104,207,223,230,240,
　257,261-2,264,279-80,290,301,383
Refugee.....71,117,131,148,211,217-18,
　307,310,348

Religion.....43,48,57-9,73,99,111,161,166,
　169,183,213,233,236,238-40,284-5,337
Renaissance.....250
Repeatability.....82,100,107,143,160,239
Repute.....37,95,155,188,204,208,217,224
Research, see Empirical research
Research program.....50-1,59,100,236,256,
　261,268,283,298
Responsibility.....20,25,32,68,114,130,
　217-18,234,251,261,263-7,269,273-4,
　278,281,286,288,290,304
Revalidation.....141,145,148,160-1
Risk.....11,61,79,113,116,134,214,265,274
Rites, see Magic
Romanticism.....23,41,47,60,97,175,178,
　252,258,260,265,278,282,285,288,
　308,383

Safeguards see Democracy
Scholasticism...47,82,149,250,289,299,300
Schools.....81,95-7,177,205,264,285
Scientism.....73,206,236,382
Secession.....284-5
Self-selection.....252,266
Sensationalism.....52,89,244,295,308,318,
　320,323,325
Sex.....36,49,74,83,95-6,145,147,168-9,
　172,180,201,311,359,366,371
Shoah.....96,133,220,260
Siblinghood of humanity.....29,32
Situational logic.....95,146,237,273,303
Slavery.....99,227-8,340-2,344
Social contract.....258
Socialism.....54,80-1,133,223,237,248,261,
　266,304
Sociology of science.....45,50,141-2,166,
　268,279
Sociologism.....263,379,383
Sophism.....31,240,344,369,382
Sovereignty.....253

Stalinism.....199,208,261,283,261
Stonewall.....23-4,250
Substance.....98,183,201,297,321,346,383
Suicide.....173-4,190
Superstition.....36,60,124,289,337
Syntax..... 322-3

Talmud.....65,121,157,189,204,353
Tautology.....94,139,236,333
Test, see Empirical Evidence
Testability.....50-1,73,99,125,137,141,181, 213,215,272,283,311-12,334-5,358,382
Textbooks.....31-2,104,129,142,159
Theology...30,34,44,58,138,213,296,321
Therapy.....124,177
Toleration.....25,46,49,57-8,91,120,167, 253,257,283-4,299,316
Totalitarianism.....127,220,261,340,343
Tradition.....15,19,34,43,50-1,72,83,97-8, 156-7,163,167,173,175-6,182,190,203, 208,223,227,234,239,250,252,258,262, 266,276,296,300,310,320,328,341,383
Traditionalism.....178,221,384
Tribalism.....99,201,240

Twin-paradox.....159
Two tiers..... 160,250,266
Tyranny...74,176,220,252-3,264,269,343-4

Unanimity.....283,299-300,316,382
Uncertainty.....317-18
Universalism.....,59,285,346,383
Universals.....11,151
Utopianism.....65,81,160,176,207,239,241, 253-4,256,258,260-1,264,268, 339-40,384

Verificationism.....50,327
Verisimilitude..... 159,247,292,319,343, 347-8

Welfare.....25,41,186,218,256,264
Whiff.....36,77,159,220,242-3
Work.....18,20,58-9,71,93,114,116,119, 127,158,173,203,225,267,273-4,305
Workshops.....107,130,179

Zionism.....45,60,168,260